SUSTAINABLE MICRO IRRIGATION DESIGN SYSTEMS FOR AGRICULTURAL CROPS

Methods and Practices

Innovations and Challenges in Micro Irrigation

VOLUME 2

SUSTAINABLE MICRO IRRIGATION DESIGN SYSTEMS FOR AGRICULTURAL CROPS

Methods and Practices

Edited by

Megh R. Goyal, PhD, PE, and P. Panigrahi, PhD

Apple Academic Press Inc.
3333 Mistwell Crescent
Oakville, ON L6L 0A2
Canada

Apple Academic Press Inc.
9 Spinnaker Way
Waretown, NJ 08758
USA

Exclusive worldwide distribution by CRC Press, a member of Taylor & Francis Group

Printed in the United States of America on acid-free paper

International Standard Book Number-13: 978-1-77188-274-3 (Hardcover)

International Standard Book Number-13: 978-1-77188-275-0 (ebook)

Library and Archives Canada Cataloguing in Publication

Sustainable micro irrigation design systems for agricultural crops : methods and practices / edited by Megh R. Goyal, PhD, PE, and P. Panigrahi, PhD.

(Innovations and challenges in micro irrigation; volume 2)
Includes bibliographical references and index.
ISBN 978-1-77188-274-3 (bound)
1. Microirrigation. 2. Sustainable agriculture. I. Goyal, Megh Raj, editor II. Panigrahi, P. (Pravukalyan), author, editor III. Series: Innovations and challenges in micro irrigation; v. 2

S619.T74S98 2015 631.5'87 C2015-903927-4

Library of Congress Cataloging-in-Publication Data

Sustainable micro irrigation design systems for agricultural crops: methods and practices.

pages cm
Edited by Megh R. Goyal.
Includes bibliographical references and index.
ISBN 978-1-77188-274-3 (alk. paper)
1. Microirrigation. 2. Irrigation farming. 3. Irrigation water. 4. Water in agriculture.
I. Goyal, Megh Raj, editor.

S619.T74S874 2015 631.5--dc23 2015020319

Apple Academic Press also publishes its books in a variety of electronic formats. Some content that appears in print may not be available in electronic format. For information about Apple Academic Press products, visit our website at **www.appleacademicpress.com** and the CRC Press website at **www.crcpress.com**

CONTENTS

LIST OF CONTRIBUTORS

Hatira Abdessatar
Faculté des Sciences de Tunis, Tunisie

Pradip K. Bora, PhD
Associate Professor (Irrigation and Drainage Engineering), College of Agricultural Engineering and Post Harvest Technology, Central Agricultural University, Ranipool, Imphal, East Sikkim, Sikkim. E-mail: kusrebharat@gmail.com; pradip66@gmail.com

Vincent F. Bralts, PhD, PE
Professor and Ex-Associate Dean, Agricultural and Biological Engineering Department, Purdue University, West Lafayette-IN 47907–2093; bralts@purdue.edu

Vishal K. Chavan, PhD
Assistant Professor and Senior Research Fellow in SWE, Agriculture University, Akola, Maharashtra, Website: www.pdkv.ac.in; E-mail: vchavan2@gmail.com

M. M. Deshmukh, PhD
Professor, Department of Irrigation and Drainage Engineering, Dr. Panjabrao Deshmukh Krishi Vidyapeeth, Krishinagar PO, Akola (MS) 444104, India

Simeon DiGennaro, MSc
Administrative Manager, The Ohio State University, Department of Agriculture, Environmental, and Development Economics, 2120 Fyffe Road, Columbus, OH 43210, USA. Mobile: +255 686037461, E-mail: simwilldig@gmail.com

G. D. Gadade, PhD
Jr. Agronomist, AICRP on Water Management, VNMKV, Parbhani – 431402, M.S

Megh R. Goyal, PhD, PE
Retired Professor in Agricultural and Biomedical Engineering, University of Puerto Rico – Mayaguez Campus; and Senior Technical Editor-in-Chief in Agriculture Sciences and Biomedical Engineering, Apple Academic Press Inc., PO Box 86, Rincon – PR – 00677 – USA. E-mail: goyalmegh@gmail.com

S. K. Gupta, PhD
INAE Distinguished Professor, ICAR – Central Soil Salinity Research Institute, Karnal – 132001 (India). Mobile: +91–9416081613, E-mail: skgupta@cssri.ernet.in

Moncef Hammami, PhD
Professor, High School of Agronomy of Mateur, 7030 Mateur, Univ. of Carthage, Tunisia, E-mail: hammami.moncef@ymail.com Tel: 72 465 074

Ben Ali Hédi, PhD
Agricultural Investment Promotion Agency, 6000 Gabès, Tunisia

Daghari Hedi, PhD
Institut National Agronomique de Tunisie, Avenue Charles Nicole, 1082 Tunis, Tunisie

A. D. Huchche, PhD
Scientist, ICAR-National Research Centre for Citrus, Nagpur-440010, Maharashtra, India

Ashok Shrikrishna Kadale, PhD
Chief Scientist, AICRP on Water Management, VNMKV, Parbhani – 431402. E-mail: kadaleashok@rediffmail.com

Ibrahim Khaleel, M.Tech.
Agricultural Engineer. E-mail: ibrahimkhaleel1075@gmail.com

David S. Kraybill, PhD
Professor, The Ohio State University, Department of Agriculture, Environmental, and Development Economics, 2120 Fyffe Road, Columbus, OH, 43210, U.S.A., Mobile: +255 757 420 308,
E-mail: kraybill.1@osu.edu

B. C. Kusre, PhD
Associate Professor (Irrigation and Drainage Engineering), College of Agricultural Engineering & Post Harvest Technology (CAEPHT), Central Agricultural University, Ranipool, East Sikkim, Sikkim. Mobile: +91–09933440970, E-mail: kusrebharat@gmail.com

Yogesh V. Mahatale, PhD
Assist Prof, College of Agricultural Engineering and Tech., Warwat Road, Jalgaon Jamod, Dist. Buldhana, Maharashtra, India. E-mail: yogeshvmahatale@gmail.com, ymmahatale@yahoo.com

A. M. Michael, PhD
Former Professor and Project Director, Water Technology Centre, IARI, New Delhi, Director, IARI, New Delhi; and Ex-Vice-Chancellor, Kerala Agricultural University, Trichur, Kerala; Present address: Dr. A. M. Michael 34/81, V.P. Marikar Road, Edappally North P.O., Kochi–682024, Kerala, India

S. Mohanty, PhD
Scientist, ICAR-Directorate of Water Management, Bhubaneswar-751023, Odisha, India

Miguel A. Muñoz-Muñoz, PhD
Ex-President of University of Puerto Rico, University of Puerto Rico, Mayaguez Campus, College of Agriculture Sciences, Call Box 9000, Mayagüez, PR. 00681–9000. Tel. 787–265–3871, E-mail: miguel.munoz3@upr.edu

B. J. Pandian, PhD
Director and Incharge, Water Technology Centre, Tamil Nadu Agricultural University, Coimbatore – 641003. Tel.: 0422–6611278, 6611478 and Mobile: +91–9443286711; E-mail: directorwtc@tnau.ac.in

Balram Panigrahi, PhD
Professor and Head, Department of Soil and Water Conservation Engineering, College of Agricultural Engineering and Technology, Orissa University of Agriculture & Technology, Bhubaneswar-751003, Odisha, India. Mobile: +91–9437882699, E-mail: kajal_bp@yahoo.co.in

P. Panigrahi, PhD
Scientist, ICAR-Directorate of Water Management, Bhubaneswar–751023, Odisha, India. E-mail: pravukalyan.panigrahi@gmail.com; pra73_nag@yahoo.co.in

G. Parthasarathi, PhD
Research Scholar, Department of Agricultural Economics, Tamil Nadu Agricultural University, Coimbatore, Tamil Nadu – 641 003

Subramanian Senthilnathan, PhD
Professor, Water Technology Centre, Tamil Nadu Agricultural University (TNAU), Coimbatore, Tamil Nadu – 641 003. E-mail: senthildemic@gmail.com

R. K. Sharma, PhD
Scientist, ICAR-Water Technology Centre, IARI, New Delhi–110012, India

A. K. Srivastava, PhD
Scientist, ICAR-National Research Centre for Citrus, Nagpur–440010, Maharashtra, India

L. Suresh, PhD
Research Scholar, Department of Management Studies, Periyar Maniammai University, Thanjavur, Tamilnadu, India

D. V. Tathod, PhD
Assist Prof., College of Agricultural Engineering and Tech., Warwat Road, Jalgaon Jamod, Dist. Buldhana, Maharashtra, India. E-mail: dnyanutathod@gmail.com, Tel. (0257) 2200829; Mobile: 9604818220

Jadhao Pandit Vishwanath
M.Tech., Plant Engineer, SPP – Maharashtra State Seeds Corporation Ltd., At Post – Dahatonda, Tal – Murtizapur, Dist. Akola (M.S.) – 444107. Mobile: +91–8275343885, 7588608593. E-mail: pvjadhao@gmail.com

Mohammed Waseem
M.Tech., Project Engineer, Irrigation Company Saudi Arabia, E-mail: waseem2075@gmail.com

Khemaies Zayani, PhD
Professor, High Institute of Environmental Sciences and Technology of Borj Cedria, BP 1003, Hammam Lif 2050, Univ. of Carthage, Tunisia, E-mail: khemaies.zayani@isste.rnu.tn

LIST OF ABBREVIATIONS

@	at the rate of
%	percent
ANOVA	analysis of variance
ASAE	American Society of Agricultural Engineers
BAW	best available water
BBF	broad bed furrow
BCR	benefit-cost ratio
BD	bulk density
BI	basin irrigation
BPF	band placement of fertilizer
CAET	College of Agriculture Engineering and Technology
CD	critical difference
cm	centimeter
cm hr^{-1}	centimeter per hour
cm^{-2}	square centimeter
COPE	cumulative open pan evaporation
CPE	cumulative pan evaporation
CRI	crown root initiation
CU	coefficient of uniformity
CUC	Christiansen's uniformity coefficient
CV	coefficient of variation
CV_m	manufacturers' coefficient of emitter variation
DAS	days after sowing
DCF	discounted cash flow
DI	deficit irrigation
DMRT	Duncan multiple range test
DPR	delivery performance ratio
dS/m	deciSiemens per meter
DSW	distillery spent wash
DU	distribution uniformity
E	evaporation
E.C.	electrical conductivity
e_a	application efficiency
EC	electrical conductivity
e_d	distribution efficiency

ET	evapotranspiration
ETc	crop evapotranspiration
EU	emission uniformity
F	flowering
FAO	Food and Agriculture Organization
FC	field capacity
FI	full irrigation
FUE	fertilizer use efficiency
FYM	farm yard manure
gm cc^{-1}	gram per cubic centimeter
gm l^{-1}	gram per liter
ha	hectare
ha cm	hectare centimeter
hp	horse power
hr	hour
ID	irrigation depth
iDE	International Development Enterprises
IRR	internal rate of return
IW	irrigation water
IWUE	irrigation water use efficiency
kg	kilogram
kg cm^{-1}	kilogram per centimeter
kg ha^{-1}	kilogram per hectare
kg ha^{-1} cm^{-1}	kilogram per hectare per centimeter
kg ha^{-1} mm^{-1}	kilogram per hectare per millimeter
kg m^{-3}	kilogram per cubic meter
kgf cm^{-2}	kilogram force per square centimeter
km hr^{-1}	kilometer per hectare
K_p	pan coefficient
Kpa	kilopascal
LAI	leaf area index
LLDPE	linear low density polyethylene
lpd	liters per day
lph	liters per hour
LWC	leaf water concentration
LWUE	leaf water use efficiency
m	meter
M ha	million hectare
m^2	square meter
MAD	management allowable depletion or deficit
MAI	moisture availability index
Mg ha^{-1}	million gram per hectare

MI	micro irrigation
MIS	micro irrigation system
mm	millimeter
MP	Madhya Pradesh
MS	Maharashtra State
MSI	micro sprinkler irrigation
MSL	mean sea level
Mt	metric ton
NPK	nitrogen, phosphorus and potassium
NPV	net present value
O&M	operation and maintenance
P.D.K.V.	Punjabrao Deshmukh Krishi Vidyapeeth
PD	pressure differential method
PRD	partial root-zone drying
PTI	prosperity through innovation
PVC	poly vinyl chloride
PWP	permanent wilting point
q ha^{-1}	quintals per hectare
RBD	randomized block design
RDF	recommended dose of fertilizer
RLWC	relative leaf water content
RPM	revolutions per minute
Rs	rupees
Rs ha^{-1}	rupees per hectare
RSC	residual sodium carbonate
SAR_{iw}	sodium adsorption ratio
SDI	subsurface drip irrigation
SE	standard error
SEM	standard error mean
SMC	soil moisture content
SPD	split plot design
T	transpiration
t ha^{-1}	tons per hectare
TAW	total available water
TDM	total dry matter
TN IAMWARM	Tamil Nadu Irrigated Agriculture Modernization and Water Bodies Restoration and Management
TSS	total soluble solid
TSSRS	reducing sugar
UC	uniformity coefficient
UCS	statistical uniformity
VC_p	pressure variation coefficient
WUE	water use efficiency

LIST OF SYMBOLS

Symbol	Description
$(e_s - e_a)$	saturation vapor pressure deficit (kPa)
/	per
%	percent
°C	degree celsius
| |	symbol for absolute value of quantity between the bars
å	symbol for summation
A	area to be fertilized (ha)
a and b	the width and depth of flow path of emitters in meters
A_s	sectional area of the emitter channel
A_s	soil bulk density of soil, g/cm^3
B	extent of area covered by foliage, fraction
B:C	benvi
efit cost ratio	
B_t	the benefits in the year t
c	maximum leaf/stem water potential measured during the study
C	concentration of nutrient in fertilizer (kg/l)
Ca	calcium
C_f	concentration of fertilizer in irrigation water in mg/liter
cm	centimeter
C_t	cost in the year t
Cu	coefficient of uniformity
CU	consumptive water use
CV	discharge coefficient of variation (%)
CV(h)	coefficient of variation of emitter flow caused by the hydraulic design (dimensionless)
CV(m)	coefficient of manufacturers variation of emitter flow (dimensionless)
D	mean tree canopy spread diameter (North-South and East-West) in meter
d	net amount of water to be applied during irrigation, cm
d_e	equivalent diameter which is equal to four times of the hydraulic radius r_h (m).
D_i	binary adoption variable (1=redeemed voucher for drip or pump equipment)
d_i	obtained drip kit with voucher (1=obtained drip kit, 0=otherwise)
d_s	effective root zone depth, cm

D_s	supplied water depth (mm)
ds/m	deciSiemens per meter
E	evaporation
E	efficiency of micro sprinkler irrigation system
e_a	actual vapor pressure (kPa)
e_a	application efficiency (%)
E_i	irrigation efficiency of drip system (90%)
Ep	cumulative pan evaporation for two consecutive days (mm)
e_s	saturation vapor pressure (kPa)
ET_o	reference evapotranspiration (mm day^{-1})
E_{UK}	emission uniformity
FC	field capacity (v/v, %)
F_r	fertilizer rate (kg/ha)
F_r	fertilizer rate per application (kg/ha)
G	soil heat flux density (MJ m^{-2} day^{-1})
H	tree canopy height (difference between tree height and stem height, meter)
h [L]	soil suction head
ha	hectare
h_e	emitter inlet pressure head (m)
hi	lateral line pressure
h_i	emitter outlet pressure head (m)
h^{max}_{req}	maximum required emitter inlet pressure head
I_0	no irrigation
I_1	irrigation at IW/CPE = 0.6
I_2	irrigation at IW/CPE = 0.8
I_3	irrigation at IW/CPE = 1.0
I_4	irrigation at IW/CPE = 1.2
I_c	irrigation at Control
IW	irrigation water, (mm)
K	potassium
k	the discharge coefficient
k_c	crop coefficient
Kc	crop factor
Kc	crop factor
kg/ha	kilogram per hectare
Kp	pan factor
M	mean application
M_{bi}	moisture content before irrigation, %
M_{fc}	moisture content at field capacity, %
Mm/day	millimeter per day

MRP_L	marginal revenue product of labor curve
N	nitrogen
n	number of days in the interval
N_c	nutrient concentration (kg/liter) in the stock solution
PAN-E	cumulative evaporation from US weather Bureau Class A pan less rain since previous irrigation
P_{ave}	the mean hydraulic pressure
p_i	obtained pump with voucher (1=obtained pump, 0=otherwise)
q	emitter discharge (l/h)
Q	quantity of water required per plant, liters;
q/ha	quintal per hectare, 1 q = 100 kg
Q_f	quantity of fertilizer to be injected (lph)
Q_{Fc}	field capacity (%)
qi	emitter discharge
q_{max}	maximum flow rate
q_{min}	minimum emitter flow rate
Q_{pwp}	permanent wilting point
q_{var}	emitter flow variation
R	number of replications
r	the rate of interest
R_e	Reynolds number
R_f	width of wetted strip (cm) measured on soil surface
R_{Max}	maximum width of the wetted area on the soil surface
R_n	net radiation at the crop surface (MJ m^{-2} day $^{-1}$)
Rs	Rupees
RSM	required soil water level 23.9 (v/v, %)
S_d	standard deviation of discharge rates of sample emitters (lph)
S_L	supply curve of labor
Sq	standard deviation
S_ψ	water stress integral (MPa day)
t	time period
T	time of injection (hr)
t/ha	tons per hectare
u_2	wind speed at 2.0 m height (ms^{-1})
u_i	error term
Us	statistical uniformity
U_{sh}	the statistical uniformity of emitter discharge rate due to pressure head
v	flow velocity
V	volume of water (liter/tree/day)
Vhs	hydraulic design coefficient of variation
V_{id}	irrigation volume (liter/tree) applied in each irrigation

viz. namely
V_{pc} tree canopy volume (m^3)
Vpf emitter performance coefficient of variation
Vqh emitter discharge coefficient of variation due to pressure head
Vqs emitter discharge coefficient of variation
W net amount of irrigation water applied during the injection period (mm)
W_f water delivered to the field in liters
Wp wetting factor
W_s water stored in root zone in liters
x emitter discharge exponent
Xi individual application amount
x_{i1} income from nonirrigated crop sources (natural log)
x_{i10} farmer crop type (1=tomatoes or rapeseed, 0=other)
x_{i11} herbicide input use (1=yes, 0=no)
x_{i12} mulch use (1=yes, 0=no)
x_{i13} land quality (1=good, 0=fair/poor)
x_{i14} land topography (1=flat, 0=gentle/steep slope)
x_{i15} farmer irrigates with pump or drip but did not redeem voucher
x_{i16} distance to nearest market for selling outputs
x_{i17} region (1=Kafue, 0=Kabwe)
x_{i2} irrigated land area (average 2008 and 2009)
x_{i3} total land area household has rights to (hectares)
x_{i4} household head's age (years)
x_{i5} household head's sex (1=male, 0=female)
x_{i6} household head's educational attainment (years of schooling)
x_{i7} size of household (number of members)
x_{i8} size of household squared
x_{i9} farmer agricultural knowledge index
Y economic yield (kg/ha)
Y economic yield
Y_i change in value of household assets
Z_d[L] and J_L[L] emitter burial depth and head loss along the lateral, respectively
Z_f the maximum wetted soil depth (cm)
z_{i1} household head's age (years)
z_{i10} attended an agricultural fair (1=yes, 0=no)
z_{i11} neighbor or friend using irrigation equipment (1=yes, 0=no)
z_{i12} land tenure status (1=land title or traditional ownership rights, 0=use rights but not ownership rights)
z_{i13} land quality (1=good, 0=fair/poor)
z_{i14} land topography (1=flat, 0=gentle / steep slope)
z_{i15} distance to market for buying inputs (km)

z_{i16} region dummy (1=Kafue, 0=Kabwe)
z_{i2} household head's sex (1=male, 0=female)
z_{i3} household head's education level attainment (years of schooling)
z_{i4} household size (number of members)
z_{i5} farmer's agricultural knowledge (an index)
z_{i6} number of technologies already adopted by household
z_{i7} household income in 2007 (logarithm)
z_{i8} distance to water source for irrigation
z_{i9} number of irrigation meetings and extension events attended
Z_{Max} the maximum wetting front depth
Zn/ha zinc per hectare
Z_r rooted soil depth (mm)
Z_r effective root zone depth, (m)

GREEK SYMBOLS

γ bulk density, (gm/cm^3)
θ $[L^3L^{-3}]$ the volumetric water content
θ_{BI} moisture content before irrigation, (%)
θ_i and θ_c initial and at field capacity soil moisture
θ_s saturated soil water content
μ fluid viscosity coefficient (kg/(m.s))
μ-mol/g micro molecular per gram
ρ fluid density (kg/m^3)
$\psi_{i, i+1}$ average midday leaf/stem water potential for any interval i and i+1 (MPa)

PREFACE

Due to increased agricultural production, irrigated land has increased in the arid and subhumid zones around the world. Agriculture has started to compete for water use with industries, municipalities, and other sectors. This increasing demand along with increases in water and energy costs have made it necessary to develop new technologies for the adequate management of water. The intelligent use of water for crops requires understanding of evapotranspiration processes and use of efficient irrigation methods.

Micro irrigation is sustainable and is one of the best management practices. I attended 17th Punjab Science Congress on February 14–16, 2014, at Punjab Technical University in Jalandhar, India. I was shocked to know that the underground water table has lowered to a critical level in Punjab. My father-in-law in Dhuri told me that his family bought 0.10 acres of land in the city for US $100.00 in 1942 because the water table was at 2 feet depth. In 2012, it was sold for US $233,800 because the water table had dropped to greater than 100 feet. This has been due to luxury use of water by wheat-paddy farmers, he said. The water crisis is similar in other countries, including Puerto Rico, where I live. We can, therefore, conclude that the problem of water scarcity is rampant globally, creating the urgent need for water conservation. The use of micro irrigation systems is expected to result in water savings and increased crop yields in terms of volume and quality.

Our planet will not have enough potable water for a population of >10 billion persons in 2115. The situation will be further complicated by multiple factors that will be adversely affected by global warming. The United Nations states that

> Water scarcity already affects every continent. Around 1.2 billion people, or almost one-fifth of the world's population, live in areas of physical scarcity, and 500 million people are approaching this situation. Another 1.6 billion people, or almost one quarter of the world's population, face economic water shortage (where countries lack the necessary infrastructure to take water from rivers and aquifers).
>
> Water scarcity is among the main problems to be faced by many societies and the World in the XXIst century. Water use has been growing at more than twice the rate of population increase in the last century, and, although there is no global water scarcity as such, an increasing number of regions are chronically short of water.
>
> Water scarcity is both a natural and a human-made phenomenon. There is enough freshwater on the planet for seven billion people but it is distributed unevenly and too much of it is wasted, polluted and unsustainably managed. (http://www.un.org/waterforlifedecade/scarcity.shtml)

Every day there news on the importance of micro irrigation appear all around the world indicating that government agencies at central/state/local level, research and educational institutions, industry, sellers, and others are aware of the urgent need to adopt micro irrigation technology that can have an irrigation efficiency of up to 90% compared to 30–40% for the conventional irrigation systems.

It is important to adopt a suitable drip irrigation system to grow agricultural crops, space plants, forest trees, landscape plants and shrubs, and garden plants because all vegetation require different water intake. For better results, one should plan and install proper irrigation systems for the land under consideration. Micro irrigation is one of the most efficient watering methods, as it can save water and give better quality of products. The trickle irrigation system can be designed and adapted to varying irrigation needs for arid, semiarid and humid regions; a wide range of crops; and different climatic and soil conditions. Drip irrigation can save our planet from the water scarcity.

The trickle irrigation design must be carried out by a professionally registered engineer who is qualified and has the necessary knowledge. This is not job for a layperson. Investment in the design phase will pay off in the long run. In November of 1979, a hydraulic technician came to my office and tried to convince me that he could design a drip irrigation system better than the engineer. One of his systems at a 500-hectare vegetable farm in Puerto Rico failed during the first crop. I helped to save this farm from total failure. We had to do the necessary modifications to the existing design and replace the necessary parts. I recommend 100% to consult an engineer to design the drip irrigation system.

Drip Irrigation Zone (http://www.dripirrigationzone.com/drip-irrigation-systems/#installing) indicates that "design is an important aspect of a drip system because the irrigation scheduling depends on it. While designing the system, keep in mind different plants with different watering needs. Also take the soil and slopes into consider-ation. Suppose the field has a heavy clay soil, one may require setting up a system with high water pressure. According to the irrigation needs in a particular situation, one should choose emitters and make sure where the lines, laterals, accessories and connectors should be best placed." According to this website, there are variations in a drip irrigation system. They have listed various types of drip systems to meet different watering requirements, such as: Toro drip irrigation system, gravity drip irrigation system, rain bird drip irrigation system, lawn drip irrigation system, automatic drip irrigation system, farm drip irrigation system, in-door drip irrigation system, dig drip irrigation system, mini-sprinkler irrigation system, drip irrigation fertilization, seasonal drip irrigation, eco drip irrigation, and Jain drip irrigation.

Jain Irrigation Systems, Ltd. (http://www.jaindrip.com/Designtechnical/design.htm) indicates that "an irrigation system is a sophisticated and complex one, in which each component plays a very important role. The reliability and efficiency of an irrigation system is a function of superior product, proper and skilled services, and professional design. Our design department has a staff of skilled engineers with experience in every aspect of irrigation systems, beginning with water source pumping and delivery systems right up to the design of the system. Our professional and cost effective design helped us to bring thousands of acre area under micro irrigation throughout India and abroad."

Micro irrigation, also known as trickle irrigation or drip irrigation or localized irrigation or high frequency or pressurized irrigation, is an irrigation method that saves water and fertilizer by allowing water to drip slowly to the roots of plants, either onto the soil surface or directly onto the root zone, through a network of valves, pipes, tubing, and emitters. It is done through narrow tubes that deliver water directly to the base of the plant. It is a system of crop irrigation involving the controlled delivery of water directly to individual plants and can be installed on the soil surface or subsurface.

The other important benefits of using micro irrigation systems include expansion in the area under irrigation, water conservation, optimum use of fertilizers and chemicals through water, and decreased labor costs, among others. The worldwide population is increasing at a rapid rate and it is imperative that food supply keeps pace with this increasing population. Micro irrigation systems are often used in farms and large gardens, but are equally effective in the home garden or even for houseplants or lawns. They are easily customizable and can be set up even by inexperienced gardeners. Putting a drip system into the garden is a great do-it-yourself project that will ultimately save the time and help the plants grow. It is equally used in landscaping and in green cities.

The mission of this book volume is to serve as a reference manual for graduate and under graduate students of agricultural, biological and civil engineering, and horticulture, soil science, crop science and agronomy. I hope that it will be a valuable reference for professionals who work with micro irrigation and water management; for professional training institutes, for technical agricultural centers, for irrigation centers, for agricultural extension services, and for other agencies that work with micro irrigation programs. I cannot guarantee the information in this book will be enough for all situations. One must consult an irrigation engineer for an optimum design.

After my first textbook, *Drip/Trickle or Micro Irrigation Management*, published by Apple Academic Press Inc., and response from international readers, AAP has published for the world community the ten-volume series on Research Advances in Sustainable Micro Irrigation, of which I am the editor.

This new book, *Sustainable Micro Irrigation Design Systems for Agricultural Crops: Methods and Practices,* is volume two of a new book series, Innovations and Challenges in Micro Irrigation. This volume is unique because it is complete and simple, a one-stop manual, with worldwide applicability to irrigation management in agriculture. This series will be a must for those interested in irrigation planning and management, namely, researchers, scientists, educators, and students. My longtime colleague, Dr. Pravukalyan Panigrahi, Scientist at Directorate of Water Management, Bhubaneswar-Odisha-India, joins me as a co-editor of this edition. His contribution to the quality of this book has been invaluable.

Volume one of the series is *Principles and Management of Clogging in Micro Irrigation,* is edited by me along with my colleagues Vishal K. Chavan, and Vinod K. Tripathi.

The contributions by all cooperating authors to this book have been most valuable in this compilation. Their names are mentioned in each chapter and in the list of contributors. This book would not have been written without the valuable cooperation of these investigators, many of whom are renowned scientists who have worked in the field of micro irrigation throughout their professional careers.

I would like to thank editorial staff, Sandy Jones Sickels, Vice President, and Ashish Kumar, Publisher and President at Apple Academic Press, Inc., for making every effort to publish the book when the diminishing water resources are a major issue worldwide. Special thanks are due to the AAP Production Staff for the quality production of this book.

We request readers offer us your constructive suggestions that may help to improve future works.

I express my deep admiration to my family for understanding and collaboration during the preparation of this book, especially my wife Subhadra Devi Goyal. With my whole heart and best affection, I dedicate this book to Dr. A.M. Michael, who taught me first undergraduate course on irrigation engineering at Punjab Agricultural University, Ludhiana, Punjab, India. One of his textbooks, titled *Irrigation: Theory and Practice,* published by Vikas Publishing House Pvt. Ltd, India, is widely used in Asia. He was my guru who taught me to love irrigation and asked me to never play a trick to the irrigation design, as it is a serious responsibility of the engineer. He also told me in one of our informal meeting that the engineer "who never sells his profession" is married for life to the engineering profession. He helped me to inherit many ethical and professional qualities. I am a mirror image of his humble and honest personality, though I will never be equal to his status. Without his advice and patience, I would not have been a "Father of Irrigation Engineering of twentieth century in Puerto Rico," with zeal for service to others. My salute to him for his irrigation legacy in India. As an educator, I offer this advice to one and all in the world: "Permit that our Almighty God, our Creator and excellent Teacher, irrigate the life with His Grace of rain trickle by trickle, because our life must continue trickling on…"

— Megh R. Goyal, PhD, PE, Senior Editor-in-Chief
August 01, 2015

FOREWORD 1

With only a small portion of cultivated area under irrigation and with the need to expand this area, which can be brought about by irrigation, it is clear that the most critical input for agriculture today is water. It is important that all available supplies of water should be used intelligently to the best possible advantage. Recent research around the world has shown that the yields per unit quantity of water can be increased if the fields are properly leveled, the water requirements of the crops as well as the characteristics of the soil are known, and the correct methods of irrigation are followed. Significant gains can also be made if the cropping patterns are changed so as to minimize storage during the hot summer months when evaporation losses are high, if seepage losses during conveyance are reduced, and if water is applied at critical times when it is most useful for plant growth.

Irrigation is mentioned in the Holy Bible and in the old documents of Syria, Persia, India, China, Java, and Italy. The importance of irrigation in our times has been defined appropriately by N. D. Gulati: "In many countries irrigation is an old art, as much as the civilization, but for humanity it is a science, the one to survive." The need for additional food for the world's population has spurred rapid development of irrigated land throughout the world. Vitally important in arid regions, irrigation is also an important improvement in many circumstances in humid regions. Unfortunately, often less than half the water applied is used by the crop—irrigation water may be lost through runoff, which may also cause damaging soil erosion, deep percolation beyond that required for leaching to maintain a favorable salt balance. New irrigation systems, design and selection techniques are continually being developed and examined in an effort to obtain high practically attainable efficiency of water application.

The main objective of irrigation is to provide plants with sufficient water to prevent stress that may reduce the yield. The frequency and quantity of water depends upon local climatic conditions, crop and stage of growth, and soil-moisture-plant characteristics. The need for irrigation can be determined in several ways that do not require knowledge of evapotranspiration (ET) rates. One way is to observe crop indicators such as change of color or leaf angle, but this information may appear too late to avoid reduction in the crop yield or quality. Other similar methods of scheduling include determination of the plant water stress, soil moisture status, or soil water potential. Methods of estimating crop water requirements using ET and combined with soil characteristics have the advantage of not only being useful in determining when to irrigate, but also enables us to know the quantity of water needed. ET estimates have not been made for the developing countries though basic information on weather data is available. This has contributed to one of the existing problems that the vegetable crops are over irrigated and tree crops are under irrigated.

Water supply in the world is dwindling because of luxury use of sources; competition for domestic, municipal, and industrial demands; declining water quality; and losses through seepage, runoff, and evaporation. Water rather than land is one of the limiting factors in our goal for self-sufficiency in agriculture. Intelligent use of water will avoid problem of seawater seeping into aquifers. Introduction of new irrigation methods has encouraged marginal farmers to adopt these methods without taking into consideration economic benefits of conventional, overhead, and drip irrigation systems. What is important is "net in the pocket" under limited available resources. Irrigation of crops in tropics requires appropriately tailored working principles for the effective use of all resources peculiar to the local conditions. Irrigation methods include border-, furrow-, subsurface-, sprinkler-, sprinkler, micro, and drip/trickle, and xylem irrigation.

Drip irrigation is an application of water in combination with fertilizers within the vicinity of plant root in predetermined quantities at a specified time interval. The application of water is by means of drippers, which are located at desired spacing on a lateral line. The emitted water moves due to an unsaturated soil. Thus, favorable conditions of soil moisture in the root zone are maintained. This causes an optimum development of the crop. Drip/micro or trickle irrigation is convenient for vineyards, tree orchards, and row crops. The principal limitation is the high initial cost of the system that can be very high for crops with very narrow planting distances. Forage crops may not be irrigated economically with drip irrigation. Drip irrigation is adaptable for almost all soils. In very fine textured soils, the intensity of water application can cause problems of aeration. In heavy soils, the lateral movement of the water is limited, thus more emitters per plant are needed to wet the desired area. With adequate design, use of pressure compensating drippers and pressure regulating valves, drip irrigation can be adapted to almost any topography. In some areas, drip irrigation is used successfully on steep slopes. In subsurface drip irrigation, laterals with drippers are buried at about 45 cm depth, with an objective to avoid the costs of transportation, installation, and dismantling of the system at the end of a crop. When it is located permanently, it does not harm the crop and solve the problem of installation and annual or periodic movement of the laterals. A carefully installed system can last for about 10 years.

The publication of this book series is an indication that things are beginning to change, that we are beginning to realize the importance of water conservation to minimize the hunger. It is hoped that the publisher will produce similar materials in other languages.

In providing this book series on micro irrigation, Megh Raj Goyal, as well as the Apple Academic Press, is rendering an important service to the farmers. Dr. Goyal, "Father of Irrigation Engineering in Puerto Rico," has done an unselfish job in the presentation of this series that is simple and thorough. I have known Megh Raj since

1973 when we were working together at Haryana Agricultural University on an ICAR research project in "Cotton Mechanization in India."

Dr. Gajendra Singh, PhD
New Delhi,
August 1, 2015

Dr. Gajendra Singh, PhD,
Former Vice Chancellor, Doon University, Dehradun, India, and
Adjunct Professor, Indian Agricultural Research Institute, New Delhi.
Ex-President (2010–2012), Indian Society of Agricultural Engineers.
Former Deputy Director General (Engineering), Indian Council of Agricultural Research (ICAR), New Delhi. Former Vice-President/Dean/Professor and Chairman, Asian Institute of Technology, Thailand

FOREWORD 2

Water is becoming increasingly a scarce resource and is limiting agricultural development in many developing and developed economies across the world. Developing infrastructure for the water resources development, conservation and management have been the common policy agenda in many economies. The water use efficiency in the agricultural sector, which still consumes over 80% of water, is only in the range of 30–40% in India, indicating that there is considerable scope for improving the existing water use efficiency. Moreover, increasing dependence on groundwater in many regions has resulted in negative externalities such as over pumping, changes in crop pattern towards more water intensive crops, well deepening, increase in well investments, pumping costs, well failure and abandonment and out migration which are increasing at a much faster rate. Deepening of existing wells and drilling new bore wells led to further decline in water table and deterioration of water quality. Similarly, productivity of crops is not comparable with other developed and even in some developing countries due to over irrigation. Therefore it is necessary to efficiently utilize water to bring additional area under irrigation so as to reduce the cost of irrigation and increase the productivity per unit area and unit quantum of water. Micro irrigation, particularly drip and sprinkler irrigation, is followed in many developed countries like USA, Austria, Germany, Israel, and Great Britain. It is in this context, the present book by Dr. Megh R. Goyal on *Sustainable Micro Irrigation Design Systems for Agricultural Crops: Methods and Practices* assumes a critical and timely one.

The book has chapters from eminent irrigation specialists across the world. The book has two major parts. The first part focuses on practices in micro irrigation. This covers the major issues of principles and theories of micro irrigation, details of technology, and evolution of micro irrigation technology. The second part deals with design methods of micro irrigation, more specifically micro irrigation. It covers different design methods for various crops.

I sincerely hope that this book is certainly a breakthrough in the field of irrigation. This book would be very useful for researchers, scholars, and development personnel, commercial firms dealing with micro irrigation equipment, nongovernment organizations, and policymakers.

D. Suresh Kumar, PhD
August 1, 2015

FOREWORD 3

Irrigation has been a vital resource in farming since the evolution of humans. Due importance to be given for irrigation was not accorded to the fact that the availability has been persistent in the past. Sustained availability of water cannot be possible in future and there are several reports across the globe where severe water scarcity might hamper farm production. Hence, in modern day farming, the most limiting input being water, much importance should be given to conservation and the judicious use of the irrigation water for sustaining the productivity of food and other cash crops. Though the availability of information on micro irrigation is adequate, its application strategies must be expanded for the larger benefit of the water-saving technology by users.

In this context under Indian conditions, the attempt made by Prof. R. K. Sivanappan, Former Dean, Agricultural Engineering College of TNAU, in collating all pertinent particulars and assembling them in the form a precious publication proves that the author is continuing his eminent service and supporting the farming community by way of empowering them in adopting the micro irrigation technologies at ease and the personnel involved in irrigation are also enriched by the knowledge on modern irrigation concepts. While seeking the blessings of Dr. R. K. Sivanappan and Dr. Megh Raj Goyal (editor of this book), I wish the publisher and authors success in all their endeavors for helping the users of micro irrigation.

B. J. Pandian, PhD

WARNING/DISCLAIMER

The goal of this compendium, *Sustainable Micro Irrigation Design Systems for Agricultural Crops: Methods and Practices,* is to guide the world community on how to manage efficiently for eco-nomical crop production. The reader must be aware that dedication, commitment, honesty, and sincerity are most important factors in a dynamic manner for complete success. This reference is not intended for a one-time reading; we advise you to consult it frequently. To err is human. However, we must do our best. Always, there is a place for learning new experiences.

The editor, the contributing authors, the publisher, and the printer have made every effort to make this book as complete and as accurate as possible. However, there still may be grammatical errors or mistakes in the content or typography. Therefore, the contents in this book should be considered as a general guide and not a complete solution to address any specific situation in irrigation. For example, one size of irrigation pump does not fit all sizes of agricultural land and work for all crops.

The editor, the contributing authors, the publisher, and the printer shall have neither liability nor responsibility to any person, organization, or entity with respect to any loss or damage caused, or alleged to have caused, directly or indirectly, by information or advice contained in this book. Therefore, the purchaser/reader must assume full responsibility for the use of the book or the information therein.

The mention of commercial brands and trade names are only for technical purposes and does not imply endorsement. The editor, contributing authors, educational institutions, and the publisher do not have any preference for a particular product.

All web links that are mentioned in this book were active on December 31, 2014. The editors, the contributing authors, the publisher, and the printing company shall have neither liability nor responsibility if any of the web links are inactive at the time of reading of this book.

OTHER BOOKS ON MICRO IRRIGATION TECHNOLOGY FROM AAP

Management of Drip/Trickle or Micro Irrigation
Megh R. Goyal, PhD, PE, Senior Editor-in-Chief

Evapotranspiration: Principles and Applications for Water Management
Megh R. Goyal, PhD, PE, and Eric W. Harmsen, Editors

BOOK SERIES: RESEARCH ADVANCES IN SUSTAINABLE MICRO IRRIGATION

Senior Editor-in-Chief: Megh R. Goyal, PhD, PE

Volume 1: Sustainable Micro Irrigation: Principles and Practices
Senior Editor-in-Chief: Megh R. Goyal, PhD, PE

Volume 2: Sustainable Practices in Surface and Subsurface Micro Irrigation
Senior Editor-in-Chief: Megh R. Goyal, PhD, PE

Volume 3: Sustainable Micro Irrigation Management for Trees and Vines
Senior Editor-in-Chief: Megh R. Goyal, PhD, PE

Volume 4: Management, Performance, and Applications of Micro Irrigation
Senior Editor-in-Chief: Megh R. Goyal, PhD, PE

Volume 5: Applications of Furrow and Micro Irrigation in Arid and Semi-Arid Regions
Senior Editor-in-Chief: Megh R. Goyal, PhD, PE

Volume 6: Best Management Practices for Drip Irrigated Crops
Editors: Kamal Gurmit Singh, PhD, Megh R. Goyal, PhD, PE, and
Ramesh P. Rudra, PhD, PE

Volume 7: Closed Circuit Micro Irrigation Design: Theory and Applications
Senior Editor-in-Chief: Megh R. Goyal, PhD; Editor: Hani A. A. Mansour, PhD

Volume 8: Wastewater Management for Irrigation: Principles and Practices
Editor-in-Chief: Megh R. Goyal, PhD, PE; Coeditor: Vinod K. Tripathi, PhD

Volume 9: Water and Fertigation Management in Micro Irrigation
Senior Editor-in-Chief: Megh R. Goyal, PhD, PE

Volume 10: Innovations in Micro Irrigation Technology
Senior Editor-in-Chief: Megh R. Goyal, PhD, PE;
Coeditors: Vishal K. Chavan, MTech, and Vinod K. Tripathi, PhD

BOOK SERIES: INNOVATIONS AND CHALLENGES IN MICRO IRRIGATION

Senior Editor-in-Chief: Megh R. Goyal, PhD, PE

Volume 1: Principles and Management of Clogging in Micro Irrigation
Editors: Megh R. Goyal, PhD, PE, Vishal K. Chavan, and Vinod K. Tripathi

Volume 2: Sustainable Micro Irrigation Design Systems for Agricultural Crops: Methods and Practices
Editors: Megh R. Goyal, PhD, PE, and P. Panigrahi, PhD

ABOUT THE SERIES EDITOR-IN-CHIEF

Megh R. Goyal, PhD, PE, is a Retired Professor in Agricultural and Biomedical Engineering from the General Engineering Department in the College of Engineering at University of Puerto Rico–Mayaguez Campus; and Senior Acquisitions Editor and Senior Technical Editor-in-Chief in Agriculture and Biomedical Engineering for Apple Academic Press Inc. He received his BSc degree in engineering in 1971 from Punjab Agricultural University, Ludhiana, India; his MSc degree in 1977 and PhD degree in 1979 from the Ohio State University, Columbus; and his Master of Divinity degree in 2001 from Puerto Rico Evangelical Seminary, Hato Rey, Puerto Rico, USA. He spent one-year sabbatical leave in 2002–2003 at the Biomedical Engineering Department at Florida International University in Miami, Florida, USA. Since 1971, he has worked as Soil Conservation Inspector (1971); Research Assistant at Haryana Agricultural University (1972–75) and Ohio State University (1975–79); Research Agricultural Engineer/Professor at the Department of Agricultural Engineering of UPRM (1979–1997); and Professor in Agricultural and Biomedical Engineering in the General Engineering Department of UPRM (1997–2012).

He was first agricultural engineer to receive the professional license in Agricultural Engineering in 1986 from College of Engineers and Surveyors of Puerto Rico. On September 16, 2005, he was proclaimed as "Father of Irrigation Engineering in Puerto Rico for the twentieth century" by the ASABE, Puerto Rico Section, for his pioneer work on micro irrigation, evapotranspiration, agroclimatology, and soil and water engineering. During his professional career of 45 years, he has received awards such as Scientist of the Year, Blue Ribbon Extension Award, Research Paper Award, Nolan Mitchell Young Extension Worker Award, Agricultural Engineer of the Year, Citations by Mayors of Juana Diaz and Ponce, Membership Grand Prize for ASAE Campaign, Felix Castro Rodriguez Academic Excellence, Rashtrya Ratan Award and Bharat Excellence Award and Gold Medal, Domingo Marrero Navarro Prize, Adopted Son of Moca, Irrigation Protagonist of UPRM, and Man of Drip Irrigation by Mayor of Municipalities of Mayaguez/Caguas/Ponce and Senate/Sec-retary of Agriculture of ELA, Puerto Rico.

He has authored more than 200 journal articles and textbooks, including *Elements of Agroclimatology* (Spanish) by UNISARC, Colombia, and two *Bibliographies on Drip Irrigation.* Apple Academic Press Inc. (AAP) has published his books, namely *Biofluid Dynamics of Human Body, Management of Drip/Trickle or Micro Irrigation, Evapotranspiration: Principles and Applications for Water Management, Sustainable Micro Irrigation Design Systems for Agricultural Crops: Practices and Theory, Biomechanics of Artificial Organs and Prostheses*, and *Scientific and Technical Terms in Bioengineering and Biotechnology.* During 2014–15, AAP is publishing his ten-volume set, Research Advances in Sustainable Micro Irrigation.

Readers may contact him at goyalmegh@gmail.com.

ABOUT THE COEDITOR

Pravukalyan Panigrahi, PhD, is working as Senior Scientist at Indian Institute of Water Management (formerly Directorate of Water Management) under the Indian Council of Agricultural Research (ICAR), Bhubaneswar in Odisha, India. He obtained his BTech and MTech degrees in Agricultural Engineering in 1996 and 1999, respectively, from Orissa University of Agriculture and Technology (OUAT), Bhubaneswar, Odisha, India. He received his PhD in Water Science and Technology from Indian Agricultural Research Institute (IARI), New Delhi, India in 2012.

He has worked on water management aspects in vegetables and fruit crops, especially in citrus. He developed the methodology to enhance water use efficiency under surface irrigation and standardized the alternate furrow irrigation and mulching in different vegetables, including okra. He has worked to standardize water harvesting, micro irrigation and fertigation, and irrigation design and scheduling in citrus. He studied the response of deficit and partial root zone drying techniques in citrus under semiarid condition and developed a methodology to forecast fruit yield under differential water stress condition in citrus during his PhD research. Presently he is working on a runoff harvesting and recycling system, partial root zone drying in mango and micro irrigation in rice-based cropping sequence in subhumid tropical region. He is also teaching on watershed management and irrigation management in India.

He has worked as an expert in different consultancy projects related to water management. He has published more than 50 peer-reviewed research publications and bulletins, and has reviewed a number of research papers for different prestigious international journals. He has also attended and presented his works in different national and international conferences. Readers may contact him at pravukalyan.panigrahi@gmail.com; pra73_nag@yahoo.co.in.

BOOK REVIEWS

"I congratulate the editors on the completion and publication of this book volume on micro irrigation design. Water for food production is clearly one of the grand challenges of the twenty-first century. Hopefully this book will help irrigators and famers around the world to increase the adoption of water savings technology such as micro irrigation."

—Vincent F. Bralts, PhD, PE, Professor and Ex-Associate Dean, Agricultural and Biological Engineering Department, Purdue University, West Lafayette, Indiana

"This book is user-friendly and is a must for all irrigation planners to minimize the problem of water scarcity worldwide. The *Father of Irrigation Engineering in Puerto Rico of twenty-first century and pioneer on micro irrigation in the Latin America*, Dr. Goyal (my longtime colleague) has done an extraordinary job in the presentation of this book."

—Miguel A Muñoz, PhD, Ex-President of University of Puerto Rico; and Professor/ Soil Scientist

"I am moved by seeing the dedication of this textbook and recalling my association with Dr. Megh Raj Goyal while at Punjab Agricultural University in India. I congratulate him on his professional contributions and his distinction in irrigation. I believe that this innovative book will aid the irrigation fraternity throughout the world."

—A. M. Michael, PhD, Former Professor/Director, Water Technology Centre – IARI; Ex-Vice-Chancellor, Kerala Agricultural University, Trichur, Kerala, India

PART I

DESIGN METHODS IN DRIP/TRICKLE OR MICRO IRRIGATION

CHAPTER 1

DRIP IRRIGATION AND INDIGENOUS ALTERNATIVES FOR USE OF SALINE AND ALKALI WATERS IN INDIA: REVIEW

S. K. GUPTA

CONTENTS

1.1 INTRODUCTION

Drip irrigation, a method of applying water directly to the plant root zone, is characterized by water application at slow rates but relatively more frequently over the crop growing season as per crop water requirement. It is most suited to vegetables crops, fruits trees and vine crops although currently its application has extended over a wide range of row crops because of its water conservation potential and higher crop productivity. As such, drip irrigation has spread over an area of around 1.9 million ha (M ha), may be the highest in the world in term of area under drip irrigation [26]. Another major advantage of drip irrigation manifests from its potential to use saline water. The low matric stress in the root zone due to frequent applications enables the plants to overcome high osmotic stress encountered in saline water irrigation. In spite of the numerous evidences generated over the last 3 decades or so not much progress has been made in expanding the use of saline/alkali water with drip irrigation. Besides, high capital cost and knowledge required to address several complex issues in drip systems, prompted the technologists to develop local alternatives of drip irrigation. Several of them not only mimic the drip irrigation but are also relatively cheap and pose less complexity in their application by resource poor farmers.

This chapter highlights evidences: (i) on the use of saline/alkali water with drip irrigation; and (ii) to understand impediments and describe some of the potential local alternatives of drip irrigation to use saline/alkali water.

1.2 WATER QUALITY SCENE IN INDIA

Ground water has been characterized and classified in India on the basis of electrical conductivity (EC_{iw}), sodium adsorption ratio (SAR_{iw}), residual sodium carbonate (RSC) and individual ion concentrations of several elements. Considering the data generated on water quality and various management options required to manage these waters, a Committee of Experts classified the water into four main classes (Table 1.1). Saline and alkali water were further subgrouped into three classes each to underline the severity of the problems.

TABLE 1.1 Classification of Poor Quality Water

Water quality	EC_{iw} (dSm^{-1})	SAR_{iw} $(mmoll^{-1})^{1/2}$	RSC ($meql^{-1}$)
A. Good	<2	<10	<2.5
B. Saline			
i. Marginally saline	2–4	<10	<2.5
ii. Saline	>4	<10	<2.5
iii. High-SAR saline	>4	>10	<2.5

TABLE 1.1 *(Continued)*

Water quality	EC_{iw} (dSm^{-1})	SAR_{iw} (mmoll^{-1})$^{1/2}$	RSC (meql^{-1})
C. Alkali water			
i. Marginally alkali	<4	<10	2.5–4.0
ii. Alkali	<4	<10	>4.0
iii. High-SAR alkali	variable	>10	>4.0
D. Toxic Water	The toxic water has variable salinity, SAR and RSC but has excess of specific ions such as nitrate, boron, fluoride, chloride, sodium, silica or heavy metals such as selenium, cadmium, lead and arsenic, etc.		

Source: Ref. [38], Toxic waters added from Ref. [23].

Central Soil Salinity Research Institute, Karnal undertook a study to compile the information, what so ever was available, and prepared a map of ground water quality for irrigation on a 1:6 million scale [22]. The map has four legends namely, good water (EC_{iw} < 2 and SAR < 10), saline water (EC_{iw} > 2 and SAR < 10), high-SAR saline water (EC_{iw} > 4.0 and SAR > 10) and alkali water (EC_{iw} variable, SAR variable and RSC > 2.5). Notwithstanding the large spatial variations, high salinity ground waters are mostly encountered in arid parts of northwest states like Rajasthan, Haryana and Punjab. Associated with salinity, the ground waters in some pockets of these states may contain toxic levels of B, F, NO_3, Se and Si, etc. The alkali waters are found mainly in the semiarid parts of India where the annual rainfall is in the range of 500–700 mm. The total area underlain with the saline ground water (EC_{iw} > 4 dS m^{-1}) is approximated as 193,438 km^2 (Table 1.2) with annual replenishable recharge of 11,765 million m^3 yr^{-1} [14].

TABLE 1.2 Estimated Area (M-ha) Underlain With Saline Ground Water (EC_{iw}>4 dS m^{-1})

State	Total area of the state (M-ha)	Area underlain with saline ground water (M ha)
Rajasthan	34.2	14.10 (41.22)
Gujarat	19.6	2.43 (12.39)
Haryana	4.4	1.14 (25.90)
Karnataka	19.1	0.88 (4.60)
Tamil Nadu	13.0	0.33 (2.53)
Punjab	5.0	0.30 (6.00)
Uttar Pradesh	2.9	0.13 (4.48)
Delhi	0.14	0.01 (7.14)
Total	125.0	19.3 (15.44)

Note: Figures in brackets express percentage.
Source: Ref. [65] .

Over the last few decades, an interesting situation is emerging in India. While the water table is declining in areas underlain by fresh water due to overexploitation, areas underlain by saline/alkali waters are experiencing rising trend in the water table. It is resulting in problems of water logging and soil salinization. Such a situation calls for exploitation of poor quality saline/alkali waters in crop production programs to turn this liability into an opportunity.

1.3 USE OF SALINE/ALKALI WATER IN CROP PRODUCTION

A large number of studies have been conducted in India to establish the potential of poor quality waters in crop production programs under various agro-climatic conditions. The generic management options that have been tried and adopted under field conditions are categorized under five groups namely crop management, soil management, irrigation water management, chemical management and rainwater management (Table 1.3) [23]. One of the options under irrigation water management is switchover to improved irrigation techniques under which sprinkler, drip and few local alternatives of drip have been tried to establish their potential to use saline/alkali water. This issue is the subject matter of this chapter.

TABLE 1.3 Management of Saline/Alkali Water [23]

Group	Technology
Crop management	Crop selection
	Exploitation of varietal differences
	Growth stage sensitivity
	Agronomic practices
Soil management	Soil specific management
	Land forming/seeding
	Fallowing
Irrigation water management	Shallow depth-high frequency irrigation
	Conjunctive use of multisource waters
	Pre-sowing irrigation
	Post sowing irrigation
	Switchover to improved irrigation techniques
	Leaching
Chemical management	Application of chemical amendments
	Application of additional nutrients
Rainwater management	*In-situ* rainwater conservation
	Rainwater harvesting and reuse

1.4 DRIP IRRIGATION FOR WATER CONSERVATION

The drip irrigation system has become quite popular in areas of acute water scarcity and places where commercial cultivation mainly of cash or horticultural crops is practiced. In line with the global experiences, numerous studies conducted in different parts of India have documented the benefits of drip irrigation in terms of increased productivity on different crops and also quantified the benefits in terms of water saving [11, 12, 15, 19–21, 34, 46, 49, 51–55, 57, 61, 68, 76, 78, 79, 81–84, 86, 87]. In order to have a realistic figure out of so many multilocation trials, Saxena and Gupta [62] compiled and prepared a new table documenting the average benefits in yield increase and water saving by taking the arithmetic average of all the reported values of a given crop (Table 1.4). Among the top 12 crops that responded relatively higher increase in yield under drip irrigation are sweet lime, carrot, beans, mango, turmeric, popcorn, baby corn, papaya, capsicum, chickpea and watermelon (Table 1.4). On the other hand, chili, coconut, radish, ridge gourd, tomato, guava, cabbage, banana, potato, beetroot and mango gave higher water use efficiency. High water saving was observed among beetroot, bitter gourd, sweet potato, papaya, radish, sweet lime, sweet lime, pomegranate, turmeric, cotton, coconut and acid lime crops.

TABLE 1.4 Average Crop Yield, Percentage Increase in Yield, Water Use Efficiency and Water Saving in Drip Over the Conventional Irrigation System for Various Crops

Sr. No.	Crop	Number of References	Yield (t ha^{-1})	Yield increase (%)	WUE (t ha^{-1} cm^{-1})	Water saving (%)
1	Acid lime	1	78.0	56.0	1.3	50.0
2	Baby corn	1	9.9	72.4	0.5	43.8
3	Banana	7	71.5	29.3	3.0	42.5
4	Bean	1	10.3	81.8	0.4	36.9
5	Beetroot	1	48.9	7.0	2.8	79.0
6	Ber	3	71.0	27.7	0.7	34.3
7	Bitter gourd	4	2.7	44.4	1.4	69.5
8	Bottle gourd	1	55.8	46.8	1.0	35.7
9	Brinjal	7	16.0	44.6	1.5	42.6
10	Cabbage	5	50.5	37.5	3.2	37.4
11	Capsicum	1	22.5	66.6	0.8	43.1
12	Carrot	1	26.3	92.3	0.8	33.6
13	Castor	2	7.3	30.2	1.7	33.0
14	Cauliflower	3	19.5	39.7	0.7	37.1
15	Chickpea	1	3.8	66.6	1.6	42.6

TABLE 1.4 *(Continued)*

Sr. No.	Crop	Number of References	Yield (t ha^{-1})	Yield increase (%)	WUE (t ha^{-1} cm^{-1})	Water saving (%)
16	Chilli	5	68.0	28.7	7.5	47.3
17	Coconut (Nos./plant)	2	181.0	7.1	6.9	50.5
18	Cotton	3	36.0	40.0	0.9	51.1
19	Cucumber	1	22.5	45.1	0.9	37.8
20	Grain Corn	1	6.5	52.9	2.2	45.0
21	Grape	5	29.9	20.9	1.0	43.0
22	Groundnut	2	3.5	62.5	1.0	32.4
23	Guava	2	25.5	63.0	3.5	9.0
24	Mango	3	19.5	80.7	2.4	28.9
25	Sweet lime, 1000 pcs	1	15.0	98.0	0.2	61.0
26	Oil Palm	1	-	-	-	21.0
27	Okra	12	20.1	20.7	1.9	44.7
28	Onion	3	17.0	42.6	1.2	36.7
29	Papaya	5	56.6	72.0	0.9	68.0
30	Pomegranate, 100 pcs	3	44.7	55.7	0.5	57.3
31	Popcorn	1	5.5	75.4	2.1	42.0
32	Potato	5	28.7	50.0	2.8	24.6
33	Radish	2	17.0	27.5	5.0	64.0
34	Ridge gourd	3	17.4	14.5	4.4	43.4
35	Round gourd	1	36.6	24.0	0.5	0.0
36	Sapota	1	-	17.2	-	21.4
37	Sweet potato	1	50.0	39.0	2.0	68.0
38	Sugarcane	6	145.9	43.6	1.2	46.7
39	Sweet lime	1	15.0	50.0	2.3	61.4
40	Tapioca	2	54.6	12.6	0.6	23.4
41	Tomato	11	36.6	46.0	3.8	37.4
42	Turmeric	2	18.4	76.3	0.6	53.1
43	Watermelon	3	46.8	64.8	2.1	46.1

WUE = Water Use Efficiency, pcs = Pieces.

1.5 DRIP IRRIGATION FOR USE OF SALINE WATER

Numerous studies have also been conducted to establish the potential of drip irrigation to use saline/ alkali water. Before making any comment, a brief review will be presented.

Agarwal and Khanna [6] conducted a trial at Hisar to grow radish crop with saline tube well water (EC_{iw} = 6.5 dS m^{-1}) and good quality canal water (EC_{iw} = 0.25 dS m^{-1}). The yield was higher with drip being maximum in subsurface than surface drip. Moreover, the yield reduction with saline water was much less in drip (10.3% in surface drip) as compared to surface irrigation (39.6% in best treatment). Recent studies have shown that subsurface drip may not be beneficial in many cases if designs fail to take care of soil, crop, salinity of the soil and water interactions. Kumar and Sivanappan [35] concluded that drip irrigation gave higher yield than any other surface irrigation method for all levels of saline water tested (EC_{iw} 0.85, 2.5, 5.0, 7.5, and 10.0 dS m^{-1}). It emerged that saline water having an EC_{iw} of 7.5 dS m^{-1} is safe to grow crops with drip irrigation. Subba Rao et al. [85] observed up to 50% decrease in yield of tomato when EC_{iw} exceeded 6 dS m^{-1}.

Jain and Pareek [31] observed lesser salt accumulation in drip irrigation when saline waters of EC_{iw} ranging from 2.7 to 9.0 dS m^{-1} were used to irrigate date palm trees. Similar results were reported by Singh et al. [77] when sodic waters containing RSC 2.1, 8.45 and 12.45 meq l^{-1} were applied to grow kinnow (*Citrus reticulata*) plantation. Singh et al. [75] compared the plant performance and soil salinity before and after three years of application of 0.4, 4.0, 8.0 and 12.0 dS m^{-1} water through drip and basin irrigation in sapota crop at Anand, Gujarat. On the basis of plant performance and accumulation of salts at the end of experiment, it was concluded that drip irrigation performed better compared to the basin method at all salinity levels. The yield of tomato decreased from 38.7 to 29.8 t ha^{-1} (about 24% less over the good quality water) as the salinity of irrigation water increased from 0.21 to 5 dS m^{-1} [33]. Jangir and Yadav [32] reported the results of Yadav [88] on the effect of saline water irrigation on cauliflower as affected by drip and surface irrigation methods. Higher yields were obtained at all salinity levels in drip than with check basin. The yield obtained at EC_{iw} of 10 dS m^{-1} in drip was about the same as obtained at EC_{iw} of 2.5 dS m^{-1} in check basin irrigation (Table 1.5). Similar results for chili crop have been reported from Agra in Table 1.6 [4].

Irrigating tomato and brinjal crops through drip using canal water and waters of 4 and 8 dS m^{-1} at three IW/CPE levels (0.75, 1.00 and 1.25) at different irrigation intervals of 2, 3 and 4 days gave better yield at higher IW/CPE (Irrigation water applied/cumulative pan evaporation) ratios [2].

TABLE 1.5 Effect of Method of Irrigation on Cauliflower Yield With Saline Water

EC_{iw} (dS m^{-1})	Drip irrigation	Check basin	Mean
0.0	17.5	13.0	15.3
2.5	17.7	13.1	15.4
5.0	15.0	10.0	12.5
10.0	13.0	7.5	10.2
Mean	15.8	10.9	

CD 5% for method of irrigation = 1.65 and EC_{iw} = 0.87.

TABLE 1.6 Effects of Saline Water and IW/CPE Ratio on Yield (t/ha) of Chili in Drip and Surface Irrigation

IW/CPE* ratio	EC_{iw} levels (dS m^{-1})			Mean	EC_{iw} levels (dS m^{-1})			Mean
	Control	4	8		Control	4	8	
	Drip Irrigation				Surface Irrigation			
0.75	15.40	9.82	9.43	11.55	10.24	6.38	5.60	7.41
1.00	15.60	9.87	9.26	11.58	10.34	6.36	4.34	7.01
1.25	14.74	9.52	8.64	10.97	10.30	5.63	4.28	6.74
Average	15.21	9.74	9.10		10.30	6.13	4.74	—

CD 5% for drip = 0.68 and for surface = 1.07; IW/CPE ratio and interaction of EC_{iw} and IW/CPE ratio were nonsignificant; and *Irrigation water applied/cumulative pan evaporation.

It was observed that if total amount of water application is constant, 13 and 33% higher yields can be obtained at irrigation intervals of 3 and 4 days compared to 2 days interval. Significant differences in cotton yield were observed at an EC_{iw} of 11 dS m^{-1} compared to 2–2.5, 5.0 and 8.0 dS m^{-1} with 37.3% decrease in yield over 2–2.5 dS m^{-1}. Subsurface drip proved inferior to surface drip in all cases [47]. The tomato yield in Punjab (India) decreased significantly with increase in salinity levels of irrigation water with 26% and 11% decrease at 6.3 and 9.1 dS m^{-1} salinity compared to 0.38 dS m^{-1} salinity [45]. The yield of okra (*Abelmoschu sesculentus* L. Moench, Var: Mahyco10 Hy) reduced by 22.9% at EC_{iw} of 8 dS m^{-1} compared to only 5.9% less at 6.0 dS m^{-1} [64]. The water use efficiencies at 0.2, 2.0, 4.0, 8.0 dS m^{-1} were 0.49, 0.49, 0.46 and 0.38 tha^{-1} cm^{-1}, respectively.

Alkali water with high RSC has also been used for crop cultivation with appropriate amendments to neutralize RSC of water. The amended water applied through drip irrigation besides giving higher yield helped to conserve water to the tune of 21–42% (Tables 1.7 and 1.8). Much higher decline in surface irrigation was experienced during the second year. Both the amendments namely gypsum and distillery spent wash (DSW) used were found to be effective ameliorants [3].

TABLE 1.7 Effect of Drip Irrigation Using Ameliorated Alkali Water on Yield and Yield Attributes of Sugarcane Crop

Treatments	Irrigation water treatments				Mean
	Drip irrigation			Furrow irrigation	
	Gypsum bed	Spent wash	Untreated		
2004–2005 Cane yield (tha^{-1})					
50% GR	102.00	72.95	70.93	71.00	79.22
No gypsum	81.00	69.00	53.00	50.62	63.41
Mean	91.50	70.98	61.96	60.81	
CD 5%	9.04 (1)	1.99 (2)	9.47 (3)	3.97 (4)	
2005–2006 Cane yield (tha^{-1})					
50% GR	71.57	53.98	59.26	42.27	56.77
No gypsum	65.74	59.77	30.00	38.47	48.50
Mean	68.66	56.88	44.63	40.37	

CD 5% for methods, for water treatments for M X S interaction and S X M interaction are 2.87, 3.17, 5.32 and 6.34, respectively. GR = Gypsum requirement.

TABLE 1.8 Amount of Water Used in Sugarcane in Drip and Furrow Irrigation

Treatment	2004–2005		2005–2006	
	Quantity of water used (mm)	Water saving (%)	Quantity of water used (mm)	Water saving (%)
Drip irrigation				
Gypsum treated water	1812	42	1783	21
Spent wash treated water	1831	41	1801	20
Untreated alkali water	1812	42	1783	21
Furrow irrigation	2580	–	2173	-

1.6 THE PROBLEMS IN THE USE OF SALINE WATER

The evidences included in the foregoing section together with many others studies have proved that drip irrigation can be used to use relatively more saline water for several crops with no or limited yield losses [35, 60, 63, 70]. Drip irrigation proved beneficial even under shallow water tale situations for cauliflower, vegetables and cotton [71, 73]. In spite of these evidences use of saline water with drip irrigation has not picked up the necessary momentum for reasons discussed in the following sections.

1.6.1 LIMITED PERIOD STUDIES

Most studies have been conducted on small scale and were in operation for limited periods. Apparently, the relevant data required to perfect the technology have not been fully documented. For example; use of saline water through drip irrigation is likely to result in accumulation of salts at the wetting front. Under such situations, it may be prudent to apply reclamation leaching before the cultivation of next crop in the sequence. For example; in one of the commercial cultivation at Indore, a heavy irrigation with good quality water proved beneficial to leach down the salts even within a single cropping season of bitter gourd. Besides, light shower during the season might push the accumulated salts into the root zone to adversely affect the production. At Agra, summer chili crop with drip failed to survive in one year and there was almost no harvestable yield because of this reason. A similar failure was experienced at Karnal, when the ridge gourd crop failed under pitcher irrigation because the salts accumulated at the surface moved in to the root zone following light showers. Moreover, under the hot sunny days commonly encountered in arid and semiarid regions of India, operational schedule of drip irrigation may require frequent irrigation with appropriate leaching requirement for salts to move farther away from the plants and deeper in the profile. Such issues have not been resolved in most studies. Although recommendations have emerged that application of saline water can be used without any adverse effect on soil, crop, environment and building up the salts in the soil In areas having good monsoon rainfall [47], yet there is a need to work out reclamation leaching requirement and leaching requirement for salt balance in saline water drip irrigated crops.

1.6.2 CLOGGING OF EMITTERS

Another major problem that might emerge with the use of saline water is due to frequent clogging of emitters. The common elements to clog drip emitters by precipitation and/or sedimentation are calcium, magnesium, iron and manganese. Water containing high levels of these elements, and having pH above 7.0, has a high potential to cause clogging of drip emitters. It may be mentioned that most saline/alkali water have pH above 8.0 and EC is of course high. Alkali waters have ***preponderance*** of bicarbonates resulting in high RSC values. It is noticed that there are not many evidences where impact of saline water on clogging has been studied along with its potential of use to grow crops. In one of the studies conducted in Gujarat, fertigation with water soluble fertilizer increased the risk of clogging, particularly when fertilizer contained P, Ca, Mg, Fe and Mn. The drippers with higher discharge rate recorded less clogging but the risk increased with increasing salinity being maximum and minimum at 8 dS m^{-1} and best available water (BAW) of 1.47 dS m^{-1}, respectively as shown in Table 1.9 [68]. Similar results were obtained by these authors in farmers' fields.

TABLE 1.9 Interaction Effect of Different Treatments on Periodical Dripper Clogging (%)

Treatments	Clogging (%)* after period (days)							
	15	30	45	60	75	90	105	120
No fertigation	6.82	11.70	15.07	17.86	20.17	22.29	24.23	26.43
Fertigation 150:50:50	7.46	12.34	15.58	18.59	21.14	23.50	25.85	28.13
NPK								
CD 5%	0.411							
Dripper discharge								
1.2 lph	8.93	13.87	17.01	19.54	22.10	24.57	26.89	29.52
2 lph	7.70	13.03	16.18	18.85	21.16	23.52	25.79	28.03
3 lph	5.08	10.38	14.10	17.39	20.09	22.16	24.54	26.49
4 lph	6.86	10.80	14.02	17.13	19.27	21.35	22.93	25.06
CD 5%	0.511							
EC_{iw}								
1.5 dS m^{-1}	7.06	11.41	14.38	17.27	19.32	21.32	23.12	24.69
4 dS m^{-1}	7.61	12.56	16.14	19.04	21.26	22.93	26.56	26.56
6 dS m^{-1}	7.15	11.71	14.78	17.54	20.25	23.05	25.48	28.15
8 dS m^{-1}	6.74	12.39	16.01	19.05	21.79	24.30	26.86	29.71
CD 5%	0.511							

*The reduction in discharge from the original discharge.

As per the Indian Standards, guidelines on clogging of drip irrigation are given in Table 1.10. The standards in respect of several parameters are more rigid than those recommended by Nakayama [44]. Since the spread of drip irrigation in India is driven by the subsidy offered by the government, most systems are installed under the BIS guidelines by the industry. If the water quality standards are not adhered to by the stakeholders as per these guidelines, industry might not service the system during the guarantee. No stakeholder would like to take such a risk. In spite of several options available to avoid clogging problem, farmers go by the manufacturer's advice. Besides, most farmers anticipate a reduced life of the system with saline water. Not much data on this issue have been generated so far. Nevertheless some policy support in terms of higher subsidy is required to encourage the farmers to use saline water.

TABLE 1.10 Water Quality and Potential For Clogging Problems in Drip Systems

Potential problem	Units	Degree of restriction on use		
Suspended solids	mgl^{-1}	< 10	10 – 100	> 100
pH*		< 7.0	7.0 – 8.0	> 8.0
Dissolved solids	mgl^{-1}	< 500	500 – 600	> 600
EC_{iw}	$dS\ m^{-1}$	0.8	0.8–3.0	>3.0
Manganese	ppm	< 0.2	0.2 – 0.4	> 0.4
Iron	ppm	< 0.1	0.1 – 0.4	> 0.4
Sulphides	ppm	< 0.1	0.1 – 0.4	> 0.4
Hardness	ppm	< 200	200–300	>300
Bacterial populations*	maximum number $(ml)^{-1}$	<10 000	10 000 – 50 000	>50 000

Source: BIS [13]; * As per Nakayama [44].

1.7 INDIGENOUS ALTERNATIVES TO DRIP IRRIGATION

Several indigenous technologies that mimic the irrigation with drip irrigation are in vogue in India. The most notable is the bamboo drip irrigation wherein bamboo pipes are used to divert and convey water from the perennial springs on the hilltops to the field plots in Meghalaya for more than 200 years [25]. The system named as bamboo trickler is so perfect that about 18–20 L of water entering the bamboo pipe system per minute at the entry point gets reduced to 20–80 drops per minute at the field plot site. This system is commonly used to irrigate the betel leaf or black pepper crops planted in areca nut orchards or in mixed orchards [27].

In the other one, porous emitters made from the local clay soil and cow dung are used to irrigate vegetable crops. These emitters are connected to a storage tank through plastic tubes. The buried emitters supply water to the soil/plants because the soil is at relatively high soil moisture tension than the emitters [25, 89]. Studies made on cabbage, cauliflower and knolkhol in Jobner, Rajasthan using earthen cups revealed that water requirement could be as low as 5–10% compared to check basin irrigation [89].

Most of these developments have emerged to conserve water. The most important developments in this respect have been the pitcher irrigation and pipe irrigation, which have been tested for their potential to conserve water as well as use of saline/alkali water for irrigation.

1.7.1 PITCHER IRRIGATION

Pitcher irrigation derives its name from the baked earthen pitchers, which are used for water storage and distribution in Indian homes. It is an ancient technique known for the last 2000 years or so in China/Africa and was used by the Romans for many centuries. It is a simple, efficient, and economic way to provide localized subsurface irrigation. It was reinvented at Central Soil Salinity Research Institute, Karnal, India [39] particularly to assess its utility with saline water, which has not been tested ever before. Currently, the technology is being used for small-scale agricultural irrigation in the arid and semiarid regions of India, China, Iran, Argentina, Brazil, Ecuador, Bolivia, Indonesia, Tanzania, Pakistan and Mexico [1, 7, 16, 17, 24, 37, 40, 50, 56, 66, 69]. Pitcher irrigation makes efficient use of water because like drip irrigation, it also delivers water directly to the plant's roots with a steady supply of moisture. The system has been used to cultivate most vegetable crops, coffee, mango, grapes, *litchi*, and mesquite showing its widespread application for crops and trees. The technique is most effective under the following conditions:

- water is either scarce or expensive;
- soils are difficult to level such as under undulating terrain;
- water is saline and cannot be normally used in surface methods of irrigation;
- in remote areas where transport of vegetables is expensive and uneconomical.

1.7.2 INSTALLATION OF THE PITCHER SYSTEM

In this method, unglazed baked porous pitchers (or pots) are buried in shallow pits. Exact number of pitchers per unit area is worked out on the basis of crop characteristics, a spreading type of crop requiring lesser number of pitchers. Places to bury the pitchers are marked, normally on a square grid. At each location, a semispherical pit of 90 cm diameter and at least 60 cm deep or in tune with the size of the pitcher is dug. Soil brought out from the pit is preserved. Clods are broken to less than 1 mm diameter and enough farmyard manure, basal dose of fertilizers (phosphorus and potash) and amendment (if soil is sodic) are mixed [43]. After the material is thoroughly mixed, a part of the mixture is placed in the pit to a height of about 30 cm. The pitcher is then placed at the center of the pit (Fig. 1.1, left). Thereafter, remaining portion of the mixture is poured around the pitcher so as to cover the whole space from the bottom to the neck of the pitcher. For heavy soils, a thin layer of sand can be placed around the pitchers. The mixture is thoroughly tapped to ensure a good contact between the soil and the pitcher. In the absence of good contact, the water will either not flow out of the pitcher or the flow will be irregular. Thereafter, pitchers are filed with clean water. At least 6–8 seeds/seedlings are sown around the pitcher (Fig. 1.1, right). Sowing is generally accomplished two days after the first filling of the pitchers with water. To facilitate germination, presoaked and germinated seeds are sown or water is added from outside to wet the seedbed. The pitchers are filled with water at predetermined intervals. Although the water is filled manually by buckets, a hose pipe can be used for convenience. A study to assess the filling schedule revealed that filling on every third day proved better than daily, alternate, four or five days filling in case of fresh water. On the other hand, alternate day filling proved better with saline water. The actual schedule may get modified depending upon the water depletion rate, which in turn depends upon the type of crop, stage of growth, and the climatic conditions. Fertilizers can be added to the water in the pitcher to ensure their uniform application in the root zone.

FIGURE 1.1 The sectional view of an installed pitcher (left) and photographic view of plants around the pitcher (right).

A number of studies have been conducted to perfect the technology. These studies have helped to develop irrigation schedules, study distribution of roots and salts especially under saline water irrigation and placement of pitchers either on the soil surface or below ground, etc. [9, 17, 24, 39, 41]. Following major observations have emerged:

The amount of water flowing out of the pitcher i.e. discharge per unit surface area per unit time, depends upon the water head at a given point being maximum at the bottom, the porosity and hydraulic conductivity of the pitcher walls, vapor-transpiration need of the plant, being greater during the day than at night and soil texture being greater in a light than a heavy textured soil. A study using HYDRUS-2D model has also shown that water distribution pattern varies significantly under various soil textures [80]. Whereas in heavy textured soils, the steady state wetting front attains the shape of the pitcher/pot, in light textured soils it is oblong taking the shape of a parabola under no crop conditions. With cropping, the wetting front even in light textured soils takes the shape of the pitcher/pot. The wetting front may also vary with placement of pots [59] and the permeability of the pot's bottom (Permeable or impermeable). Many people recommend that bottom of the pitcher can be made impermeable to save water. The experiments conducted at CSSRI, Karnal do not support this contention because fine roots make a thick mat over the whole pitcher in due course of time. It leaves little opportunity for deep percolation of the water.

For a pitcher of 30 cm diameter and 10 L water holding capacity, wetted surface area at the soil surface is around 0.7 m^2. It is sufficient to grow at least four plants of any vegetable crop. The moisture content of the soil was about the field capacity of the soil near the pitcher walls that reduces with distance from the wall. Due to limited hydraulic conductivity of the pitcher wall, moisture levels above these were not observed. Some change might occur during nights because of low evapotranspiration demand.

TABLE 1.11 Estimated Irrigation Water Use Efficiency in Various Irrigation Systems

Irrigation method	WUE (t ha^{-1} cm^{-1})[1]	WUE (t ha^{-1} cm^{-1})[2]
Closed furrow (basin)	0.7	0.7
Sprinkler	0.9	0.9
Drip	1 – 2.5	1.4
Porous capsule (pressure)	1.9 +	1.9
Porous capsule (no pressure)	2.5 +	2.5
Buried clay pot	2.5 – 7	2.5–6.0
Wick	-	4.0

[1]Bainbridge [8]

[2]http://www.pakissan.com/english/newtech/pitcher [28].

The water saving potential of the technology has been worked out. Depending upon the number of pitchers, crop duration and filing schedule, water equivalent to 1–2 irrigations with surface irrigation method may suffice for the whole crop growing season. For example for alternate day filling schedule, with 800 pitchers each of 10 L capacity, it may require only 4.8 cm of water per ha for a crop of 120 days duration. Bainbridge [8] made elaborate calculations and showed that water use efficiency is the highest in the buried clay pot system than any other irrigation system (Table 1.11). Same results with some additions/modifications have also been reported by other workers (Table 1.11).

Besides water saving, yield benefits have also been documented in India and elsewhere. Studies at CSSRI have shown that good yields of many vegetable crops can be obtained with this method (Table 1.12). Kurian et al. [36] used buried clay pot irrigation to grow *Prosopis* (*Prosopis* spp.) seedlings. Trees irrigated with clay pots were more than three times taller than rain fed trees and 70% taller than surface irrigated trees. Reddy and Rao [58] found that the dry weight of weeds in crops irrigated by buried clay pots was only 13% compare to weeds in control plots irrigated by basin irrigation. Batta [10] reported the application of pitcher irrigation to cultivate *litchi* (*Litchi chinensis*) crop using four porous cups per plant (Fig. 1.2). The application of pot irrigation for coffee plants during low rainfall months to overcome water stress has been documented [18].

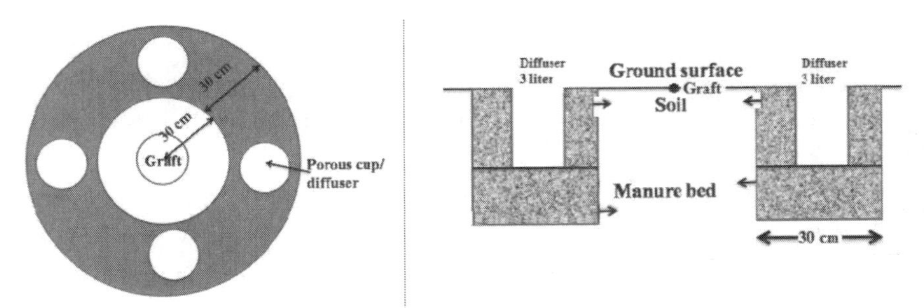

FIGURE 1.2 A schematic view of porous cup irrigation for horticultural crops.

TABLE 1.12 Yield of Various Crops With Fresh Water (with EC_{iw}= 0.5 dS m^{-1})

Crop	Crop yield (kg $pitcher^{-1}$)	Crop	Crop yield (kg $pitcher^{-1}$)
Water melon	11.3	Tomato	5.8
Muskmelon	7.4	Cauliflower	5.2
Bottle gourd	21.5	Brinjal	5.1
Bitter gourd	7.5	Cabbage	4.8
Ridge gourd	4.5	Radish	8.0
Cucumber (*Kakri*)	14.0	Grapes	3.5

1.7.3 USE OF SALINE WATER

Evidences have been generated to show that pitcher irrigation helps to use relatively high salinity irrigation water to grow salt sensitive vegetables and horticulture crops. It is attributed to the fact that the principle involved in this irrigation technique is similar to the drip irrigation. The irrigation water salinity values reported in Table 1.13 for pitcher irrigation gave same yield per pitcher as was obtained with fresh water reported in Table 1.12. Although the data reported is not generated under similar set-ups, it is apparent that that highly saline water can be used to cultivate even sensitive vegetable crops at which yield decline in any surface irrigation method may be around 50% or more (Table 1.13). Since the most active plant roots remain around the pitcher wall, these are not subjected to salt stress at the soil surface or at the wetting front boundaries.

TABLE 1.13 Salinity of Irrigation Water Under Pitcher At Which Crop Yield Is Same As With Fresh Water and Relative Yields Under Surface Irrigation

Crop	EC_{iw} for relative yield			EC_{iw} under pitcher for same yield with fresh water
	90%	75%	50%	
Tomato#	2.4	4.1	6.9	5.7
Brinjal#	2.3	4.1	7.1	9.8
Cauliflower	0.9–2.7*	-	-	15.0
Ridge gourd	-	-	-	3.2
Cabbage**	1.9	2.9	4.6	9.7
Watermelon@	2.4	3.8	-	9.0
Muskmelon	-	-	-	9.0
Grapes@	1.7	2.7	-	4.0

Minhas and Gupta [38], – Data not available; and
** <http://www.fao.org/docrep/003/t0234e/T0234E03.htm> [29] @http://archive.agric.wa.gov.au/objtwr/imported_assets/content/lwe/water/irr/fn2007_h2osalinity_jburt.pdf [30].

A comparison between pitchers placed on the soil surface and beneath the ground revealed that salts accumulate on the pitcher walls when placed above ground because the evaporating surface in this case is the pitcher wall. On the other hand, in pitchers placed beneath the ground salts do not accumulate on the walls because salts accumulation zone is the wetting front far away from the pitcher walls. As a result, the life of the pitcher is much more than surface placed pitchers, which becomes brittle and looses porosity earlier because of salt accumulation. As such, baked clay pitchers even with saline water have been continuously used for 3–6 years (6–12 seasons). To prolong the life of the pitcher, the following precautions should be taken:

- Always keep the mouth of the pitcher covered. This will minimize evaporation losses as well as prevent sunlight to enter the pitcher to minimize algal information and growth.
- Only clean water should be used for filing the pitcher. Muddy runoff from rainfall should be used after it has been passed through a sand filter.
- Before storing the pitchers, wash them with good quality water so as to remove the earth and the salts that might be sticking on the pitcher walls.

Although, there is little fear of clogging of the pitcher walls, leaching water may be required to leach the accumulated salts after the cropping season. In the Indian context if the cropping system is planned well, monsoon rainfall can take care of the leaching at least for one season.

1.8 POLYVINYL CHLORIDE (PVC) PIPES FOR IRRIGATION

The perforated PVC pipes have been used for quite some time in small-scale irrigation as a means of spot irrigation for widely spaced row crops to conserve water. The same system has been used in India to apply alkali water to horticultural crops namely sapota and *ber* at Indore, India [4]. The system comprises of PVC pipes of 100 mm diameter of about 40 cm lengths and embedded to 30 cm in the soil (Fig. 1.3, left). The perforations are made in the pipes and oriented towards crop plants. The embedded pipes are wrapped with locally available coir material to avoid soil particles to clog the perforations/pipes. The tests revealed that only 400 L of water per year per plant was required in this method compared to 1600 L in check basin method, resulting in a saving of 75% of the irrigation water over the check basin. To use alkali water, DSW was used to lower the RSC of the water. Earlier experimental evidences have shown that about 1 L of DSW can reclaim about 250 L of alkali water available in the region. Thus, the exact quantity of DSW required for each irrigation was calculated and accordingly added to water before its application. The performance of both the plants in terms of height, girth and productivity was better with embedded pipes using fresh or treated alkali water. It was followed by drip and basin irrigation. The performance treated alkali water (EC 0.9 dS m^{-1}, SAR 7.3 (mmol $l^{-1})^{½}$ and RSC nil) was even better than the fresh water (EC 0.5 dS m^{-1}, SAR 1.1 (mmol $l^{-1})^{½}$ and RSC nil) in basin irrigation [5].

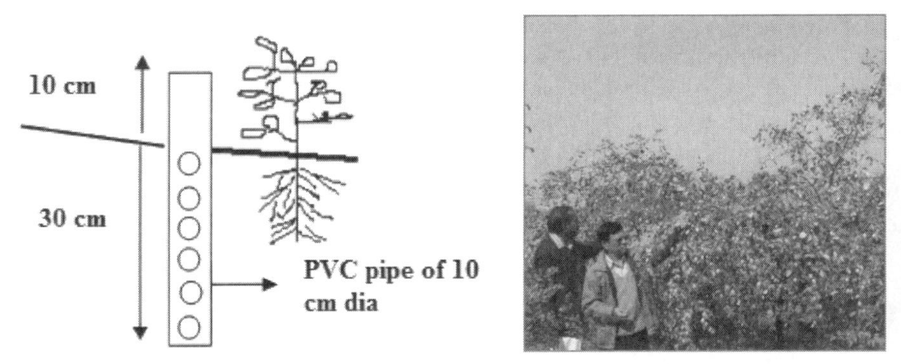

FIGURE 1.3 Perforated pipe irrigation method and crop of ber irrigated with this method.

1.9 SUMMARY

Anticipating the ever increasing water shortage, scientific community in India is engaged in developing technologies to achieve 'more crop per drop' and to exploit every drop of water including naturally occurring saline/alkali ground and recycled waters. The research results paved the way for a major shift from traditional surface irrigation methods to drip irrigation, drip irrigation being extended to cover an area of 1.9 million ha. This chapter summarizes the benefits of drip irrigation in terms of increase in crops yield and water savings in 42 crops. It also includes a comprehensive review of evidences that highlights the use of this technology in using saline/alkali ground waters. The limitations of these evidences and impediments in extending drip irrigation with saline/alkali water are included. The indigenous methods that mimic the drip irrigation have been reviewed. Amongst these, pitcher irrigation and pipe irrigation methods are described in more detail. The results suggest that pitcher irrigation has the potential to conserve water as well as allow the use of relatively more saline water than any other surface irrigation method. Application of pipe irrigation to use alkali water for two horticultural crops is described. If necessary policy support and subsidies are provided by the government, drip irrigation and indigenous alternatives can prove to be useful tools to make use of saline/alkali ground waters in crop production programs.

KEYWORDS

- **alkali water**
- **amendments**
- **bamboo trickler**
- **clogging**
- **crop cultivation**
- **distillery spent wash**
- **drip irrigation**
- **gypsum**
- **India**
- **pipe irrigation**
- **pitcher irrigation**
- **porous cups**
- **pot irrigation**
- **saline water**
- **water quality**
- **water saving**
- **water use efficiency**

REFERENCES

1. Alemi, M. H. (1981). Distribution of water and salt in soil under trickle and pot irrigation regimes. Agricultural Water Management, 3:195–203.
2. Anonymous, (2000a). Tolerance of vegetable crops to saline irrigation under drip and surface irrigation systems. Biennial Report (1998–2000). AICRP on Management of Salt Affected Soils and Use of Saline Water in Agriculture. Coordinating Unit, CSSRI, Karnal, pp. 61–71.
3. Anonymous, (2000b). Evaluation of irrigation schedules for sugarcane using alkali water through drip irrigation. Biennial Report (1998–2000). AICRP on Management of Salt Affected Soils and Use of Saline Water in Agriculture. Coordinating Unit, CSSRI, Karnal, p. 72.
4. Anonymous, (2007). Biennial Report (2004–2006). AICRP on Management of Salt Affected Soils and Use of Saline Water in Agriculture. Coordinating Unit, CSSRI, Karnal.
5. Anonymous, (2011). Biennial Report (2008–2010). AICRP on Management of Salt Affected Soils and Use of Saline Water in Agriculture. Coordinating Unit, CSSRI, Karnal, 208p.
6. Agarwal, M. C., Khanna, S. S. (1983). Efficient Soil and Water Management in Haryana. Bulletin HAU, Hisar, 118 pp.
7. Bainbridge, D. A. (1986). Pitcher Irrigation. University of California, Dry Lands Research Institute, Riverside, CA. 3 pp.
8. Bainbridge, D. A. (2001). Buried clay pot irrigation: a little known but very efficient traditional method of irrigation. Agricultural Water Management, 48:79–88.
9. Balakumaran, K. N., Mathew, J., Pillai, G. R., Varghese, K.,1982. Studies on the comparative effect of pitcher irrigation and pot watering in cucumber. Agricultural Research Journal (Kerala), 20:65–67.
10. Batta, R. K. (2011). Micro-irrigation for water conservation and high productivity. In: Salinity Management for Sustainable Agriculture in Canal Commands (Dey and Gupta, Eds.). Central Soil Salinity Research Institute, Karnal, pp. 34–42.
11. Batta, R. K., Singh, S. R. (2000). Drip irrigation to sugarcane: Indian experience. Proceedings of the National Seminar on Micro-Irrigation Research in India: Status and Perspectives for the twenty-first century, WTCER, Bhubaneswar. July 27–28, (1998). Institution of Engineers (India), pp. 143–147.
12. Bendale, S. K., Chauhan, H. S., Shukla, K. N. (1993). Field study of drip irrigation of banana around Sangli (Maharashtra). Proc. Workshop on Sprinkler and Drip Irrigation Systems held at Jalgaon, Maharashtra, 8–10 December 1993, CBIP, New Delhi, pp. 137–141.
13. BIS, (2000). Indian Standard: Prevention and Treatment of Blockage Problem in Drip Irrigation System–Code of Practice: IS 14791:2000. Bureau of Indian Standards, New Delhi.
14. CGWB, (1997). Inland Groundwater Salinity in India. Government of India, Ministry of Water Resources, Central Ground Water Board, Faridabad, 62 pp.
15. Chandaragiri, K. K. (2002). Water management for fruit crops. *In:* Recent Advances in Irrigation Management for field crops by Chinnamuthu et al. Eds., Center of Advance Studies in Agronomy, Tamil Nadu Agric. Univ., Coimbatore, pp. 198–212.
16. Dubey, S. K., Gupta, S. K., Mondal, R. C. (1990). Response of ridgegourd to saline irrigation water applied through earthen pitchers. Transaction of the Indian Society of Desert Technology, 15:53–61.
17. Dubey, S. K., Gupta, S. K., Mondal, R. C. (1991). Pitcher irrigation technique for arid and semiarid zones. Dryland Resources & Technology, 6:137–177.
18. Elavarasan, K., Govindappa, M., Hareesh, S. B. (2014). Clay pot irrigation of young coffee seedling under Pulney hills of Tamil Nadu, India. Journal on New Biological Reports, 3:91–96.
19. Garg, M. L. (1995). Water Dynamics and Crop Response under Different Methods of Micro Irrigation for Cabbage and Cauliflower. M.Tech. (Agril. Engg.) Dissertation, CCS HAU, Hisar.

20. Gidnavar, V. S., Adder, G. D. (1993). A review on drip irrigation system in field crops-case study. Proc. Workshop on Sprinkler and Drip Irrigation Systems, Jalgaon, Maharashtra, 8–10 December 1993, CBIP, New Delhi, pp. 147–149.
21. Goyal, Megh R., (ed.), (2015). *Research Advances in Sustainable Micro Irrigation, Volumes 1 to 10.*Oaksville, ON, Canada: Apple Academic Press Inc.,
22. Gupta, R. K., Singh, N. T., Sethi, M. (1994). Ground Water Quality for Irrigation in India. Tech. Bull No. 19. Central Soil Salinity Research Institute, Karnal, pp. 13.
23. Gupta, S. K. (2011). Irrigation with saline and alkali waters: management strategies. In: Salinity Management for Sustainable Agriculture in Canal Commands by Dey and Gupta, Eds. Central Soil Salinity Research Institute, Karnal. Pp. 69–81.
24. Gupta, S. K., Dubey, S. K. (2001). Pitcher irrigation for water conservation and use of saline water in vegetable production. Indian Farming, 51:33–34, 36.
25. Gupta, S. K., Gupta, I. C. (1987). Management of Saline Soils and Waters. Oxford and IBH Publishing Co. Pvt. Ltd. New Delhi, 339 pp.
26. http://bhuwanchand.wordpress.com/2012/01/20/drip-irrigation-huge-scope-for-adoption-in-indian-farms/ opened on 12.02.2015.
27. <http://kiran.nic.in/publications/Efficient_Water_Management_NorthEast.pdf>.
28. http://www.pakissan.com/english/newtech/pitcher.irrigation.a.water.shtml (Opened on 27.02.2015).
29. http://www.fao.org/docrep/003/t0234e/T0234E03.htm
30. http://archive.agric.wa.gov.au/objtwr/imported_assets/content/lwe/water/irr/fn2007_h2osalinity_jburt.pdf (opened on 26.02.2015).
31. Jain, B. L., Pareek, O. P. (1989). Effect of drip irrigation and mulch on soil and performance of date palm under saline water irrigation. Annals of Arid Zones. 28:245–248.
32. Jangir, R. P., Yadav, B. S. (2011). Management of saline irrigation water for enhancing crop productivity. Jour. of Scientific and Industrial Research. 70:622–627.
33. Kadam, J. R., Patel, K. B. (2001). Effect of saline water through drip irrigation system on yield and quality of tomato. J. Maharashtra Agric. Univ. 26:8–9.
34. Kailasham, C. (2002). Water management for sugarcane. In: Recent Advances in Irrigation Management for Field Crops by Chinnamuthu, et al., Eds. Centre of Advance Studies in Agronomy, Tamil Nadu Agric. Univ., Coimbatore. 173–181.
35. Kumar, V., Sivanappan R. K. (1983). Utilization of salt water by the drip system. Proceedings Second National Seminar on Drip Irrigation. March 5–6, 1983, Tamil Nadu Agricultural University, Coimbatore: 47–55.
36. Kurian, T., Zodape, S. T., Rathod, R. D. (1983). Propagation of *Prosopis juliflora* by air layering. Transactions Indian Society of Desert Technology and University Centre of Desert Studies. 8:104–108.
37. León, B. (1995). Pottery Irrigation on Peru's Arid Coast. IDRC Reports, International Development Research Centre (IDRC) of Canada. 8–9.
38. Minhas, P. S., Gupta, R. K. (1992). Quality of Irrigation Water–Assessment and Management. Publications and Information Division, Indian Council of Agricultural Research, *KrishiAnusandhan Bhawan*, Pusa, New Delhi. 123 p.
39. Mondal, R. C. (1974). Farming with pitcher: a technique of water conservation. World Crops. 26:91–97.
40. Modal, R. C. (1978). Pitcher farming is economical. World Crops. 30:124–127.
41. Mondal RC. (1983). Salt tolerance of tomato grown around earthen pitchers. Indian Journal of Agricultural Science. 53:380–382.
42. Mondal R. C., Dubey, S. K., Gupta, S. K. 1992.Use pitcher when water for irrigation is saline. Indian Horticulture. 36:13–15.

43. Mondal, R. C., Gupta, S. K., Dubey S. K., Barthwal, H. K. (1987). Pitcher Irrigation. Central Soil Salinity Research Institute, Karnal, India. 11p.
44. Nakayama F. S. (1982). Water analysis and treatment techniques to control emitter plugging. Proc. Irrigation Association Conference, 21–24 February 1982. Portland, Oregon.
45. Nangare, D. D., Singh, K. G., Kumar, S. (2013). Effect of blending fresh-saline water and discharge rate of drip on plant yield, water use efficiency (WUE) and quality of tomato in semi arid environment. African Journal of Agricultural Research. 8:3639–3645.
46. Narayanamoorthy, A. (2004). Drip irrigation in India: can it solve water scarcity? Water Policy. 6:117–130.
47. Nasrabad, G. G., Rajput, T. B. S., Patel, N. (2013). The effect of saline water under subsurface drip irrigation on cotton. Indian journal of Agricultural Sciences. 83:81–84.
48. Nasrabad, G. G., Rajput, T. B. S., Patel, N., Sehgal, V. K. (2011). Soluble salts distribution under saline water condition in drip irrigation. Asian Journal of Microbial Biotech Env. Sci. 13:569–575.
49. NCPA. (1990). Status, Potential and Approach for Adoption of Drip and Sprinkler Irrigation Systems. National Committee on the Use of Plastics in Agriculture, Pune, India.
50. Oswal, M. C., Singh, K. (1975). Pitcher farming of vegetables under dry lands: a new dimension in water harvesting. Haryana Agricultural Journal of Research. 5:351–353.
51. Padmakumari, O., Sivanappan, R. K. (1985). Drip irrigation for cotton. In: Drip/Trickle Irrigation in Action. Proceedings, third International Drip/Trickle Irrigation Congress, November 18–21, 1985. ASAE. Publication No. 10–85. Vol. I: 262–267.
52. Padmakumari, O., Sivanappan, R. K. (1989). Drip irrigation for papaya. Proc. of the 11th International Congress on Agril. Engg., Dublin, Ireland. September 4–8, 1989. 134–136.
53. Parikh, M. M., Shrivastava, P. K., Savani, N. G., Raman, S. (1993). Feasibility study of drip in sugarcane. Proc. Workshop on Sprinkler and Drip Irrigation Systems, Jalgaon, Maharashtra, 8–10 December 1993, CBIP, New Delhi: 124–127.
54. Patel, R. M., Patel, Z. B., Vyas, H. N. (1993). Large scale adoption of drips in fruit crops- a case study. Proc. Workshop on Sprinkler and Drip Irrigation Systems, Jalgaon, Maharashtra, 8–10 December 1993, CBIP, New Delhi: 142–146.
55. Praveen Rao, V. (2002). Drip irrigation and its application in farmers' fields of India. In: Recent Advances in Irrigation Management for Field Crops by Chinnamuthu, et al., Eds. Centre of Advance Studies in Agronomy, Tamil Nadu Agric. Univ., Coimbatore. 182−185.
56. Power, G. (1985). Porous pots help crops in NE Brazil. World Water. 10:21–23.
57. Raman, S. (1999). Status of research on microirrigation for improving water use efficiency in some horticultural crops. Proceedings of the National Seminar on Problems and Prospects of Micro-Irrigation–A Critical Appraisal, Bangalore, Nov 19–20, 1999. Institution of Engineers (India): 31–45.
58. Reddy, S. E., Rao, S. N. (1980). Comparative study of pitcher and surface irrigation methods on snake gourd. Indian Journal of Horticulture. 37:77–81.
59. Saleh, E., Setiawan, B. I. (2010). Numerical modeling of soil moisture profiles under pitcher irrigation application. Agricultural Engineering International (ejournal). 12:1–13.
60. Samuel, J. C., Singh, H. P. (1999). Micro-irrigation in Indian horticulture–Retrospect and prospects. Proceedings of the National Seminar on Micro-Irrigation Research in India: Status and Perspectives for the twenty-first century, Bhubaneswar. July 27–28, 1998. Institution of Engineers (India): 1–11.
61. Savani, N. G., Parikh, M. M., Srivastava, P. K., Solia, B. M., Zalawadia, N. M., Raman, S. (1999). Drip technologies for south Gujarat. In: Water Management Research in Gujarat. SWMP Pub. 10. Soil and Water Management Research Unit, Gujarat Agricultural University, Navsari Campus, Navsari, Gujarat: 87–94.
62. Saxena, C. K., Gupta, S. K. (2004). Drip irrigation for water conservation and saline/sodic environments in India: a review. In Proceedings of International Conference on Emerging

Technologies in Agricultural and Food Engineering. IIT, Kharagpur. December 14–17, 2004. Natural Resources Engineering and Management and Agro-Environmental Engineering. Amaya Publishers. New Delhi. 234–241.

63. Saxena, C. K., Gupta, S. K. (2006). Effect of soil pH on the establishment of litchi (*Litchi chinensis*) plants in an alkali environment. The Indian Journal of Agricultural Sciences. 76:547–549.
64. Saxena, C. K., Gupta, S. K., Purohit, R. C., Bhaka, S. R., Upadhyay, B. (2013). Performance of okra under drip irrigation with saline water. Journal of Agricultural Engineering. 50:72–75.
65. Sharma, S. K. (2000). Ground water management. In: Conf. Proc. 3rd Water Asia 2000. Interads Ltd. New Delhi.
66. Shiek'h, M. T., Shah, B. H. (1983). Establishment of vegetation with pitcher irrigation. Pakistan Journal of Forestry. 33:75–81.
67. Shinde, P. P., Deshmukh, A. S., Jadhav, S. B. (2000). Field evaluation of drip irrigation in sugarcane. Proceedings of the National Seminar on Micro-Irrigation Research in India: Status and Perspectives for the twenty-first century, July 27–28, (1998). Institution of Engineers (India), WTCER, Bhubaneswar: 233–236.
68. Shinde D. G., Patel K. G., Solia B. M., Patil R. G., Lambade B. M., Kaswala A. R. (2012). Clogging behavior of drippers of different discharge rates as influenced by different fertigation and irrigation water salinity levels. Journal of Environmental Research and Development. 7:917–922.
69. Silva, A. de S., Silva, D. A., Gheyi, H. R. (1983). Viability of irrigation by porous capsule method in arid and semiarid regions:1. International Commission on Irrigation and Drainage, 12th Congress. 753–764.
70. Singh, Pratap. (2000). Pressurized Irrigation systems for enhanced water use efficiency with saline waters. Lecture notes–training course on Use of saline water for Irrigation (26th June–16th July 2000) IDNP, CSSRI, Karnal: 299–308.
71. Singh, Pratap and Kumar, R.1994. Effect of irrigation methods for cauliflower grown in heavy soils with shallow water table. J. Agril. Engg. 31:36–43.
72. Singh, Pratap and Kumar, R. (1989). Comparative performance of trickle irrigation for tomato. J. Agril. Engg., ISAE, New Delhi. 26:39–48.
73. Singh, Pratap and Kumar, R. (2000). Comparative performance of drip irrigation for vegetable and cotton in heavy soils with shallow water table. Proc. International Conference on Micro and Sprinkler Irrigation Systems, Jalgaon. 87–92.
74. Singh, P., Kumar, R., Agarwal, M. C., Mangal, J. L. (1990a). Performance of drip and surface irrigation for tomato in heavy soils. Proc. XI International Convention on Use of Plastics in Agriculture, NCPA, New Delhi. 67–72.
75. Singh, R., Das, M., Kundu, D. K., Kar, G. (2000). Prospect of using saline water in fluventiceutrochrept through drip irrigation. Proceedings of the National Seminar on Micro-Irrigation Research in India: Status and Perspectives for the twenty-first century, Bhubaneswar, July 27–28, 1998; 255–257.
76. Singh, S. R., Patil, N. G., Islam, A. (2002). Drip irrigation system for efficient water and nutrient management. Bulletin No. 2. ICAR Research Complex for Eastern Region, Patna, Bihar. 39p.
77. Singh, V. P., Samra, J. S., Gill, H. S. (1990b). Use of poor quality water through drip system in kinnow orchards. Proc. XI International Congress on the Use of Plastics in Agriculture. Oxford & IBH Pub. Co. Pvt. Ltd., New Delhi. 165–175.
78. Sivanappan, R. K. (1999). Status and perspectives of microirrigation research in India. Proceedings of the National Seminar on Micro-Irrigation Research in India: Status and Perspectives for the twenty-first century, Bhubaneswar. July 27–28, 1998. Institution of Engineers (India): 17–29.

79. Sivanappan, R. K., Kumar, V. (1982). Status and development of drip irrigation in India. Proceedings Second National Seminar on Drip Irrigation. March 5–6, 1983, Tamil Nadu Agricultural University, Coimbatore: 3–18.
80. Siyal, A. A., van Genuchten, M.Th., Skaggs, T. H. (2009). Performance of Pitcher Irrigation System. Soil Science. 174:312–320.
81. Srinivas, K. (1997a). Growth, yield and quality of banana in relation to N fertigation. Tropical Agric. 74:260–264.
82. Srinivas, K. (1997b). Growth, yield and water use of papaya under drip irrigation. Ind. J. Horti. 53:19–22.
83. Srinivas, K. (1999). Yield and water use of Anab-e-Shahi grapevines under drip and basin irrigation. Indian J. Agric. Sci. 69:21–23.
84. Srinivasan, K. (2002). Water management for vegetables. *In:* Recent Advances in Irrigation Management for Field Crops by Chinnamuthuet al. Eds. Centre of Advance Studies in Agronomy, Tamil Nadu Agric. Univ., Coimbatore. 173 −181.
85. Subba Rao, N., Subbiah, G. V., Ramaiah, B. (1987). Effect of saline water on tomato yield and soil properties. J. Indian Soc. Soil Sci. Research. 5:407–409.
86. Tiwary, K. N., Kannan, N., Mal, P. K. (2000). Enhancing productivity of horticultural crops through drip irrigation. Proceedings of the National Seminar on Micro-Irrigation Research in India: Status and Perspectives for the twenty-first century, July 27–28, 1998. Institution of Engineers (India), WTCER, Bhubaneswar: 17–29.
87. Veeraputhiran, R., Kandasamy, O. S. (1999). Water and nitrogen use efficiency through drip irrigation in hybrid cotton. National Seminar on problems and prospects of micro irrigation–A critical appraisal. November 19–20. Institution of Engineers(India) and Micro Irrigation Society of India, Bangalore. India.
88. Yadav, B. R. (2006). Status of management of saline and alkali water for crop production in arid and semiarid regions. In: Enhancing Water Use Efficiency in Arid and Semiarid Areas for Sustainable Agriculture (Yadav, B. S. Ed.). Agricultural Research Station, Sriganganagar. 236–241.
89. Yadav, R. K. (1983). Clay-drip irrigation-an automatic system. Transaction of the Indian Society of Desert Technology and University Centre of Desert Studies, 8:18–22.

CHAPTER 2

HYDRAULIC PERFORMANCE OF DRIP IRRIGATION SYSTEM: REVIEW

B. C. KUSRE and PRADIP K. BORA

CONTENTS

2.1 INTRODUCTION

Irrigation always plays a catalytic role in enhancing production and productivity. Research studies across different countries have confirmed that irrigation plays a paramount role in achieving food security through enhanced use of inputs, increasing crop intensity and productivity [13, 34, 38]. With the decline of freshwater availability and need for higher food production to feed ever increasing global population, irrigation engineers are compelled to look for greener pasture in water management techniques. Agriculture consumes more than 70% of available fresh water resources. Demand of water has further increased with high yielding varieties and hybrid crops that are basically input intensive. On the other hand, for the economical development of any country, which is closely linked with industrialization, the demand for fresh water in the other competing sectors (such as, industries, power generation, domestic consumption, recreation, etc.) has increased exponentially. Agriculture is the affected sector in this scenario, which is parting with its major share. It has also been reported that creating newer water sources for irrigation has doubled in India, Pakistan, Indonesia, Sri Lanka and Philippines during 1970–1990 [14, 33]. In such water stress situations, efficient use of available water or use of recycled water is one of alternate solutions available to irrigation engineers. Micro irrigation techniques uses small quantity of water that is generally equal to daily crop evapotranspiration. Application of exact quantity of water to meet evapotranspiration demand not only minimizes unproductive water loss through deep percolation and runoff, but also increases crop productivity and minimizes water foot prints. Micro irrigation techniques include drip or trickle and micro-jet irrigation. Drip irrigation system offers number of advantages over conventional irrigation and sprinkler irrigation systems.

2.2 ADVANTAGES AND DISADVANTAGES OF MICRO IRRIGATION

Advantages as well as certain disadvantages of micro irrigation must be considered and understood before adopting this technology. Goyal [16, 17] reported that the advantages are: water conservation, reduced deleterious water quality impacts due to high application efficiencies, automation capabilities, improved or increased yields, ease of chemical applications, and potential sustainability. Disadvantages include a high potential for emitter plugging, high system costs and required high levels of management.

2.2.1 ADVANTAGES

Various studies have been undertaken to assess the impact of micro irrigation (MI). The common results of these studies have shown encouraging impacts on crop productivity and production. The studies have also indicated that micro irrigation can bring diversification of crop production from rainfed to high value horticultural crops and development of cultivable waste lands. Micro irrigation saves up to 40–65% of water among horticultural crops and 30–40% water in vegetable crops (Table 2.1). The drip irrigation also results in labor saving in the farm practices, such as: irrigation application, weeding, fertilization, harvesting, etc. Further it also aids in eliminating drudgery in farm management (irrigating crops during irregular and odd hours of power supply). It also results in saving of energy use in agriculture due to reduced hours of pumping. The adoption of MI has also resulted in enhanced crop yield. However, the extent varied from crop to crop as compared to conventional irrigation. Some improvement in crop quality (such as uniform pod filling in case of groundnut, better shine in sweet orange, uniform bigger sized bananas) has also been reported leading to higher output realization from the produce (Table 2.2). The improved quality and reduction in labor involvement have resulted in enhancing gross value of output. The magnitude of these changes has, however, varied from crop to crop and region to region. Impact of adopting micro irrigation has shown an increase of financial return to the extent of 30–50%. The payback period on investment under MI varied from 0.5 to 1.17 years in groundnut, potato, cotton while it was somewhat higher in horticultural crops.

TABLE 2.1 Water Savings and Crop Yield by Drip Irrigation System [17]

Crop	Water savings (%)	Yield increase (%)
Banana	45	52
Chili pepper	63	45
Grape	48	23
Pomegranete	45	45
Sugarcane	56	33
Sweet lime	61	50
Tomato	31	50
Water melon	36	88

Source: INCID (1994) Drip irrigation in India [17].

TABLE 2.2 Yield Increase Under Drip Irrigation Over Conventional Irrigation [17]

Crop	Yield (tons/ha)		
	Conventional	**Drip**	**% increase**
Banana	57.5	87.5	52
Chillies	4.2	6.1	44
Cotton	2.3	2.9	26
Grapes	26.4	32.5	23
Okra	15.3	17.7	16
Papaya	13.4	23.5	75
Pomegranate	55.0	109.0	98
Sugarcane	128.0	170.0	33
Sweet lime	100.0	150.0	50
Sweet Potato	4.2	5.9	39
Tomato	32.0	48.0	50
Water Melon	24.0	45.0	88

Source: INCID (1994) Drip irrigation in India [17].

2.2.2. DISADVANTAGES [16, 17]

a. **Clogging of emitters**: Emitter orifice diameter is between 0.5–1 mm and is therefore vulnerable to clogging by root penetration, sand rust, micro organisms or other impurities. The clogging reduces flow and may lead to non-uniform distribution of water.
b. **Salt accumulation at the periphery of water front**: It causes serious problems in subsequent crops if irrigated by a method other than drip or in arid areas where rainfall is insufficient to leach the accumulated salts.
c. **Lack of proper root growth**: Due to localized application of water near the root zone, limited growth of roots has been reported. Such restricted growth weakens the trees to withstand external impacts of storms. This is important particularly in orchard crops.
d. **Damage due to rodents**: The plastic components of drip irrigation system can attract rodents such as rats, squirrels, etc. These rodents can damage the system by puncturing the system.
e. **Operational constraints**: The drip irrigation system operation requires skills for operation and maintenance (O&M). The O&M steps may decide the duration of operation, scheduling of sub – sections, knowledge of impacts of using chemicals on the components of micro irrigation, etc. Proper training and awareness is required for judicious operation of the system.

f. **High cost**: The drip irrigation system involves assembly of sub components. These parts and sub parts involve cost, which many a times are beyond the reach of poor farmers. The involvement of high cost discourages adoption of drip irrigation.

2.3 UNIFORMITY OF MICRO IRRIGATION

The drip irrigation system is known for its high water distribution uniformity compared to conventional system. Evaluation of distribution uniformity on land surface is accepted as one of the key criteria for monitoring irrigation performance [42]. The uniformity of application is mostly dependent on the physical and hydraulic characteristics of the various components of drip system [1]. Mizyed and Kruse [30] enumerated the factors that affect the uniformity of water application uniformity. These factors are: (i) manufacturing variations in emitters and pressure regulators, (ii) pressure variations caused by elevation changes, (iii) friction head losses throughout the pipe network, (iv) emitter sensitivity to pressure and irrigation water temperature changes, and (v) emitter clogging. Barragan et al. [4] stated that among various factors affecting uniformity of micro irrigation, emitter plugging is most significant followed by grouping of emitters for low-density crops and spacing for high-density crops. Capra and Sciolone [11] mentioned that variation in uniformity of drip emitter is due to emitter design, the material used to manufacture the drip tubing, and precision. Wu [40, 41] stated that both the hydraulic design and the manufacturing variations, provided they are designed within a specified range, are less significant. Bralts et al. [7] mentioned the following equation for variation of discharge of emitters due to both hydraulic and manufacturing variation:

$$CV^2 = CV_{(m)}^2 + CV_{(h)}^2 \quad (1)$$

where, CV(m) is the coefficient of manufacturers variation of emitter flow (dimensionless), and CV(h) is the coefficient of variation of emitter flow caused by the hydraulic design (dimensionless).

Hezarjaribi et al. [18] conducted hydraulic performance analysis of various emitters at different operating pressures. They assessed the manufacturing variation coefficient, emitter discharge coefficient and emitter discharge exponent in order to establish flow sensitivity to pressure and compared these with manufacturer's specifications. Their results indicated that design should be based on reliable test data, not on manufacturer's specifications. It can be interpreted that using manufacturer's data will lead to nonuniformity of discharge throughout the system. Ozekici, and Sneed [32] reported that observed values of coefficient of manufacturing variation were higher than those specified by the manufacturer. High coefficients of manufacturing variation could result in low emission uniformities. Designs based on supplied data may deliver too little water to some plants and too much water to others. Therefore, designs should be based on reliable test data, not on manufacturer's specifications.

2.4 EMISSION UNIFORMITY FOR DESIGN OF MICRO IRRIGATION

Emission uniformity (E_u) has been one of the most frequently used design criteria for microirrigation since the development of drip irrigation [4]. ASAE also recommended evaluation of system uniformity of micro irrigation by the ASAE Standards [2, 3, 27]. Emission uniformity expresses the emitter flow variation of micro irrigation that is affected by hydraulic variation, manufacturer's variation and emitter grouping.

Generally two concepts of emission uniformity are being accepted: manufacturers' variation and hydraulic variation. The estimation of uniformity of application can be achieved by number of methods such as from simple range of maximum to minimum, or minimum to mean to the statistical terms, such as uniformity coefficient and coefficient of variation. Some of the methods for estimating uniformity coefficient are discussed in this section.

2.4.1 *UNIFORMITY COEFFICIENT PARAMETERS*

Various parameters are used for measuring emitter discharge uniformity. These parameters are emitter flow rate variation (q_{var}), ratio of maximum to minimum discharge, ratio of minimum discharge to average discharge, coefficient of variation (CV), uniformity coefficient (UC), and distribution uniformity (DU). These parameters are described below:

1. Flow rate variation [39]

$$q_{var} = \frac{q_{max} - q_{min}}{q_{max}} \tag{2}$$

where, q_{max} is the maximum flow rate, and q_{min} is the minimum emitter flow rate.

2. Ratio of maximum to minimum discharge [19]

$$U_c = \frac{q_{\min(h)}}{q_{\max(h)}} \tag{3}$$

3. Keller and Karmeli [22] defined emission uniformity as the ratio of minimum emitter flow to mean emitter flow:

$$E_{UK} = \frac{q_{min}}{\bar{q}} \tag{4}$$

where, E_{UK} is the emission uniformity by Keller and Karmeli [22], dimensionless; q_{min} is average discharge from emitters in the lowest 25% of emitter flow rate in lph; and is the average of all emitter flow rate in lph.

4. Coefficient of variation was introduced by Keller and Karmeli [22, 23, 24] as statistical measure of manufacturers' variation, and is described below:

$$CV = \frac{S_d}{\dot{q}} \tag{5}$$

where, CV is the discharge coefficient of variation (%); S_d represents the standard deviation of discharge rates of sample emitters (lph), $\overline{q}$ is the mean emitter flow rate (lph). Several guidelines have been suggested for classifying the values of CV by International Standard Organization [21] and ASAE [2, 3]. The values recommended by both the organizations are shown in Tables 2.3 and 2.4.

TABLE 2.3 Recommended Classification of Manufacturer's Coefficient of Variation (CVm) by ASAE [2, 3]

S. No	CV_m (%)	Classification
1	<5	Excellent
2	5–7	Average
3	7–11	Marginal
4	11–15	Poor
5	>15	Unacceptable

TABLE 2.4 Classification of Coefficient of Variation Values According to ISO Standards [21]

Category	CV	Details	Classification
A	0 to ±%5	Higher uniformity of emission rate and smaller deviations from the specified nominal emission rate	Good
B	±5 to ±10%	Medium uniformity of emission rate and medium deviations from the specified nominal emission rate	Medium
C	>10%	Lower uniformity of emission rate and greater deviations from the specified nominal emission rate	Poor

5. UC is defined by Christiansen [12] and is modified to express it in percentage:

$$UC = 100\left[1 - \frac{\frac{1}{n}\sum_{i=1}^{n}\left|q_i - \overline{q}\right|}{\overline{q}}\right] \tag{6}$$

where, n is the number of emitters under considerations.

6. DU is defined in the following equation by Kusre [26]:

(7)

where, $\overline{q}_{lq}$ is the mean of lowest one fourth of emitter.

7. ASAE EP458: ASAE [2] also suggested estimation of uniformity coefficient of drip irrigation system. ASAE Engineering Practice EP458 was adopted by American Society of Agricultural Engineers as an approved practice to evaluate the micro irrigation systems in the field [2]. The method was revised in 1997 to bring clarity and to reduce complexity of the evaluation [27]. The method uses statistical uniformity, emitter discharge variation, hydraulic variation, and emitter performance variation to evaluate drip irrigation systems in the field. Confidence limits (95%) for calculated uniformity parameters were determined using the procedure by Bralts and Kesner [8], because confidence limits were not included in EP458 for the number of emitters under considerations [24]. Most of the uniformity values require the determination of mean emitter discharge rate, $\overline{q}$, and standard deviation, *Sq*, which were calculated using the equations:

$$\overline{q} = \frac{1}{n}\sum_{i=1}^{n} q_i \tag{8}$$

$$S_q = \sqrt{\frac{\sum_{i=1}^{n} q_i^2 - \frac{1}{n}\left(\sum_{i=1}^{n} q_i\right)^2}{n-1}} \tag{9}$$

The emitter discharge coefficient of variation, *Vqs*, and statistical uniformity, *Us*, are described by following equations:

$$V_{qs} = \frac{S_q}{q} \tag{10}$$

$$U_s = 100(1 - V_{qs}) \tag{11}$$

The mean hydraulic pressure, P_{ave}, and hydraulic design coefficient of variation, *Vhs*, can be determined using Eqs. (6) to (8), respectively, by substitution of lateral line pressure, *hi*, with emitter discharge, *qi*, while all other variables are previously described in these equations. The emitter discharge coefficient of variation due to pressure head, *Vqh*, can be calculated using the equation:

$$V_{qh} = xV_{hs} \tag{12}$$

where, x is the emitter discharge exponent. The statistical uniformity of emitter discharge rate due to pressure head, U_{sh}, can be calculated as follows:

$$U_{sh} = 100(1 - V_{qh}) \tag{13}$$

The emitter performance variation is a measure of emitter discharge variability due to water temperature, emitter manufacturers' variation, emitter wear, and emitter plugging. The emitter performance coefficient of variation, Vpf, was calculated using the previously determined emitter discharge coefficient of variation, Vqs, the emitter discharge coefficient of variation due to pressure head, Vqh, and the equation:

$$V_{pf} = \sqrt{V_{qs}^2 - V_{qh}^2} \tag{14}$$

For the case where flow was adjusted for a constant pressure, Vhs and Vqh are set equal to zero, and $Ush = 100$. As a result, we have $Vpf = Vqs$.

Wu [41] stated that the three statistical parameters (uniformity of coefficient (UCC) in Eq. (6), the CV in Eq. (5), and statistical uniformity (UCS) in Eq. (11)) are interrelated. Further, it has also been reported that the emitter flow variation, q_{var}, also shows high correlation between UCC and CV.

2.4.2 EMITTER EXPONENT AND FLOW REGIME

Apart from the above-mentioned criteria, the relationship between emitter discharge and pressure variation also indicates performance of drip irrigation system. In the field, it is expected that an emitter should uniformly discharge water at wide ranges of pressure. Technically the emitter discharge is a function of operating pressure [20, 23]:

$$q = kP^x \tag{15}$$

where, k is the discharge coefficient, and x is the flow exponent that is a characteristic of the emitter flow regime and may be used to characterize hydraulic performance of any given emitter [6]. These two coefficients can be calculated by using the following equations [23]:

$$\log k = \frac{\sum logq_i \sum (\log P_i)^2 - \sum (\log q_i \log P_i) \sum P_i}{m \sum (\log P_i)^2 - \left(\sum \log P_i\right)^2} \tag{16}$$

$$x = \frac{m\sum \log q_i \log P_i - \sum \log q_i \sum \log P_i}{m\sum (\log P_i)^2 - \left(\sum \log P_i\right)^2} \tag{17}$$

The relationship between the flow velocity (v) and discharge (q) can be described by the following equation:

$$v = \frac{q}{3.6 \times 10^6 A_s} \tag{18}$$

where, A_s = Sectional area of the emitter channel. Substituting Eq. (15) into Eq. (18), we get:

$$v = \frac{1}{3.6 \times 10^6 A_s} kH^x \tag{19}$$

The Eq. (19) shows that the v-H relationship is determined by the flow exponent in the relationship of q-H. The Reynolds number can be calculated by the following equation [15]:

$$R_e = \frac{\rho v d_e}{\mu} \tag{20}$$

where, R_e = Reynolds number, ρ = fluid density (kg/m^3), μ = fluid viscosity coefficient (kg/(m.s)), d_e = equivalent diameter which is equal to four times of the hydraulic radius r_h (m). In case of a rectangular cross section, d_e can be calculated as follows:

$$r_h = \frac{ab}{2(a+b)} \tag{21}$$

$$d_e = 4r_h = \frac{2ab}{(a+b)} \tag{22}$$

where, a and b are the width and depth of flow path of emitters in meters. Substituting Eqs. (19) and (22) into Eq. (20), we get:

$$R_e = \frac{\rho ab}{1.8 \times 10^6 A_s \mu (a+b)} kH^x \tag{23}$$

where, A_s = area = *ab* for a rectangular. Then, Eq. (23) reduces to:

$$R_e = \frac{\rho As}{1.8 \times 10^6 \mu (a+b)} kH^x \tag{24}$$

The Eq. (24) shows that the relationship between Reynolds number and operating pressure head is characterized by the flow exponent.

2.4.3 FLOW REGIMES

Four flow regimes were defined as a function of Reynolds number by Mane et al. [29]:

- Laminar flow regime, when $R_e \leq 2000$.
- Unstable flow regime (Transition or critical), when $2000 < Re \leq 4000$.
- Partially turbulent flow regime, when $4000 < Re \leq 10{,}000$.
- Fully turbulent flow regime, when $Re > 10{,}000$.

For the discharge to be least sensitive to pressure variations, the flow exponent x should be equal to zero but practically such emitters do not exist in the market. Typically, the value of x varies from 0.1 to 1.0, depending on the make/ model/ type and design of an emitter. At $x = 1$, the flow is fully laminar; and at $x = 0$, the flow is turbulent. Generally under fully laminar flow regime, emitters must be very sensitive to changes in pressure. This implies that the variation of pressure head will be proportional to the variation of discharge, in a fully laminar flow. It has been reported that most noncompensating emitters are always fully turbulent with an $x = 0.5$, and a pressure variation of 20% will result in a flow variation of approximately 10%. Whereas the pressure compensating drippers are insensitive to pressure variation [25]. The value of emitter coefficient x generally varies between 0.1 to 0.4. Ideally a pressure compensating dripper should have x equal to zero [5, 9, 18, 37]. The variation of discharge with change in pressure can be attributed to friction, elevation, accidental restriction resulting in nonuniform water application [7, 9]. The emitters with turbulent flow cause less plugging or clogging according to Goyal [17].

2.5 RANGE FOR ACCEPTING EMITTER UNIFORMITY OF DISTRIBUTION

Ortega et al. [31] conducted a study on emission uniformity (EU), pressure variation coefficient (VC_p), and flow variation coefficient per plant (VC_q) at localized systems. They reported that systems with $VC_q > 0.4$ are unacceptable, while $VC_q < 0.1$ are most acceptable. Apart from pressure variation along irrigation tape, variation in emitter structure or emitter geometry has been known to cause poor uniformity of emitter discharge [25, 40].

Qualitative classification standards for the production of emitters, according to the manufacturers' coefficient of emitter variation (CV_m), have been developed by ASAE. CV_m values below 10% are acceptable and > 20% are unacceptable [3]. The emitter discharge variation rate (q_{var}) should be evaluated as a design criterion in drip irrigation systems; $q_{var} < 10\%$ may be regarded as good and $q_{var} > 20\%$ as unacceptable [10, 39]. The acceptability of micro irrigation systems has also been classified according to the statistical parameters, U_{qs} and EU. Namely, EU = 94–100% and Uqs = 95–100% are excellent; and EU < 50% and $U_{qs} < 60\%$ are unacceptable [2].

Comparison of uniformity through traditional approach and ASAE methods indicated both methods as suitable. However, the values for the uniformity are slightly lesser in ASAE methods as compared to other methods [35]. Safi et al. [35] reported that the ASAE showed slightly lower uniformity in both unused and used tapes (1.6% and 3.65%, respectively). Camp et al. [10] suggested using EPA-458 in estimating uniformity coefficient particularly for used emitters. In general, higher values of uniformity parameters are more desired in high value crops.

2.6 UNIFORMITY MEASUREMENT

Ideally it is necessary to measure the discharge of all emitters in the field for obtaining the above described uniformity parameters. However, practically it is not feasible to measure the discharge of all the emitters. Smajstrla et al. [36] recommended measurement of minimum of 18 emitters to accurately determine uniformity parameters. Computations will be simplified if the number of emitters under consideration is a multiple of six. The statistical coefficient of variation is then calculated from these data points. Similar observations were also made by Bralts and Kesner [8], as well as procedure adopted by EPA 458 [10]. Lamm et al. [28] stated that the random 18-sample size survey may cause a maximum of ± 36% variation in the coefficient of variation. They further stated that increase in number of emitters will yield slight improvement in uniformity coefficient values, thus sample size of 18 emitter was recommended.

2.7 CONCLUSIONS

Micro irrigation systems are promoted to achieve a higher water savings and improved uniformity of water application. The uniformity of a micro irrigation system may be explained in terms of system uniformity or uniformity in discharge in individual emitters. It is a result of hydraulic design, manufacturers' variation, grouping of individual emitters, plugging of emitter opening, etc. The performance of micro irrigation system should be monitored at regular intervals to ensure designed uniformity. This approach offers a check, and deviation from the normal conditions will require system adjustments to bring back to designed performance.

In the present scenario, the flow variation of emitter is accepted as 10–20%, while manufacturer's variation is accepted in the range of 2–20%. It has also been suggested in some studies to adopt grouping of emitters to improve the uniformity of application. Plugging has been reported by number of researchers as the principal reason for deviation from the optimum uniformity. It has also been reported that a 10% plugging may result 30% variation in coefficient of variation of all the emitter discharge and more than 10% variation in UCC on spatial uniformity. To improve the uniformity of water application, closer spacing can be effective. Spacing less than 0.5 m can achieve more than 80% in uniformity coefficient.

For estimation of uniformity of water application of micro irrigation emitter, any of the method can be used. However, the studies indicated that ASAE method gives slightly lesser value.

2.8 SUMMARY

Efficient use of water resources for producing crop has always been a challenge to irrigation engineers. Recently, number of technological interventions has been adopted worldwide for improving water use efficiency in agriculture. Drip or trickle irrigation is one of such interventions. The primary objective of drip irrigation is to improve application efficiency of water. It has been a driving force in increasing crop productivity in water scarce situation. "More crops per drop" is the hallmark of this irrigation system. Water application uniformity is also greatly achieved in a suitably designed system of drip irrigation. It has been reported that productivity of crop is related to application uniformity. On the other hand, poorly designed system can lead to nonuniform water application. There are number of reasons for non-uniform application *viz.*, non-standardized equipment and inadequate design considerations. The non-uniformity is generally expressed as emission uniformity (E_u) that is used as design criteria for drip irrigation. Emission uniformity expresses the emitter flow variation of a micro irrigation system affected by hydraulic variation, manufacturers' variation and emitter grouping. Several standards are available for evaluation of system uniformity. In this chapter, authors have discussed most frequently used E_u parameters used in drip irrigation system design. The information in this chapter can be helpful to irrigation professionals to improve the design of drip irrigation system.

KEYWORDS

- **application uniformity**
- **ASAE**
- **clogging**
- **design considerations**
- **drip irrigation**
- **emission uniformity**
- **emitter**
- **field uniformity**
- **flow variation**
- **hydraulic variation**
- **irrigation**
- **micro irrigation**
- **system uniformity**
- **trickle irrigation**
- **water quality**
- **water use efficiency**

REFERENCES

1. Al-Amound, A. I. (1995). Significance of energy losses due to emitter connections in trickle irrigation lines. *J. Agric. Eng. Res.*, 60:1–5.
2. ASAE Standards, (1996). EP458: Field evaluation of micro irrigation systems. *St. Joseph, MI, ASAE*, 43:756–761.
3. ASAE Standards, (2005). EP405.1: Design and installation of micro irrigation systems. *ASAE, St. Joseph, MI.*
4. Barragan, J., Bralts, V. F., Wu, I. P. (2006). *Assessment of emission uniformity for micro irrigation design. Biosystems Engineering (ASAE),* 93(1):89–97.
5. Boswell, M. J. (1985). Design characteristics of line-source drip tubes. *Proceedings of the Third International Drip/Trickle Irrigation Congress, vol I.* California, USA, pp. 306–312.
6. Bralts, V. F., Wu, I. P. (1979). Emitter flow variation and uniformity for drip irrigation. ASAE Paper No. 79–2099. St. Joseph, MI.
7. Bralts, V. F., Kesner, C. (1982). Drip irrigation field uniformity estimation. *Transactions of the ASAE*, 24:1369–1374.
8. Bralts, V. F., Wu, I. P., Gitlin, H. M. (1981). Manufacturing variation and drip irrigation uniformity. *Transactions of the ASAE*, 24:113–119.
9. Braud, H. J., Soon, A. M. (1980). Trickle irrigation design for improved application uniformity. ASAE and CSAE joint meeting, paper no. 79–2571, Winnipeg, Canada.
10. Camp, C. R., Sadler, E. J., Busscher, W. J. (1997). *A comparison of uniformity measure for drip irrigation systems. Transactions of the ASAE*, 40:1013–1020.
11. Capra, A., Scicolone, B. (1998). Water quality and distribution uniformity in drip/trickle irrigation systems. *J. Agric. Eng. Res.*, 70:355–365.
12. Christiansen, J. E. (1942). Hydraulics of sprinkler irrigation system. *Trans. Amer. Soc. Civ. Eng.*, 107:221–239.
13. Dhawan, B. D. (1988). Role of Irrigation in raising intensity of cropping. *Journal of Indian School of Political Economy*, 3(4, October–December):632–671.
14. Dinar, A., Subramanian, A. (1997). Water pricing experiences: An international perspective. World Bank Technical Paper No. 386. Washington, D.C
15. Ferziger, J. H., Peric, M. (1996). Computational Methods for Fluid Dynamics. Springer, New York.
16. Megh R. Goyal (2014). *Management of Drip/Trickle or Micro Irrigation.* Oakville, ON, Canada: Apple Academic Press Inc.
17. Megh R. Goyal (2015). *Research Advances in Sustainablr Micro Irrigation.* Oakville, ON, Canada: Apple Academic Press Inc.
18. Hezarjaribi, A., Dehghani, A. A., Helghi, M. M., Kiani, A. (2008). Hydraulic performances of various trickle irrigation emitters. *Journal of Agronomy*, 7:265–271.
19. Howell, T. A., Hiler, E. A. (1974). Trickle irrigation lateral design. *Transactions of the ASAE*, 17:902–908.
20. Howell, T. A., Barinas, F. A. (1980). Pressure losses across trickle irrigation fittings and emitters. *Trans. ASAE*, 23:928–933.
21. International Standard, (1991). Agricultural irrigation equipment: emitters specification and test methods. *ISO 9260*, Acc No. 319301.
22. Keller, J., Karmeli, D. (1974). Trickle Irrigation Design Parameters. *Transactions of the ASAE*, 17(4):678–684.
23. Keller, J., Bliesner, R. D. (1990). *Trickle Irrigation Design.* Van Nostrand Reinhold, New York.
24. Keller, J., Blisner, R. D. (1990). *Sprinkler and Trickle Irrigation.* Van Nostrand Reinhold, New York.

25. Kirnak, H., Dogan, E., Demir, S., Yalcin, S. (2004). Determination of hydraulic performance of trickle irrigation emitters used in irrigation systems in the Harran Plain. *Turkish Journal of Agriculture and Forestry*, *28*(4):223–230.
26. Kruse, E. G. (1978). Describing irrigation efficiency an uniformity. *J. Irrigation and Drainage Division, ASCE*, 104(IR):35–41.
27. Lamm, F. R., Storlie, C. A., Pitts, D. J. (1997). Revision of EP-458: Field evaluation of micro irrigation systems. *ASAE Paper 97–2070*, pp. 20.
28. Lamm, F. R., Ayars, J. E., Nakayama, F. S. (2007).*Microirrigation for Crop Production: Design, Operation and Management*. Elsevier Publications. 608 pp.
29. Mane, M. S., Ayare, B. L., Magar, S. S. (2008). *Principles of Drip Irrigation System*. 2nd Edition. Jain Brothers, New Delhi–India.
30. Mizyed, N., Kruse, E. G. (1989). Emitter discharge evaluation of subsurface trickle irrigation systems. *Transactions of the ASAE*, 32:1223–1228.
31. Ortega, J. F., Tarjuelo, J. M., Juan, J. A. (2002). Evaluation of irrigation performance in localized irrigation system of semiarid regions (Castilla–La Mancha, Spain). *Agricultural Engineering International: CIGR Journal of Scientific Research and Development*, 4:1–17.
32. Ozekici, B., Sneed, R. E. (1995). Manufacturing variation for various trickle irrigation on-line emitters. *Applied Engineering in Agriculture (USA)*, 11:235–240.
33. Rosegrant, M. W., Svendsen, M. (1993). Asian food production in the 1990 s: Irrigation investment and management policy. *Food Policy*, 18 (February):13–32
34. Rosegrant, M. W., Cal, X., Cline, S. A. (2005). World Water and Food in 2025: Dealing with scarcity. *International Food Policy Research Institute*, Washington, D.C.
35. Safi, B., Neyshabouri, M. R., Nazemi, A. H., Massiha, S., Mirlatifi, S. M. (2007). Water application uniformity of a subsurface drip irrigation system at various operating pressures and tape lengths. *Turkish Journal of Agriculture & Forestry*, 31(5):275–285.
36. Smajstrla, A. G., Boman, B. J., Haman, D. Z., Pitts, D. J., Zazueta, F. S. (2012). *Field evaluation of microirrigation water application uniformity*. Bulletin 265, Agricultural and Biological Engineering Department, Florida Cooperative Extension Service, Institute of Food and Agricultural Sciences, University of Florida.
37. Solomon, K., Bezdek, J. C. (1980). Simulated flow rate requirements for some flushing emitters. ASAE and CSAE joint meeting, Paper no. 79–2571, Winnipeg, Canada.
38. Vaidyanathan, A., Krishnakumar, A., Rajagopal, A., Varatharajan, D. (1994). Impact of irrigation on productivity of land. *Journal of Indian School of Political Economy*, 6(4, October–December):601–645.
39. Wu, I. P., Gitlin, H. M. (1974). Drip irrigation design based on uniformity. *Transactions of the ASAE*, 17(3):157–168.
40. Wu, I. P., Gitlin, H. M. (1979). *The Manufacturer's coefficient of variation of emitter flow for drip irrigation*. University of Hawaii at Manoa and U.S.D.A., pp. 1–3.
41. Wu, I. P. (1993). Microirrigation design for trees. Paper No. 93–2128 at the ASAE Summer Meeting. Spokane, Washington.
42. Yavuj, M. Y., Demırel, K., Erken, O., Bahar, E., Devecıler, M. (2010). *Emitter clogging and effects on drip irrigation systems performances. African Journal of Agricultural Research*, 5(7):532–538.

CHAPTER 3

MOISTURE DISTRIBUTION UNDER DRIP IRRIGATION

D. V. TATHOD, Y. V. MAHATALE, and V. K. CHAVAN

CONTENTS

3.1 INTRODUCTION

Indian economy is mainly dependent on agriculture, and water is the major agriculture input for crop production. Therefore, irrigation plays an important role to increase crop yield. Agriculture accounts for 48% of the national income and 72% Indians depend on agriculture and its production in India.

The scientific utilization of agricultural water resource involves consideration of adopting advanced irrigation methods [5, 6]. In traditional irrigation methods, there is a fluctuation in the water content, temperature, and soil aeration, which results in plant stress. Drip irrigation is becoming increasingly popular in areas with water salinity and salt problems. Drip irrigation is a method of watering plants frequently with a volume of water approaching the consumptive use of plant, thereby minimizing losses due to deep percolation, runoff and soil water evaporation. The system applies water slowly to keep soil moisture with in desired range for plant growth. Drip irrigation can achieve 90% or more application efficiency, which can hardly be achieved by other method.

The temporal and spatial soil moisture distribution is considerable factor for healthy plant growth. The water dropping on to the ground surface enters the soil profile and percolates downward and laterally. The size and shape of wetting pattern depends on the discharge of dripper, the duration of application and type of soil. The design of drip irrigation requires the knowledge of water distribution pattern for various discharge rates. The knowledge of moisture distribution pattern will determine the effectiveness of drip irrigation. The design of drip irrigation system mainly involves the measurement of lateral and emitter spacing, which is a function of wetted area of crop root zone. It is essential to know distribution pattern with relation to discharge rate of dripper after water application.

Therefore, this study was conducted to evaluate temporal and spatial moisture distribution patterns due to 4 lph and 8 lph emitters under different soil depths.

3.2 METHODOLOGY

This study was carried out at the research farm of Department of Irrigation and Drainage Engineering at College of Agricultural Engineering & Technology, Jalgaon (Jamod) – India. The field was thoroughly investigated to select a suitable block for the experimentation. The field capacity at the site was 24.25%. Soil properties are shown in Table 3.1.

TABLE 3.1 Mechanical Analysis of Soil

Bulk Density	1.42 gm/cc
Field Capacity	24.25%
Sand	25.58%
Silt	32.25%
Clay	42.17%

FIGURE 3.1 An emitter on a 16 mm lateral.

The moisture distribution pattern was determined under an emitter of 4 and 8 lph that were installed on 16 mm diameter of lateral as shown in Fig. 3.1. The irrigation system was operated for one hour. Then soil samples were collected with an auger at in three soil depths (0–10, 10–20, 20–30 cm) and at three lateral locations from the emitter (0, 4 and 8 cm). Soil samples for moisture determination were taken an elapsed time of 0, 6, and 24 h after irrigation. The sample was oven dried for 24 h at 105 °C, to determine soil moisture content (%). The soil moisture distribution patterns were plotted by using surfer software.

3.3 RESULTS AND DISCUSSION

The Table 3.2 presents the soil moisture distribution data at different locations and elapsed time. The data was plotted as shown in Figs. 3.2 and 3.3. The Fig. 3.2 shows soil moisture distribution patterns under an emitter of 4 lph for an elapsed time of 0, 6, and 24 h after irrigation. The Fig. 3.3 shows soil moisture distribution patterns under an emitter of 8 lph for an elapsed time of 0, 6, and 24 h after irrigation.

TABLE 3.2 Soil Moisture Distribution An Emitter on a 16 mm Lateral

Soil depth	Soil moisture content, %								
	Just after irrigation at			6 h after irrigation at			24 h after irrigation at		
	Lateral distance, cm			Lateral distance, cm			Lateral distance, cm		
cm	0 cm	4 cm	8 cm	0 cm	4 cm	8 cm	0 cm	4 cm	8 cm
Under a 4 lph emitter									
Surface	28.34	26	26.25	26.66	18.9	17.57	23.57	18.45	15.22
0–10	28.9	28.36	24.32	25.19	24.37	20.16	21.12	18.72	16.5
10–20	27.81	27.97	22.22	26.44	25.07	22.25	23.27	22.7	23.07
20–30	20.60	17.33	16.78	23.5	23.57	19.77	24.57	25.8	23.2
Under a 8 lph emitter									
Surface	33.65	30.76	21.85	28.29	24.69	20.81	26.37	21.94	19.25
0–10	31.98	30.94	28.33	20.28	23.18	22.26	18.98	21.21	20.98
10–20	31.5	26.18	24.48	24.6	23.17	23.17	25.25	23.7	23.71
20–30	16.9	14.8	13.6	20.6	22.7	20.3	25.5	24.07	24.3

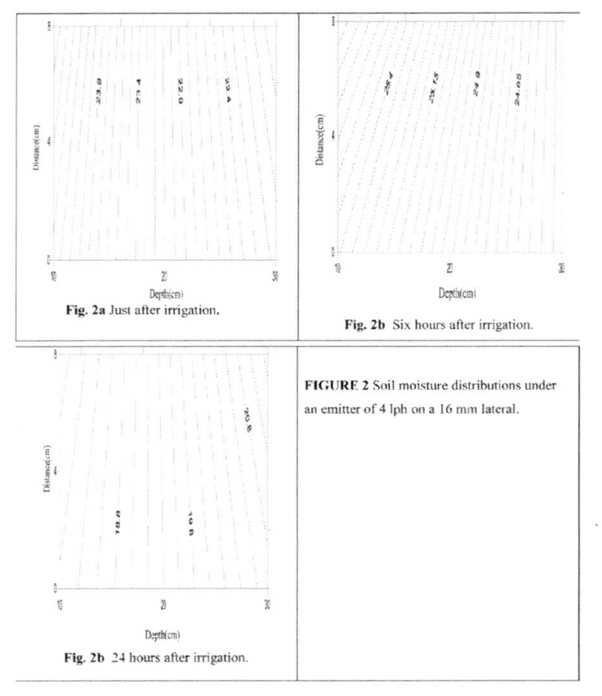

FIGURE 3.2 Soil moisture distributions under an emitter of 4 lph on a 16 mm lateral. (a) Just after irrigation. (b) Six hours after irrigation. (c) 24 h after irrigation.

Figure 3.2a shows moisture distribution pattern for "just after irrigation." It can be concluded that soil moisture content was 22.6% at 10–20 cm that was close to field capacity depth. Soil moisture content was 19.6% at 20–30 cm that was less than field capacity. Figure 3.2b shows soil-wetting pattern for 6 h after irrigation. It can be observed that moisture content from surface was decreased and was maintained up to field capacity in the root zone depth of the crop. Figure 3.2c shows moisture distribution pattern for 24 h after irrigation. Graph reveals that moisture content was decreased from surface to 20 cm depth indicating 21% moisture in 10–20 cm depth and near field capacity was in root zone.

Figure 3.3a shows soil-wetting pattern for "just after irrigation." It can be concluded that the moisture content at 10–20 cm depth was less than field capacity but higher than 20–30 cm depth. Mean soil moisture content was 20% 10–20 cm depth compared to 17.5% for 20–30 cm depth. Figure 3.3b shows moisture distribution pattern for 6 h after irrigation. The graph show that moisture was moves from surface to root zone depth. Nearly same moisture level was maintained in 10–20 cm (22.84%) and 20–30 cm (22.74%) depths. Both values are almost close to field capacity. Figure 3.3c shows moisture distribution pattern for 24 h after irrigation. It can be observed that moisture content was decreased from the surface and nearly constant moisture content was maintained in the root zone depth. There is also horizontal movement of moisture up to 8 cm distance away from point source.

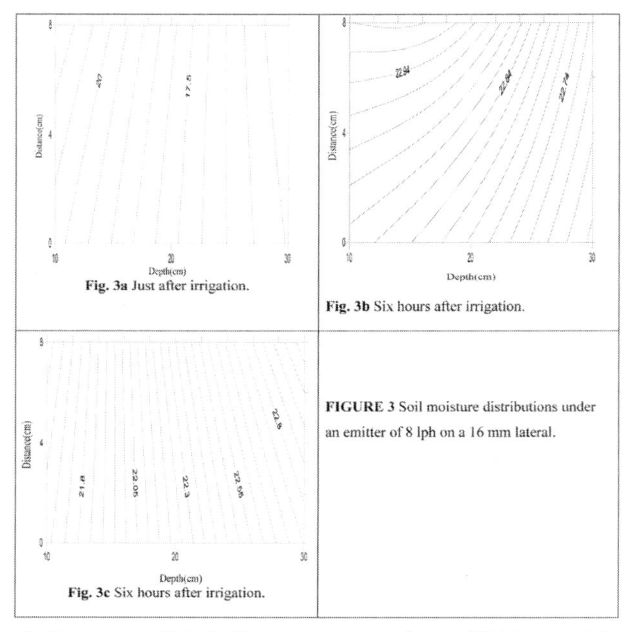

FIGURE 3.3 Soil moisture distributions under an emitter of 8 lph on a 16 mm lateral. (a) Just after irrigation. (b) Six hours after irrigation. (c) Six hours after irrigation.

From Fig. 3.3, it can be concluded that the soil moisture content was higher on the surface at just after irrigation but there was movement of moisture from surface to root zone depth. Nearly constant moisture up to field capacity was maintained in the root zone.

3.4 CONCLUSIONS

For 4 lph dripper with 15 cm spacing on 12 mm lateral, the soil moisture was moved on surface and was greater than field capacity. As the elapsed time after irrigation, the moisture was concentrated in root zone depth and was near field capacity. The moisture content was maintained to field capacity at surface as well as in the root zone for just after irrigation. For 24 h after irrigation, soil moisture was rapidly decreased from surface to root zone depth for 8 lph dripper on 16 mm lateral.

For 4 lph dripper with 15 cm spacing on 16 mm lateral, the soil moisture was higher for just after irrigation on surface, while for 24 h of irrigation moisture was maintained at field capacity in root zone depth of crop. The moisture content was more on surface just after irrigation but there was movement of moisture from surface to root zone depth and nearly constant moisture was maintained in root zone. Moisture was uniformly distributed for 4 lph dripper with 15 cm spacing on 12 mm lateral. For 16 mm lateral in both drippers, moisture was maintained near field capacity.

3.5 SUMMARY

The study was conducted to evaluate soil moisture distribution patterns under an emitter, at the research farm of Irrigation and Drainage Engineering at College of Agricultural Engineering and Technology, Jalgaon – Maharashtra. Authors studied effects of depth, distance and elapsed time after irrigation (just after irrigation, after 6 h and 24 h) under 4 lph and 8 lph dripper on 16 mm diameter laterals at 0–10, 10–20, 20–30 cm soil depths on soil moisture distribution. The field capacity, bulk density and permanent wilting point were 24.25%, 1.42 gm/cm^3 and 10.10%, respectively. The results showed that soil moisture distribution varied with elapsed time, depth and lateral distance from the emitter. The drip irrigation saved water and directed it to root crop zone. The moisture content was higher near soil surface just after irrigation, but there was movement of moisture from surface to root zone depth and moisture was nearly constant (i.e., up to field capacity was maintained in root zone). With 16 mm lateral with both drippers, moisture content was maintained near field capacity.

KEYWORDS

- **clay loam**
- **drip irrigation**
- **dripper**
- **emitter**
- **field capacity**
- **India**
- **micro irrigation**
- **micro sprinkler**
- **moisture distribution**
- **root zone**
- **soil moisture**
- **trickle source**
- **wetted soil volume**

REFERENCES

1. Acar, B., Topak, R., Milkaisoy, R. (2008). Effect of applied water and discharge rate on wetted soil volume in loam or clay loam soil from an irrigated trickle source. *Afr. Agric. Res. J.*, 4(1):49–54.
2. Fasinmirin, J. T., Oguntuase, A. M. (2008). Soil moisture distribution pattern in *Amaranthus cruentus* under drip irrigation system. *Afr. Agric. Res. J.*, 3(7):486–493.
3. Fieke, N. N., Salunke, D. S. (1992). Soil moisture movement in microsprinkler irrigation. *J. Agril. Eng. Today*, 15(1):52–55.
4. Goel, A. K., Gupta, R. K., P. Kumar, (1993). Effect of drip discharge rate on soil moisture distribution pattern. *J. Water Agric. Management*, 1(1):50–51.
5. Megh R. Goyal, (Ed.), (2013). *Management of Drip/Trickle or Micro Irrigation.* Oakville–ON, Canada: Apple Academic Press Inc., pp. 1–408.
6. Megh R. Goyal, (Ed.), (2015). *Research Advances in Sustainable Micro Irrigation, volumes 1–10.* Oakville–ON, Canada: Apple Academic Press Inc.

CHAPTER 4

WETTED ZONE BEHAVIOR UNDER MICRO IRRIGATED CROPS[1, 2]

HAMMAMI MONCEF, DAGHARI HÉDI, and HATIRA ABDESSATAR

CONTENTS

[1]Modified and printed from: Hammami Moncef, Daghari Hédi and Hatira Abdessatar, 2011. An empirical concise approach to predict the maximal wetted soil depth under trickle irrigated crops. American-Eurasian J. Agric. Environ. Sci., 11(3):334–340, idosi@idosi.org. Open Source Access.

[2]Authors are thankful to the European Union for the financial support through project "QUALIWATER: Diagnosis and control of salinity and nitrate pollution in Mediterranean irrigated agriculture (Project N°: INCO-CT-2005-015031)."

4.1 INTRODUCTION

Trickle irrigation has widely extended throughout the world. Indeed, the irrigated area under trickle irrigation has increased by 330% during the 1990s. During 2000, more than 3 million hectares were expected to be trickle irrigated worldwide [4]. In most arid countries where water resource is limiting factor, using trickle irrigation to sustain irrigated agriculture is a must. In fact, this system enables to increase the crop yield, and to reduce water losses up to 50% as compared to furrow or basin irrigation [3, 22].

The principal mission of the trickle irrigation is to supply water directly in the rhizosphere and then to keep the rooted soil volume within prescribed humidity thresholds. Consequently:

- The wetted area on the soil surface is to be reduced: Thus, water losses by evaporation from surface are significantly reduced.
- The wetted soil volume of onion shape is limited to beneath emitters: Thus, deep percolation and nutrient losses are substantially reduced.

To achieve maximum profits from these opportunities, the water distribution network and trickle irrigation management must be designed so that the wetted soil volume is matched with the rooting zone. To achieve this objective, the shape and the dimensions of the wetted soil volume behavior is to be known [4, 5].

Several analytical and numerical models for predicting water infiltration into the soil have been proposed. Because of the computational simplicity, the general insights and the direct link among the inputs and outputs, the analytical solutions are useful tools for design of trickle irrigation network and management. But most of these solutions remain valid only for steady state flow, in homogeneous and uniform soil conditions [20, 24].

Many numerical models have been proposed to simulate soil water redistribution pattern beneath point and/or linear surface sources [1, 2, 17]. Although these models are powerful in solving complexity of nonlinear soil problems, yet they are less practical because of their complexity and the saturated zone's extension on the surface remain difficult to be accurately reproduced. Moreover, only few of these models allow for water uptake by plant. In 1974, Keller and Karmelli [13] presented a table linking the soil texture (coarse, medium or fine), the emitter spacing and the emitter discharge rate to the wetted soil fraction (P) induced by 40 mm water depth. Empirical expressions have been adjusted [6, 12, 23] to allow reproducing bulb's extension. Hammami et al. [11] proposed a compact physical based approach for predicting the wetted soil depth $Z_f(t)$, beneath an emitter on the soil surface. Comparison with measured and theoretical results revealed that this approach is more reliable [11]. Because of the simplicity and feeless, some of these models remain useful.

This chapter proposes a new empirical approach that enables to predict the maximum wetted soil depth $Z_f(t)$ under trickle irrigated sweet melon and tomato.

4.2 MATERIALS AND METHODS

4.2.1 CLIMATIC DATA AND FIELD TRIAL SITE

The trials were carried out at two private plots in Kalaât Landalous district located in the north-eastern region (latitude: $37°02' \leq \alpha \leq 37°06'$ N; longitude: $10°05' \leq \varphi \leq 10°10'$ E and $0 \leq$ AMSL ≤ 5 m) of Tunisia. It is one of the widest (2905 ha) irrigated land in the country.

Environmental conditions are favorable for trickle irrigation management (shortage of water resources, the fertile soil depth did not exceed 1 m in the major parts of district, orchards and vegetables are the most irrigated summer crops). More than 85% of the average annual rainfall (497 mm) occurs between October and April (Table 4.1).

Because of the acute imbalance between annual precipitation and potential evapotranspiration (1344 mm), irrigating crops in summer is a must. The average temperature ranges from 11 °C (January) to 27 °C (August) (Table 4.1). The soil texture is a loamy-clay loam. Tables 4.2 and 4.3 show a quite uniform textured soil profile with relatively high bulk densities D_b for plots with sweet melon and tomato. Medjerda River is the main water source with a salinity ranging between 1 g/l (in winter) and 2.5 g/l (in summer).

4.2.2 MEASUREMENTS

The soil samples were taken by an auger–hole method at three random locations in each plot to determine the physical characteristics such as: particles size partition, bulk density [8], saturated soil water content and hydraulic conductivity [16].

The data were taken on two private trickle irrigated fields. In the first plot, the tomato seedlings were transplanted on March 24th of 2009. In the second plot, the melon seedlings were transplanted on April 4th of 2009. For both plots, each crop row was irrigated by a single lateral equipped with in-line emitters at 30 cm apart.

Emitter discharges (Q) were monitored using valves that were installed on the laterals upstream. Identical experimental devices were used. However, in the tomato plot, trials were performed with two different emitter discharge rates and three initial water suction (H_i, mb) values. Contrary in the melon plot, irrigation measurements were taken with three different emitter discharge rates, but the average initial water suction (H_i, mb) was similar. Soil water suction in each plot was measured with the sensors (Fig. 4.2) that were installed as shown in Fig. 4.1.

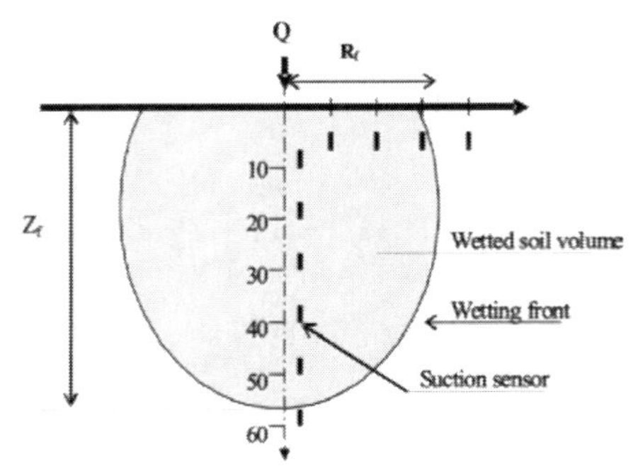

FIGURE 4.1 Schematic description of the wetted cross-section and the location of sensors to measure soil water suction at the experimental site (Fig. 4.2). R_f = Radius of the wetted surface, Z_f = Depth of the wetted bulb, and Q = Discharge rate from the emitter.

FIGURE 4.2 Measurement of wetted front advance: Visual and tensiometric (sensors).

TABLE 4.1 Climatic Data: Average Monthly Temperature T (°C), Rainfall P (mm) and Potential Evapotranspiration ETP (mm) Values

	Values, mm												
	Sep	**Oct**	**Nov**	**Dec**	**Jan**	**Feb**	**Mar**	**Apr**	**May**	**Jun**	**Jul**	**Aug**	**Total**
T	23.8	20.0	15.7	12.3	11.0	11.2	13.0	14.7	19.5	23.0	26.0	27.0	—
P	46.7	35.6	67.7	86.5	75.0	63.7	38.0	40.4	24.1	13.4	2.8	3.4	497
ETP	133	115	84	69	71	80	98	106	126	136	161	165	1344

TABLE 4.2 Soil Characteristics in the Tomato Plot: Soil Texture, Bulk Density (D_b), Saturated Soil Moisture (θ_s) and Saturated Hydraulic Conductivity (K_s)

Soil depth	Particle size distribution			Soil properties			
	Sand	Loam	Clay	Texture class	D_b	θ_s	K_s
cm	%	%	%	—	$g.m^{-3}$	%	$cm.h^{-1}$
0–20	38.5	40.5	21.0	Loam	1.47	0.45	2.40
20–40	48.0	34.0	18.0	Loam	1.50	0.44	1.65
40–60	32.5	46.5	21.0	Loam	1.48	0.46	1.20

Each observation is an average of three soil samples.

TABLE 4.3 Soil Characteristics in the Melon Plot: Soil Texture, Bulk Density (D_b), Saturated Soil Moisture (θ_s) and Saturated Hydraulic Conductivity (K_s)

Soil depth	Particle size distribution			Soil properties			
	Sand	Loam	Clay	Texture class	D_b	θ_s	K_s
cm	%	%	%	—	$gm.cm^{-3}$	%	$cm.h^{-1}$
0–20	22.0	43.5	33.5	Clay loam	1.48	0.46	2.10
20–40	20.0	40.0	38.0	Clay loam	1.51	0.45	1.50
40–60	20.0	41.0	39.0	Clay loam	1.50	0.43	1.52

Each observation is an average of three soil samples.

Each value of Q is the average of four observations for two adjacent emitters at the beginning and the end of each irrigation event. However, Hi value corresponds to the average of suction readings made just before irrigation on five sensors placed at 10, 20, 30, 40 and 50 cm depth (Fig. 4.1). Supplied water depths (Ds) were calculated as follows:

$$D_s = [Z_r(\theta_c - \theta_i)] \quad (1)$$

where: D_s = Supplied water depth (mm), Z_r = rooted soil depth (mm), θ_i and θ_c are initial and at field capacity soil moisture (determined using soil water suction sensors). Irrigations were initiated as soon as soil water suction reached the previously fixed H_i value (= 200, 400 and 600 mb). The following variables were recorded:

- The average width of the wetted area, R_f (t) (cm), was measured visually on the soil surface at elapsed times [19, 20, 21]. Each R_f(t) value is an average of three observations on three consecutive emitters at each trial site (Fig. 4.2).
- The maximum depth of the wetted bulb, Z_f(t), at 5 cm parallel to the symmetrical axis, determined using the soil water suction sensors: The wetting

front depth was recorded once a water suction reduction was observed on the tensiometer placed at the same point [22].

4.3 RESULTS AND DISCUSSION

4.3.1 CLIMATE AND SOIL CHARACTERISTICS

The long period (1970–2010) climatic data in Table 4.1 reflect an acute imbalance between precipitation and potential evapotranspiration especially for summer crops (vegetables and orchards).

The Tables 4.2 and 4.3 indicate a homogeneous loamy textured soil in the tomato plot and homogeneous clay loam textured soil in the melon plot. The relatively higher saturated hydraulic conductivity of the topsoil layer results from the frequent soil cropping activities.

4.3.2 HORIZONTAL AND VERTICAL WETTING FRONT ADVANCES

Recorded $R_f(t)$ and $Z_f(t)$ values for the elapsed time are plotted in Figs. 4.3 and 4.4. These curves are similar to those reported by several researchers [1, 6, 11]. In fact, the higher emitter discharge rates result in faster horizontal wetting front advance. The effect of such flow rates is not so clear on the vertical wetting front velocity (advance). On the other hand, it seems that the drier initial soil moisture conditions result in slower wetting front advance. This behavior is due to the fact, that under constant flux source with initial drier soil profile, the same amount of water should wet an increasing volume of soil pores, which would result in a decrease in the wetting front advance rate.

Experimental $Z_f(t)$ values as function of the corresponding $R_f(t)$ data are plotted in Figs. 4.5 and 4.6. In all cases, $Z_f(t)$ is strongly correlated with $R_f(t)$ ($r > 0.92$). The corresponding ($R_f(t)$, $Z_f(t)$) data observations are scattered on an exponential shaped curves identical to those reported by Keller and Bliesner [14] and by Hammami et al. [11]. This exponential form is as follows:

$$Z_f = a.[\exp(b.R_f)] \quad (2)$$

where: Z_f is the maximum wetted soil depth (cm); R_f refers to the width of wetted strip (cm) measured on soil surface and a and b are exponential regression coefficients. The values of these nonlinear regression coefficients were determined using nonlinear regression analysis. As a rule of thumb, the following boundary conditions must be satisfied:

$$R_f \rightarrow 0, Z_f \rightarrow 0 \quad (3a)$$

$$R_f \rightarrow R_{Max}, Z_f \rightarrow Z_{Max} \quad (3b)$$

where: at the end of irrigation, Z_{Max} is the maximum wetting front depth; and R_{Max} is the maximum width of the wetted area on the soil surface. Then, substituting the regression constants, a and b, in Eq. (2) and rearranging yields:

$$Zf = (ZMax).exp^{[(Rf - RMax) \div Rf]} \quad (4)$$

It is clear that equation (4) satisfies the physical boundary conditions (Eqs. 3a and 3b). The fitting parameters (Z_{Max} *and* R_{Max}) must be adjusted for in-situ cropping conditions. The Z_{Max} value is previously fixed equal to the maximum rooted depth and R_{Max} is fixed equal to the shaded width or the canopy lateral spread. However, these parameters are strongly dependent on soil properties and irrigation conditions. In fact, in the same textured soil and initial water content, increased R_{Max} value results with higher emitters' flow rates. However, with the same emitter flow irrigation times, lower R_{Max} and higher Z_{Max} values appear in the coarse textured soil.

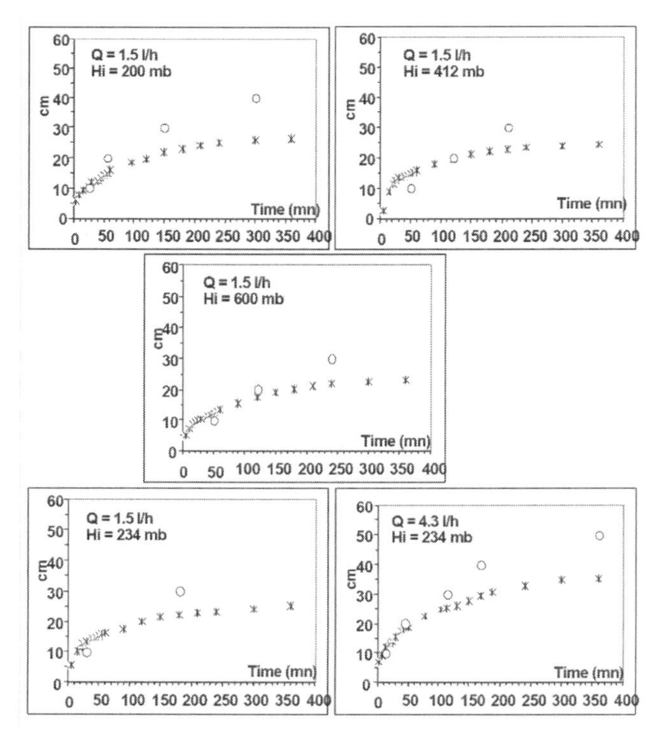

FIGURE 4.3 Tomato plot: Vertical (data shown by circles) and horizontal (data shown by crosses) wetting front advances versus elapsed time for two emitters' discharge rates (Q) and varying initial water suctions (H_i).

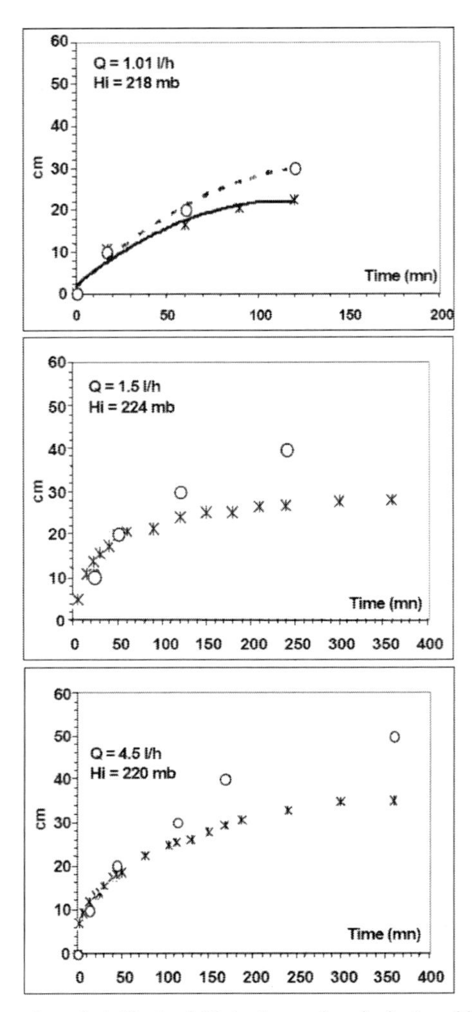

FIGURE 4.4 Sweet melon plot: Vertical (data shown by circles) and horizontal (data shown by crosses) wetting front advances for three emitters' discharge rates (Q) and initial water suctions (H_i)

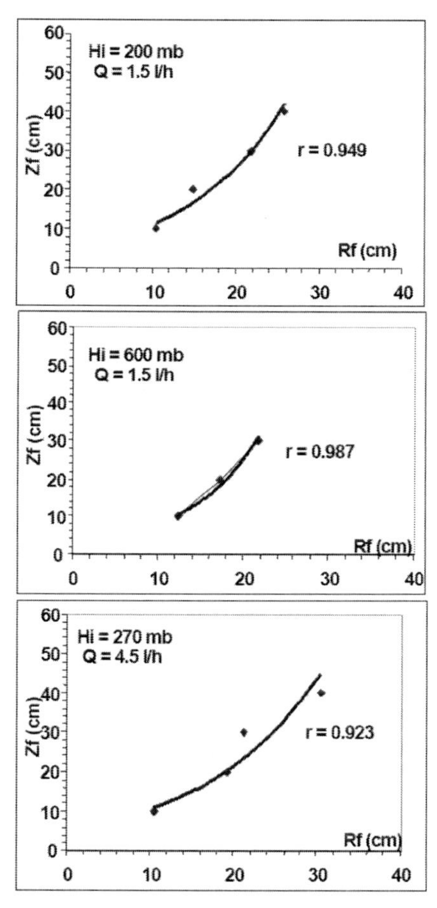

FIGURE 4.5 Tomato plot: Z_f as a function of R_f for two emitters' discharge rates (Q) and three initial water suctions (H_i).

Then using the same R_{Max} value in Eq. (4), the resulted wetting front (Z_f) will be deeper in coarse textured soils than in fine textured soils. These results agree those reported by several investigators [1, 5, 6, 11, 17]. The parameters in Eq. (4) are based on the experimental data for the two-cropped plots, distinguished emitter discharge rates and different initial water contents. It satisfies the physical boundary conditions and is in agreement with the published results on this topic.

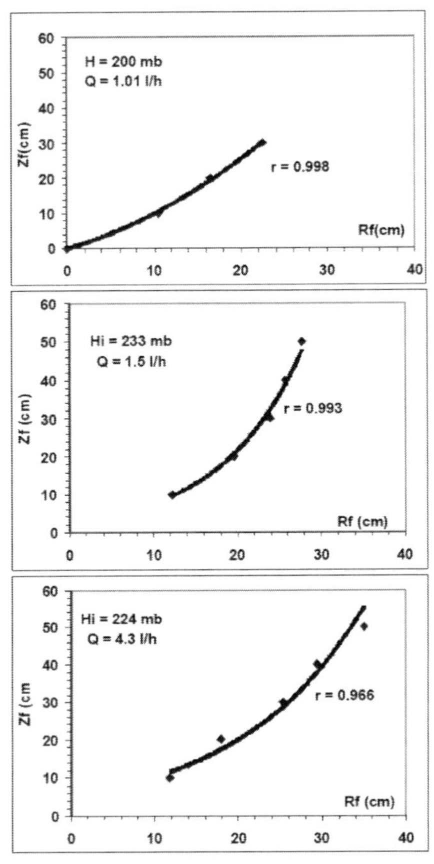

FIGURE 4.6 Melon plot: Z_f as a function of R_f for three emitter discharge rates (Q) and varying initial water suctions (H_i).

Then Eq. (4) can be a practical helpful tool to predict the wetting front depth under trickle irrigated crops, although it is valid only for the infiltration phase.

4.4 CONCLUSIONS

Using horizontal and vertical wetting front advance data, an empirical equation for predicting the maximum wetted soil depth was obtained for trickle-irrigated crops. The proposed equation was established using data recorded on two cropped plots, with different emitters' discharge rates and distinguished initial soil water contents. Based on the measurements of surface wetted area width, the proposed equation enables to compute the corresponding depth of wetting front. The fitting parameters (Z_{Max} and R_{Max}) values are strongly dependent on the local soil and irrigation management conditions. Thus these must be always adjusted for *in situ* conditions. Because of its simplicity, this approach can be helpful tool for deep percolation and fertilizer-leaching control in trickle irrigated crops. But further trials are needed to test the relevance of the proposed approach for wide range of trickle irrigated crops though it remains valid for only the watering phase.

4.5 SUMMARY

An easy, empirical and reliable new approach for predicting the wetted soil depth for the trickle irrigated crops is proposed. The approach was adjusted using field measurements of the maximum wetting front depth $Z_f(t)$ and lateral spread $R_f(t)$ in both tomato and sweet melon plots. Within each plot, measurements were made for different initial water contents and three emitter discharge rates. For all cases, results showed that $Z_f(t)$ is strongly correlated ($r > 0.92$) with $R_f(t)$. An empirical exponential relationship was inferred. Knowing the lateral wetting front spread (in situ conditions), the proposed approach enables to predict the correspondent maximum wetted soil depth. The only two empirical parameters were easily fitted to the *in situ* measurements. Because of its simplicity, the proposed approach is a practical tool for trickle irrigation management, deep-water percolation and fertilizers leaching control.

KEYWORDS

- **deep water percolation**
- **emitter**
- **fertigation**
- **irrigation**
- **leaching**
- **maximum wetted depth**
- **melon**
- **point source**
- **tomato**
- **trickle irrigation**
- **Tunisia**
- **wetted strip width**
- **wetting front**

REFERENCES

1. Ababou, R. (1981). Modeling of hydraulic conductivity of irrigated soil (Modélization des transferts hydriques dans le sol en irrogation). PhD Dissertation in Engineering. Polytechnic Institute of Grenoble.
2. Brandt, A., Bresler, E., Diner, N., Ben-Asher, I., Heller, J., Godelber, D. (1971). Infiltration from a trickle source: I–Mathematical models. Soil Sci. Soc. Am. Proc., 35:675–682.
3. Clyder, R., Timothy, K., Carlos, N. (1989). Comparison of subsurface Trickle and furrow irrigation on plastic-mulched and bare soil for tomato production. J. Am. Hor. Sci., 114:40–43.
4. Coelho, F. E., Or, D. (1997). Applicability of analytical solutions for flow from point sources to drip irrigation management. Soil Sci. Soc. Am. J., 61:1331–1341.
5. Elmaloglou, S., Diamantopoulos, E. (2007). Wetting front advance patterns and losses by deep percolation under the root zone as influenced by pulsed drip irrigation. Agric. Water Management, 90:160–163.
6. Elmaloglou, S., Malamos, N. (2006). A methodology for determining the surface and vertical components of the wetting front under a surface point source, with root water uptake and evaporation. Irrig. and Drain, 55:99–111.
7. Fernandez-Galvez, J., Simmonds, L. P. (2006). Monitoring and modeling the three-dimensional flow of water under drip irrigation. Agric. Water Management, 83:197–208.
8. Gee, G. W., Bauder, J. W. (1986). Particle-size analysis. In: *Methods of Soil Analysis–Part 1*. Agronomy Monograph 9. American Society of Agronomy, Madison, WI, pp. 383–409.
9. Hammami, M. (2001). New approach to determine the soil wetting volume under an emitter (*Nouvelle approche pour déterminer le volume de sol humidifié par un goutteur*). PhD Dissertation for the Faculty of Physical Sciences, Tunisia.
10. Hammami, M., Maalej, M. (1999). Prediction of soil wetting depth under an emitter (*Prédiction de la profondeur de sol humidifiée sous goutteur*). Rev. Sci. Eau., 12(2):273–284.
11. Hammami, M., Daghari, H., Balti, J., Maalej, M. (2002). Approach for predicting the wetting front depth beneath a surface point source: Theory and numerical aspect. Irrig. Drain, 51:347–360.
12. Healy, W., Warrick, A. W. (1988). A generalized solution to infiltration from surface soil point source. Soil Sci. Soc. Am. J., 52:1245–1251.
13. Jiusheng, L., Jiangjun, Z., Minjie, R. (2004). Wetting patterns and nitrogen distributions as affected by fertigation strategies from a surface point source. Agric. Water Management, 67:89–104.
14. Keller, J., Bliesner Ron, D. (1990). Sprinkler and Trickle irrigation. An AVI Book Van Nostrand Reinhold, New York.
15. Keller, J., Karmelli, D. (1974). Trickle irrigation design parameters. Transactions of the ASAE, 17(4):678–684.
16. Klute, A., Dirksen, C. (1986). Hydraulic conductivity and diffusivity–laboratory methods, In: *Methods of soil analysis, Part 1*. Agronomy Monograph 9. American Society of Agronomy, Madison, WI.
17. Lafolie, F., Guennelon, R., Van Genuchten, Th. M. (1989). Analysis of water flow under trickle irrigation: I. Theory and numerical solution. Soil Sci. Soc. Am. J., 53:1310–1318.
18. Lubana Singh, P. P., Narda, N. K. (1998). Soil water dynamics model for trickle irrigated tomatoes. Agric. Water Management, 37:145–161.
19. Michelakis, M., Vougioucalou, E., Clapaki, G. (1993). Water use, wetted soil volume, root distribution and yield of avocado under drip-irrigation. Agric. Water Management, 24:119–131.
20. Philip, J. R. (1985). Steady absorption from spheroid cavities. Soil Sci. Soc. Am. J., 49:828–830.

21. Raats, P. A. C. (1971). Steady infiltration from point sources cavities and basins. Soil Sci. Soc. Am. J., 35:689–694.
22. Satpute, G., Bendales, S., Kausal, A. (1992). Water requirement of tomato crop under drip and furrow irrigation. KPV Res. J., 16:83–87.
23. Schwar Tzmass, M., Zur, B. (1985). Emitter spacing and geometry of wetted soil volume. Journal of Irrigation and Drainage Engineering (ASCE), 112:242–253.
24. Warrick, A. W. (1974). Time-dependent linearized size infiltration. I. Point sources. Soil Sci. Soc. Am. J., 38:384–386.

CHAPTER 5

FERTIGATION TECHNOLOGY IN INDIAN AGRICULTURE

S. KADALE and G. D. GADADE

CONTENTS

5.1 INTRODUCTION

Efficient utilization of water and fertilizer is necessary for achieving sustainable agricultural production [1 to 10]. Among the several inputs used in agricultural production, water and fertilizer are becoming costliest day by day. These inputs play important role in enhancing the crop productivity. Use of the conventional methods of irrigation not only results in considerable loss of water but is also responsible for development of widespread salinity, water logging and leaching of nutrients [1, 10]. Fertilizers applied under traditional methods of irrigation are not efficiently used by the crops.

Micro irrigation is becoming more popular throughout the world, as water is becoming a scare commodity day by day. A micro irrigation system not only conserves water but it also allows more effective management of water and fertilizer applications compared to the traditional methods of irrigation [3].

Fertigation is defined as simultaneous application of irrigation and fertilizer through micro irrigation system to the plants, and is a new concept in Indian agriculture. Water soluble solid or liquid fertilizers are injected with irrigation water through micro irrigation system [1, 2]. Enormous growth of micro irrigation system in India has made the concept viable. Under micro irrigation only a portion of the soil volume around each plant is usually wetted. Crop root growth in essentially restricted to this volume of soil and nutrient reserves within that volume can become depleted by crop uptake/or leaching below the root zone and develop nutrient deficiencies. The depletion of nutrients from the rooting zone of microirrigated crops necessitates the continuous replenishment of nutrient reserves. The obvious way to achieve such a goal is to fertilize through the micro irrigation system, in this way it is possible to more or less control or at least influence the nutrient composition of the soil solution.

This chapter discusses scope of fertigation in Indian agriculture and fertigation technology.

5.2 ADVANTAGES OF FERTIGATION

There are several advantages of fertigation through micro irrigation. Fertilizers are precisely applied in the restricted wetted volume, where active roots are concentrated. Uniform application of fertilizers through micro irrigation prevents nutrient deficiencies that can develop because of limited soil volume explored by roots. It also minimizes loss of N due to leaching because of frequent application of soluble fertilizers in small quantities to the soil. Application of fertilizer in small quantities to the soil at any given time: Improves fertilizer use efficiency (FUE), helps to maintain nutritional balance and nutrient concentrations at optimum level, saves energy and labor, and provides opportunity to apply nutrients at critical stages of crop growth. It also provides a flexibility of fertilization since fertilizers can be added into the root zone as needed. Also hazard of ground water pollution is minimized due to nitrate leaching as compared to conventional practice of fertilizer application. Fertigation can achieve FUE as high as 60–70%. Thus for the same yield level, fertilizer economy up to 25–30% can be effected. This single advantage may even overcome the main constraint of high initial cost of equipment in adoption of micro irrigation system and makes micro irrigation economically viable.

5.3 SELECTION OF FERTILIZER

Fertilizer must be selected on the basis of following criteria:

1. It must not corrode and clog any component of the system.
2. It must be safe for field use.
3. It should increase or at least not decrease crop yield.
4. It must be water soluble or emulsifiable.
5. It should not react adversely with salts or other chemicals contained in the irrigation water.

5.4 FERTIGATION METHODS

Fertilizers and other chemicals can be injected into micro irrigation system using following methods [1, 2, 10].

1. Pressure differential.
2. Venturi system.
3. Metering pump

The pressure differential method (PD) is based on pressure head difference in the system. PD can be developed by valves, venturi, elbows or pipe friction. Fertilizer tank are often used in PD. Fertilizer tank must withstand the pressure of the irrigation system. The main advantage of PD applicators is the absence of moving parts. They are simple in operation and require no electric, gasoline, or water powered pumps. They can operate, whenever water is flowing and where a pressure drop is present. The primary disadvantage of PD units is that the rate of application is not constant and changes continuously with time, thus, a uniform concentration of a nutrient cannot be maintained.

Injecting fertilizers or chemicals solutions by means of pump is probably the more precise way of metering chemicals into an irrigation system. The solution is normally pumped from a pressurized tank. The pump may be driven by an internal combustion engine or an electric motor or tractor powers take off. However, a power supply must be near the injection point. With the pump, fertilizers may be fertigated at more or less constant rate. The pumping rate and the concentration of the stock solution can be adjusted to attain the desired level of fertilizer. However, the water flow and fertilizer flow are independently controlled. Changes in water flow, power failure or mechanical failure may cause serious deviations from the planned concentrations. Another disadvantage is the need for an external power source and the relatively high cost of the system. The use of hydraulic motor, operated by line pressure avoids these difficulties.

5.5 INJECTION RATE OF FERTIGATION

The injection rate of fertilizer into the system for a desired application rate to an area is determined by:

$$Q_f = [F_r \, x \, A]/[N_c \, x \, T] \tag{1}$$

where, Q_f = quantity of fertilizer to be injected (lph), F_r = fertilizer rate per application (kg/ha), A = area to be fertilized (ha), N_c = Nutrient concentration (kg/liter) in the stock solution, and T = time of injection (hr).

After the projected Q_f injection rate has been calculated, it must be evaluated for concentration of nutrients in the irrigation water. This can be determined by:

$$C_f = [K \, x \, F_r]/[W] \tag{2}$$

where, C_f = Concentration of fertilizer in irrigation water in mg/liter, K = Conversion constant = 100 for metric units, F_r = Fertilizer rate (kg/ha), and W = net amount of irrigation water applied during the injection period (mm).

When the desired concentration of nutrient in the irrigation water C_f has been selected, the rate of injection can be determined from flow rate in the system, density and percentage of nutrient in the fertilizer in solution.

$$Q_f = [K \, x \, C_f x \, Q]/[\rho_f \, x \, Y] \tag{3}$$

where: Q_f = quantity of fertilizer to be injected (lph), K = conversion constant = 0.36 for metric units, C_t = volume or rate of flow (lps), ρ_f = density of the fertilizing solution (kg/l), and Y = percentage of fertilizer in solution without decimals.

5.6 FERTILIZER TANK CAPACITY

For a pressure differential injection system, fertilizer tank should have an adequate capacity for a complete application. This requires a tank capacity C_t as determined below:

$$C = F_r x \, A \tag{4}$$

where, C = concentration of nutrient in fertilizer (kg/l), Fr = fertilizer rate per application (kg/ha), and A = area to be fertilized (ha).

5.7 FERTILIZERS SUITABLE FOR INJECTION

5.7.1 NITROGEN

Nitrogen is most commonly applied through micro irrigation system, because it causes few precipitation and clogging problems. Nitrogen can be applied in several forms such as [1, 2]: Anhydrous ammonia, aqua ammonia, ammonium sulfate, ammonium phosphate, ammonium nitrate, potassium nitrate, calcium nitrate and urea. Injection of anhydrous ammonia or aqua ammonia raises the pH of the irrigation water with possibility that insoluble salts of calcium and magnesium could precipitate. Ammonium salts (ammonia sulfate) are fairly soluble in water and generally cause few problems. The use of ammonium phosphate can cause problems if calcium and magnesium are present in the irrigation water in large quantities, since precipitation of calcium and magnesium phosphate is possible.

The prolonged use of ammonium containing fertilizer in lateral lines can however have very detrimental effects on soil fertility in the wetted soil volume. This is because nitrification of the applied NH^+_4 causes soil acidification. When all the fertilizer ammonium is applied to a restricted volume of soil, which contains the bulk of the root mass, such acidification can become a serious problem.

Urea is well suited for fertigation since it is highly soluble and dissolves in not-ionic form so that it does not react with other substances in the water. Thus, it is not likely to cause precipitation problems. Indeed, since the transit time though the trickle irrigation system is fast, urea is unlikely to be hydrolyzed to ammonium to a significant degree in irrigation system even if the urease enzyme is present. Urea has the advantage that it has half the potential acidifying effect (per unit N) of ammonium containing fertilizers.

Nitrate salts are characteristically soluble and are well suited for use in irrigation systems.

5.7.2 PHOSPHORUS

Phosphorus is not usually recommended for fertigation because of possible precipitation of phosphate salts. Where irrigation water is high in calcium and magnesium, the precipitation of insoluble Di-calcium phosphate and Di-magnesium phosphate in irrigation pipes and emitters is likely.

Phosphoric acid is most suitable form of phosphorus for use in trickle irrigation systems. If irrigation water is low in calcium and magnesium, few problems should be encountered in applying phosphoric acid. In such situations, ammonium or potassium di-hydrogen orthophosphate can also be used. If irrigation water is high in calcium and magnesium, it may be possible to inject high concentration of phosphoric acid in pulses, which will keep the pH of irrigation water low enough for most phosphate salts to remain soluble.

5.7.3 POTASSIUM

The common sources of potassium (potassium sulfate, potassium chloride and potassium nitrate) are readily soluble in water and will cause few precipitation or clogging problems in lateral lines and emitters.

5.7.4 MICRO NUTRIENTS

Micro nutrients (iron, manganese, zinc and copper) can be fertigated in chelated form, without causing precipitation problems. Nevertheless, if such micro nutrients are added as inorganic forms, they could possibly react with salts in the irrigation water and cause precipitates.

5.8 PROBLEMS ASSOCIATED WITH FERTIGATION

5.8.1 CLOGGING

Micro irrigation systems are prone to clogging because of the low operating pressures and small orifice sizes of emitters. Physical, chemical and biological agents in water are primary causes of clogging.

Clogging can occur when dissolved chemicals present at high concentrations precipitate out and eventually form encrustations that can restrict water movement. Calcium and magnesium carbonates and hydroxides or sulfides of iron and manganese are among the most troublesome compounds. Some injected fertilizer materials may also react directly with dissolved substances in water to form insoluble precipitates. Common problems are precipitates of calcium and magnesium carbonates and phosphates. Periodic injections of acid (HCl or H_2So_4) have been shown to partially dissolve such precipitates and hence improve emitter performance.

When injecting fertilizers, precipitation of applied chemical is a critical problem and great care must be taken to prevent partial or complete clogging. If in doubt about the mixing compatibility of chemicals, the lines should be flushed thoroughly before applying different chemicals.

Injection of fertilizer nutrients into water can induce the increase in biological activity and sizes of microbial populations in the irrigation water. Increased biological activity can, itself result, in clogging. This occurs when bacterial, algae, fungi or other organisms produce precipitates, mucus or slime products or produce or acts as flocculants for other materials in irrigation water. Injection of chlorine (usually supplied as sodium hypochlorite) at low concentrations (e.g., 1 ppm) continuously or as slug treatments (e.g., 10–20 ppm), at intervals as necessary, inhibits most biological activity in lateral lines and appears to have minimal or no effect on crop plants.

5.8.2 UNEVEN DISTRIBUTION OF NUTRIENTS

An uneven distribution of nutrients within the root zone can occur under fertigation since immobile nutrients such as phosphate become concentrated. For example, phosphate can become concentrated around the emitter while mobile ions such as nitrate and potassium move downward and outward with the wetting front and accumulate at the periphery of the wetted soil volume plants. However, it appears to have the ability to adopt to spatial variability of available nutrients in soils through two major mechanisms [2]. Firstly, the rate of nutrient uptake per unit weight or length of roots in the nutrient enriched area can be increased and secondly, localized root proliferation can occur in the zone of soil high in nutrients. The effect of an uneven nutrient distribution under micro irrigation fertigation is probably not very serious, particularly when reasonably high rates of nutrients are being applied. Although some-micro irrigation systems are operated continuously, in most cases, water is supplied in cyclical fashion so that soil directly below the emitter can be saturated during irrigation cycles but begin to dry between cycles. In the zones immediately below the emitter, temporary localized anaerobic conditions can develop when the soil remains at or near saturation during irrigation. Anaerobic conditions can result in death of roots in the center portion of the wetted volume and proliferation of roots at periphery. The cyclical pattern of moisture and the production of anoxic sites in the wetted zone directly below the emitter can have important implications to nutrient availability.

5.9 SCOPE FOR FERTIGATION RESEARCH

At present, there is not enough information available regarding nutrient requirements of crops at different stages of development to take advantage of micro irrigation technology [2]. The optimum rates and times of fertilizer injection for various crops, soils and climatic conditions are virtually unknown and substantially more research is required in this area.

Movement and transformations of nutrients in the wetted soil volume also require further study. For instance, the movement of urea from the emitted and its subsequent conversions to ammonia and then to nitrate have not been fully studied while the extent and form of gaseous losses of nitrogen under trickle irrigation are not well known. The movement or fixation of micronutrients in the soil, when supplied in chelated form through the irrigation system, has not been reported either. Research into the extent and significance of the cyclic release of native and applied soil phosphate during wetting cycle is required. The effect and significance of the uneven distribution of nutrients in the root zone, which occurs following fertigation, also warrants further study.

Thus, there is ample scope for further research in both the practical and detailed aspects of fertigation through micro irrigation systems.

5.10 SUMMARY

This chapter presents fertigation technology that includes advantages, disadvantages, limitations, methods, rates and duration, fertilizer types, problems, and future research needs of fertigation in Indian agriculture.

KEYWORDS

- **clogging**
- **drip irrigation technology**
- **economics**
- **fertigation technology**
- **fertilizer use efficiency**
- **IARI**
- **ICID**
- **India**
- **micro irrigation**
- **nutrients**
- **tank capacity**
- **trickle irrigation**
- **water scarcity**
- **Water Technology Center India**
- **water use efficiency**

REFERENCES

1. Megh R. Goyal, (Ed.), (2013). *Management of Drip/Trickle or Micro Irrigation.* Oakville–ON, Canada: Apple Academic Press Inc., pp. 1–408.
2. Megh R. Goyal, (Ed.), (2015). *Research Advances in Sustainable Micro Irrigation, volumes 1–10.* Oakville–ON, Canada: Apple Academic Press Inc.
3. http://www.icid.org/sprin_micro_11.pdf
4. IARI, (2009). *Development of micro irrigation technology*. WTC, IARI, New Delhi.
5. Rajput, T. B. S., Patel, N. (2007). *Micro irrigation manual.* WTC-IARI, New Delhi.
6. Sivanappan, R. K. (2015). Advanced technologies for sugarcane cultivation under micro irrigation, Tamil Nadu. Pp. 1:121–146, In: *Megh R. Goyal, (Ed.), (2015). Research Advances in Sustainable Micro Irrigation, volume 1.* Oakville–ON, Canada: Apple Academic Press Inc.
7. Sivanappan, R. K. (2015). Micro irrigation potential in fruit crops, India. Pp. 3:79–94, In: *Megh R. Goyal, (Ed.), (2015). Research Advances in Sustainable Micro Irrigation, volume 4.* Oakville–ON, Canada: Apple Academic Press Inc.

8. Sivanappan, R. K. (2015). Micro irrigation potential in India. Pp. 1:87–120, In: *Megh R. Goyal, (Ed.), (2015). Research Advances in Sustainable Micro Irrigation, volume 1.* Oakville–ON, Canada: Apple Academic Press Inc.
9. Sivanappan, R. K. (2015). Research advances in micro irrigation in India. Pp. 4:33–42, In: *Megh R. Goyal, (Ed.), (2015). Research Advances in Sustainable Micro Irrigation, volume 4.* Oakville–ON, Canada: Apple Academic Press Inc.
10. Sivanappan, R. K., Padmakumari, O. (2015). Principles of drip/micro or trickle irrigation. Chapter 3, In: *Megh R. Goyal, (Ed.), (2015). Research Advances in Sustainable Micro Irrigation, volume 9.* Oakville–ON, Canada: Apple Academic Press Inc.

CHAPTER 6

DESIGN OF BURIED MICRO-IRRIGATION LATERALS BASED ON SOIL WATER RETENTION

HAMMAMI MONCEF, ZAYANI KHEMAIES, and HÉDI BEN ALI

CONTENTS

Edited version of Hammami Moncef, Zayani Khemaies, and Hédi Ben Ali. Required Lateral Inlet Pressure Head For Automated SDI Management. International Journal of Agronomy, Volume 2013, 6 pages. Open Access article, http://www.hindawi.com

6.1 INTRODUCTION

In subsurface drip irrigation (SDI), water emits from the buried drippers into the soil and spreads out in the rhizosphere due to capillary and gravity forces [20, 25]. Thus, SDI system permits direct application of water to the wetted soil volume and maintaining dry the nonrooted topsoil. This pattern has advantages such as minimizing soil evaporation, deep percolation, weeds growth and thus affects evapoconcentration phenomenon. The SDI improves the water application uniformity, increases the laterals and emitters longevity, reduces the occurrence of soil-borne diseases, and infestation of weeds. Several field trials have revealed relevant profits due to adequate management of SDI for crop production. Nevertheless, the appropriate depth of buried laterals remains debatable [10, 14, 21, 28]. Comparing evaporation from surface and subsurface drip irrigation systems, Evett et al. [7] reported a saving of 51 mm and 81 mm irrigation depth with drip laterals buried at 15 cm and 30 cm, respectively. Neelam and Rajput [20] recorded maximum onion yield (25.7 t per ha) with drip laterals buried at 10 cm depth. They reported maximum drainage with drip laterals at 30 cm depth. Several investigators have analyzed the effects of soil properties on the discharge of SDI emitters and water distribution uniformity [1, 17, 23]. The analytical method by Sinobas et al. [25] predicted reasonably well the soil water suction and the pressure head distribution in the laterals and SDI units [26]. The water oozes out from the buried emitters due to inlet lateral pressure head and the soil water suction. Therefore, the emitter discharge is high at the beginning of irrigation due to dry root zone. Gradually, as the soil pore space in the vicinity of the dripper outlet is filled with water, a positive pressure head develops, which may cause a decrease in dripper discharge [24]. If the discharge is greater than the soil infiltration capacity, the resulting overpressure near the nozzle tends to reduce the flow rate [17, 30].

6.2 BASICS OF ANALYTICAL METHOD

The pressurized irrigation systems are customarily designed so that the mean pressure head throughout the pipe is equal to the nominal pressure head. On the other hand, irrigation management is based on the replenishment of the soil holding capacity. Hence, the soil moisture should range between predetermined and minimum allowable soil moisture. It is assumed that the average pressure head is equal to the emitter operating pressure head. The emitter discharge equation is defined below [15]:

$$q = KH^{x} \tag{1}$$

where: q [L^3T^{-1}] and H [L] are emitter discharge and the emitter pressure head; K [$L^{3-x}T^{-1}$] and x are nonlinear regression coefficients. Equation (1) is valid for a pressure head $\geq$ 5.0 m. It is worth pointing out that most long-path turbulent flow and pressure-compensating emitters require an operating pressure head fulfilling this condition. For buried emitters, the emitter pressure head is lumped with the water suction near the outlets, as shown below:

$$H = h_e - h_i \quad (2)$$

where: h_e and h_i refer to the pressure heads [L] at the inner and outer of the emitter, respectively. For emitters in surface drip irrigation, h_i is the atmospheric pressure. Conversely, for buried emitters, h_i is a spatial-temporal variable dependent on the prevailing soil water content. Hereinafter, we will consider the sigmoid retention curve of Van Genuchten [27] given below:

$$\theta = \{\theta_r + (\theta_s - \theta_r)[1 + (\alpha h)^n]^m\} \quad (3)$$

where: θ [L^3L^{-3}] and h [L] refer to the volumetric water content and to the soil suction head, respectively. The residual water contents are denoted as θ_r, α [L^{-1}]. The constants n and m are nonlinear regression coefficients that are found by fitting the curve to the scattered data (θ, h) according to Eq. (3); and θ_s refers to the saturated soil water content. The dimensionless parameters n and m are expressed by the Mualem [19] as shown below:

$$m = [1 - (1/n)] \quad (4)$$

The soil capillary capacity C [L^{-1}] is derived straightforwardly by differentiating Eq. (3) with respect to the suction head h as follows:

$$C = d\theta/dh = - \{mn\alpha\ (\theta_s - \theta_r)\ (\alpha h^{\ n-1})[1 + (\alpha h)^n]^{m+1}\} \quad (5)$$

Equation (5) shows that additional increase in the suction head produces an additional release of water from the soil. Besides, the value of C is the highest if the second derivative of the soil moisture content with respect to the suction head is zero. Under these conditions, the crops absorb the maximum water from the root zone for the same additional energy increment. Further analysis indicates that the coordinates of the inflection point of the retention curve as well as the maximum capillary capacity are as follows:

$$h_{op} = -m^{\ 1/n}/\alpha$$

$$\theta_{op} = \theta_r + (\theta_s - \theta_r)/(1 + m)^m$$

$$C_{max} = nm^{\ m+1}\alpha(\theta s - \theta r)/(1 + m)^{m+1} \quad (6)$$

where: h_{op}, θ_{op}, and C_{max} refer to the optimal water suction, optimal soil water content, and maximum capillary capacity, respectively. Therefore, the design of SDI systems should ascertain a suction head at the emitter outlet that matches the optimal water status within the root zone. Combining Eqs. (1) and (2) yields the following:

$$q = K(h_e - h_{op})^x \tag{7}$$

Equations (6) and (7) reveal the dependence of the emitter discharge on the pressure heads at the inner and outer tips of the nozzle. In as much as the soil is more or less dry at the beginning of the irrigation, the discharge decreases with the elapsed time. Incidentally as the soil becomes wetter, the soil pressure head increases and the emitter discharge stabilizes to a minimum value. Gil et al. [9] found that the decrease of the flow rate is steeper in loamy than in sandy soils. Yao et al. [30] recorded that the wetted soil volume in medium loam and sandy loam is virtually invariant as the inlet pressure head was increased from 60 to 150 cm. This increase of pressure head may lead to the back-pressure development. Yao et al. [30] recommended that the emitter discharge should be matched to the soil conditions so that back-pressure occurrence is avoided. According to Ben-Gal et al. [2] and Lazarovitch et al. [17], one of the main issues with SDI systems is the soil saturation. This phenomenon induces temporary asphyxia of crops and may stop the emitter discharge even though the moistened bulb is not yet spatially well extended. Based on Eqs. (1) and (2), the emitter discharge is null whenever the outlet pressure head (h_i) matches the predetermined inlet one (h_e). Afterwards, the redistribution process provides drier rooted soil profiles. Subsequently, the pressure near the emitter (h_i) decreases until the pressure differential between the outlet tips overtakes a minimum

The threshold value Δh_{min} is required for the emitter operation. The Δh_{min} is dependent on the structural form, dimension, and material of the emitter pathway. For any emitter model, Δh_{min} may be inferred from the emitter discharge-pressure head relationship provided by the manufacturer. Thus, the next irrigation is automatically triggered, once the following inequality is fulfilled:

$$[h_e - h_i] = [h_e - h_{op}] \geq \Delta h_{min} \tag{8}$$

Therefore, the required minimum pressure head at the emitter inlet h^*_{min} should comply with:

$$h^*_{min} \geq [h_{op} + \Delta h_{min}] \tag{9}$$

It is emphasized that the suction head at the vicinity of the emitter cannot be maintained constant and equal to h_{op}. Unavoidable fluctuations of the suction head are expected owing to evapotranspiration and water redistribution processes. For the sake of convenience, the suction head in the root zone should be circumscribed within a prescribed interval $[(h_{op} + \Delta h_{op})$ and $(h_{op} - \Delta h_{op})]$. Therefore, the minimum required emitter inlet pressure head h^{min}_{req} is given by:

$$h^{min}_{req} = [h_{op} - \Delta h_{op} + \Delta h_{min}] \tag{10a}$$

whereas: the maximum required emitter inlet pressure head h^{max}_{req} is given by:

$$h^{max}_{req} = [h_{op} + \Delta h_{op} + \Delta h_{min}] \qquad (10b)$$

The magnitude of the interval $[h_{op} \pm \Delta h_{op}]$ should account for the sensitivity of the crop to the water stress. As a matter of fact, for tomato crop, the reduction of the water requirement by 20% resulted in 20% increase in yield [6]. However, the decrease of the onion water requirement by 20% resulted only 2% decrease in yield [21]. It should be highlighted that these yield reductions are more or less significant according to the physiological stages.

TABLE 6.1 Tolerable soil pressure head variations for selected crops.

Crop	Pressure range, cm		Reference
	Upper limit	Lower limit	
Grape	- 2	- 1000	[12]
Grass	- 25	- 800	[3]
Soybean	- 25	- 800	[3]
Spring wheat	- 25	- 1000	[18]
Tomato	- 2	- 800	[8, 12]

6.3 REQUIRED LATERAL PRESSURE HEAD

For a buried lateral equipped with N identical emitters, the inlet discharge Q will vary within the following limits:

$$Nq_{min} \leq Q \leq Nq_{max} \qquad (11)$$

where: q_{max} and q_{min} are the maximum and minimum emitter average discharge, respectively. For design purpose, only the maximum average emitter discharge is considered. Therefore, the lateral inner diameter is designed to allow the conveyance of the upper bound of the discharge. Consequently, the minimum pressure head required at the upstream end of nontapered flat lateral is:

$$h_{Lm} = [Z_d + J_L + \Delta h_{min} + h_{op} - \Delta h_{op}] \qquad (12a)$$

whereas: the maximum pressure head required at the upstream end of the lateral is:

$$h_{LM} = [Z_d + J_L + \Delta h_{min} + h_{op} + \Delta h_{op}] \qquad (12b)$$

where: Z_d[L] and J_L[L] are emitter burial depth and head loss along the lateral, respectively. By convention, the gravitational potential Z_d is computed negatively downwards.

According to the aforementioned basics, the design procedure of SDI systems should lead to the automation of micro irrigation. Indeed, the irrigation events are triggered, whenever the mean pressure head within the root zone is reduced to the minimum prescribed value of $(h_{op} - \Delta h_{op})$. The events are automatically ended, once the pressure head within the root zone exceeds the maximum value of $(h_{op} + \Delta h_{op})$. From theoretical standpoint, a self-regulation of the flow rate by soil water properties and moisture conditions should prevail. Moreover, the variations in emitter discharge due to the head losses are offset by soil pressure head gradients. Accordingly, the irrigation events as well as the uniformity of the flow rates are controlled by the soil suction head at the depth of burial of emitters. These results agree with Gil et al. [9] who indicated higher variability in the flow rates with surface emitters than with the buried emitters.

Tolerable variations in the soil pressure head for some crops are summarized in Table 6.1. It is worth pointing out that the abovementioned approach remains valid regardless of the used soil water-retention relationship. The following steps summarizes for the proposed design procedure of SDI laterals.

6.3.1 *DESIGN STEPS*

Step 1: Carry out simultaneous in situ field measurements of soil moisture and suction heads.

Step 2: Fit the experimental dataset (θ, h) in accordance with the appropriate soil water-retention curve (for example: equation (5)).

Step 3: Derive twice the moisture content with respect to the suction head and infer h_{op}.

Step 4: Select the proper interval of the soil suction head Δh_{op} for a particular crop (for example, data provided in Table 1).

Step 5: For the emitter type under consideration, calculate the minimum inlet pressure head h^*_{min} using equation (9).

Step 6: Calculate the minimum and maximum required emitter inlet pressure heads using equations (10a) and (10b), respectively.

Step 7: Using equation (11), calculate the required lateral inlet discharge.

Step 8: Determine the minimum and maximum required lateral inlet pressure heads, using Eqs. (12a) and (12b), respectively.

6.4 EXAMPLE

Determine the minimum and maximum required lateral inlet pressure heads, for the following data:

Length of polyethylene nontapered flat pipe: 100 m.

In-line emitter spacing: 40 cm equally spaced.

Emitter depth, according to Patel and Rajput [21]: 15 cm

Crop: Tomato
Soil texture: homogeneous sandy soil.

6.4.1 PROCEDURE

Step 1: Simultaneous in situ measurements of the soil moisture and suction heads were performed [11] on three randomized ~~points~~ locations during water redistribution. In each soil profile, suction heads were measured using three tensiometers installed at 10, 30, and 50 cm soil depth. Soil cores sampled at the same depths were used to determine gravimetrically the corresponding soil moisture. For each depth, the average of the three observations was considered.

Step 2: Experimental data were fitted in accordance with Van Genuchten [27] model [11]. Scattered and fitted data are shown in Fig. 6.1. The inferred fitting parameters (θ_r, m, n, and α) are summarized in Table 6.2.

TABLE 6.2 Fitting Parameters For Van Genuchten's Equation For the Sandy Soil

θ_s (cm^3/cm^3)	θ_r (cm^3/cm^3)	α (cm^{-1})	n	R^2
0.38	0.02	0.05	1.70	0.991

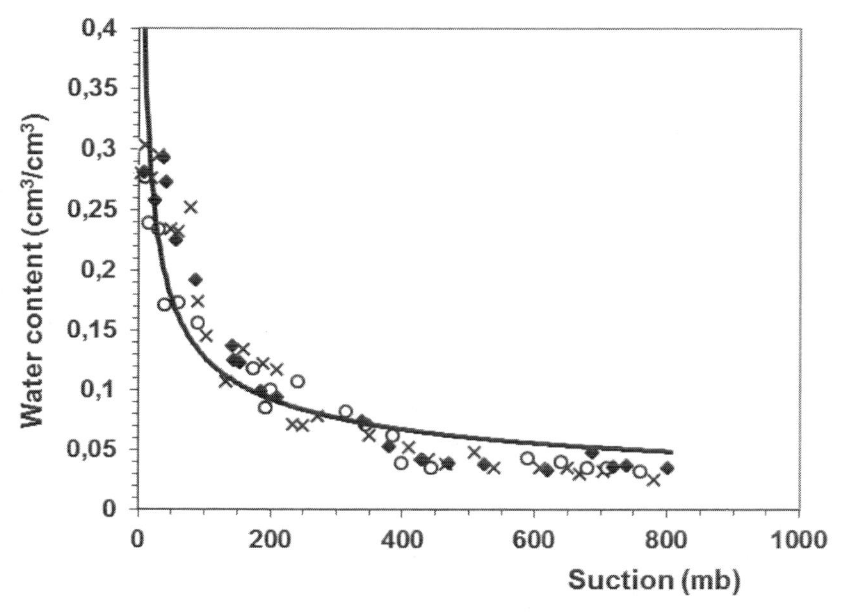

FIGURE 6.1 Soil Water retention curve and measured data at different depths. Legend: 10 cm = xx; 30 cm = oo; and 50 cm = ■■.

Step 3: Using Eq. (6), the optimum suction head h_{op} is approximately −12 cm. This value is within the optimal range of the suction head for tomato crop (See Table 6.2, [8, 12]). To prevent asphyxia risk or relative water stress at upper (−2 cm) and lower (−800 cm), tolerable pressure heads, $\Delta h_{op} = 400$ cm is acceptable.

Step 4: Therefore, the prescribed soil pressure head limits for tomato crop are determined as follows:

$$h_{op} - \Delta h_{op} \approx -12 - 400 = -412 \text{ cm, and}$$

$$h_{op} + \Delta h_{op} \approx -12 + 400 = 388 \text{ cm} \tag{13}$$

In order to avoid eventual backpressure development, the suction head should be maintained within [−412 and 0.0] cm.

Step 5: A trapezoidal labyrinth long-path emitter is used with a minimal differential operating pressure head of $\Delta h_{min} = 500$ cm. The discharge-pressure head relationship of these emitters is shown below [22]:

$$q = [0.752(h_e - h_i)^{0.478}] \tag{14}$$

where: q = emitter discharge (l/h), h_e = emitter inlet pressure head (m), and h_i = the emitter outlet pressure head (m).

Step 6: Using equations (10a) and (10b), the required emitter inlet pressure h_{req} should comply with:

$$[(-12 - 400 + 500) = 88] \leq h_{req} \text{ (cm)} \leq [(0 + 500) = 500] \tag{15}$$

To maintain an optimal suction head within the root zone (−12 cm) and to compensate the minimum differential operating pressure head (Δh_{min} = 500 cm), the optimal required emitter inlet pressure should be h_{oreq} = (−12 + 500) = 488 cm. Compared with the pressure heads customarily required for on-surface drippers (approximately 1000 cm), the obtained value underlines an outstanding energy saving with SDI systems. Therefore, according to Eq. (14), the corresponding emitter discharge q is given as:

$$\{0.752[0.88 - 0.00]0.478 = 0.707\} \leq q \text{ (l/h)} \leq \{0.752[5.00 - (-4.12)]\ 0.478 = 2.163\} \tag{16}$$

As long as the lowest differential pressure head (0.88 m) is less than the minimum differential operating pressure head (Δh_{min} = 500 cm), the emitter discharge vary within the interval [0.00, 2.163]. Nevertheless, the optimal required emitter discharge matching the optimal soil suction head q_{op} will be:

$$q_{op} = \{0.752[4.88 - (-0.12)]0.478\} \approx 1.623 \text{ l/h} \tag{17}$$

Step 7: The number of emitters along the lateral equals 100 m/0.4 m = 250. According to Eq. (11), the optimal required discharge at the lateral inlet tip is:

$$Q_{op} = 250 \times 1.623 = 405.75 \text{ l/h} \quad (18)$$

The head loss gradient j may be estimated by Watters and Keller's formula [29] as follows:

$$j = \alpha Q^{\beta} D^{-\gamma} \quad (19)$$

where: Q and D are the discharge and the lateral inside diameter, respectively. For j (m/m), Q(l/h), and D(mm), the parameters in Eq. (19) are $\beta = 1.75$, $\gamma = 4.75$, and $\alpha = 14.709598\upsilon$, where: υ (m^2 s^{-1}) is the kinematic viscosity of water. At 20° C, α is equal to 0.4655. Considering Eq. (19) and an inner diameter of 16 mm, the head loss throughout the lateral J_L is given [29] below:

$$J_L = \alpha Q^{\beta}_{max} D^{-\gamma}/(1+\beta)\text{L, or}$$

$$J_L = 0.4655(405.75)^{1.75}(16.0)^{-4.75}100/(1+1.75) = 1.184 \text{ m} \quad (20)$$

This value is doubled if we take into consideration the head losses due to emitters' connection as computed by Juana et al. [13] method.

Step 8: Using Eqs. (12a) and (12b) and accounting for emitters' connection head losses, the required pressure head at the inlet tip of the lateral will be:

$$\{(-15+2\times118.4+500-12-400)=309.8\} \le h_L \text{ (cm)} \le \{(-15+2\times118.4+500+0)=721.8\} \quad (21)$$

In the same way, the optimal required pressure head h_{Lo} at the lateral inlet will be:

$$h_{Lo} = (-15 + 2 \times 118.4 - 12 + 500) = 709.8 \text{ cm} \quad (22)$$

Therefore, it is possible to ensure a complete automation of the SDI system via the installation of an overhead basin with a constant water level.

6.5 CONCLUSIONS

Besides savings in water, energy and labor-input, the SDI system offers the opportunity to fully automate the micro irrigation and to include best management practices in agriculture. In fact, the adequate control of variation of soil moisture in the vicinity of emitters is a milestone in the management of subsurface drip irrigation. The rationale is that the flow rate of buried drippers is a function of pressure head at the soil depth of subsurface drip lines. Therefore, the temporal variation of the flow rate is dependent on soil water redistribution and water uptake by roots. The design procedure developed in this chapter provides appropriate emitter discharge and inlet lateral pressure head that fit the water uptake by plant roots. Knowing soil retention curve and water uptake, the procedure provides guidelines to design SDI laterals. The main objective of the design is to ascertain optimal suction head within the installation depth of emitters so that irrigation events are automatically controlled based on the soil moisture variations. The case study showed that soil moisture can be circumscribed within an interval suitable for plant growth. This approach can be a helpful tool for the optimum design of SDI system and the best irrigation management. However, it is worthwhile to note that the current approach completely overlooks the effects of burial drippers on clogging.

6.6 SUMMARY

SDI is based on small and frequent water application near the root zone. Since emitter lines are buried in the SDI, the emitter discharge is dependent on the soil moisture status in the vicinity of the emitters. This chapter includes design of subsurface laterals based on the soil water-retention characteristics and water uptake by the roots. The approach in this chapter permits systematic triggering and cut-off of irrigation events based on fixed water suctions in the rhizosphere. Therefore, the soil moisture is maintained at an optimal threshold value to ensure the best plant growth. The method in this chapter is a helpful tool for the optimum design of the SDI system and appropriate water management. Knowing the soil water-retention curve, the appropriate water suction for the plant growth, and the emitter discharge-pressure head relationships were developed. The method by authors allows the computation of the required hydraulics of the laterals (e.g., inlet pressure head, inside diameter, etc.). An illustrative example is presented for the design of SDI laterals in tomato.

KEYWORDS

- **buried dripper**
- **clogging**
- **deep percolation**
- **drip line**
- **dripper**
- **emitter**
- **evapoconcentration**
- **Hydrus-2d**
- **irrigation design**

- **lateral**
- **micro irrigation**
- **plant growth**
- **pressure head**
- **soil evaporation**
- **soil moisture**
- **soil water**
- **subsurface drip irrigation**
- **suction head**
- **tomato**
- **water distribution**
- **water management**
- **water use efficiency**

REFERENCES

1. Ayars, J. E., Phene, C. J., Hutmacher, R. B. (1999). Subsurface drip irrigation of row crops: a review of 15 years of research at the Water Management Research Laboratory. *Agricultural Water Management*, 42(1):1–27.
2. Ben-Gal, A., Lazorovitch, N., Shani, U. (2004). Subsurface drip irrigation in gravel-filled cavities. *Vadose Zone Journal*, 3(4):1407–1413.
3. Clemente, R. S., De Jong, R., Hayhoe, H. N., Reynolds, W. D., Hares, M. (1994). Testing and comparison of three unsaturated soil water flow models. *Agricultural Water Management*, 25(2):135–152.
4. Clothier, B. E., Green, S. R. (1997). Roots: the big movers of water and chemical in soil. *Soil Science*, 162(8):534–543.
5. Colaizzi, P. D., Schneider, A. D., Evett, S. R., Howell, T. A. (2004). Comparison of SDI, LEPA, and spray irrigation performance for grain sorghum. *Transactions of the American Society of Agricultural Engineers*, 47(5):1477–1492.
6. Doorenbos, J., Kassem, A. H., Bentverlsen, C. L. M. (1987). Yield Response to water. FAO Paper 33.
7. Evett, S. R., Howell, T. A., Schneider, A. D. (1995). Energy and water balances for surface and subsurface drip irrigated corn. *Proceedings of the 5th International Micro irrigation Congress*, Orlando, Florida, USA, April, pp. 135–140.
8. G¨arden¨as, A. I., Hopmans, J. W., Hanson, B. R., Sim˚unek, J. (2005). Two-dimensional modeling of nitrate leaching for various fertigation scenarios under microirrigation. *Agricultural Water Management*, 74(3):219–242.
9. Gil, M., Sinobas, L. R., Juana, L., Sanchez, R., Losada, A. (2008). Emitter discharge variability of subsurface drip irrigation in uniform soils: effect on water-application uniformity. *Irrigation Science*, 26(6)451–458.

10. Grabow, G. L., Huffman, R. L., Evans, R. O., Jordan, D. L., Nuti, R. C. (2006). Water distribution from a subsurface drip irrigation system and drip line spacing effect on cotton yield and water use efficiency in a coastal plain soil. *Transactions of the ASABE*, 49(6):1823–1835.
11. Hammami, M., Zayani, K. (2009). Effect of trickle irrigation strategies on tomato yield and roots distribution. *World Journal of Agricultural Sciences*, 5:847–855.
12. Hanson, B. R., Sim°unek, J., Hopmans, J. W. (2006). Evaluation of urea-ammonium-nitrate fertigation with drip irrigation using numerical modeling" *Agricultural Water Management*, 86(1):102–113.
13. Juana, L., Sinobas, L. R., Losada, A. (2002). Determining minor head losses in drip irrigation laterals. I: methodology. *Journal of Irrigation and Drainage Engineering*, 128(6):376–384.
14. Kandelous, M. M., Sim°unek, J. (2010). Numerical simulations of water movement in a subsurface drip irrigation system under field and laboratory conditions using HYDRUS-2D. *Agricultural Water Management*, 97(7):1070–1076.
15. Khemaies, Z., Moncef, H. (2009). Design of level ground laterals in trickle irrigation systems. *Journal of Irrigation and Drainage Engineering*, 135(5):620–625.
16. Lazarovitch, N. (2008). *The effect of soil water potential, hydraulic properties and source characteristic on the discharge of a subsurface source,'' (PhD thesis)*, Faculty of Agriculture of the Hebrew University of Jerusalem.
17. Lazarovitch, N., Sim°unek, J., Shani, U. (2005). System-dependent boundary condition for water flow from subsurface source. *Soil Science Society of America Journal*, 69(1):46–50.
18. Li, K. Y., De Jong, R., Boisvert, J. B. (2001). An exponential root water-uptake model with water stress compensation. *Journal of Hydrology*, 252(1–4):189–204.
19. Mualem, Y. (1976). A new model for predicting the hydraulic conductivity of unsaturated porous media" *Water Resources Research*, 12(3):513–522.
20. Neelam, P., Rajput, T. B. S. (2008). Dynamics and modeling of soil water under subsurface drip irrigated onion. *Agricultural Water Management*, 95(12):1335–1349.
21. Patel, N., Rajput, T. B. S. (2009). Effect of subsurface drip irrigation on onion yield. *Irrigation Science*, 27(2):97–108.
22. Qingsong, W., Yusheng, S., Wenchu, D., Gang, L., Shuhuai, H. (2006). Study on hydraulic performance of drip emitters by computational fluid dynamics" *Agricultural Water Management*, 84(1–2):130–136.
23. Safi, B., Neyshabouri, M. R., Nazemi, A. H., Massiha, S., Mirlatifi, S. M. (2007). Water application uniformity of a subsurface drip irrigation system at various operating pressures and tape lengths. *Turkish Journal of Agriculture and Forestry*, 31(5):275–285.
24. Shani, U., Xue, S., Gordin-Katz, R., Warrick, A. W. (1996). Soil limiting flow from subsurface emitters. I: pressure measurements. *Journal of Irrigation and Drainage Engineering*, 122(5):291–295.
25. Sinobas, L. R., Gil, M., Juana, L., Sanchez, R. (2009). Water distribution in laterals and units of subsurface drip irrigation, I: Simulation. *Journal of Irrigation and Drainage Engineering*, 135(6):721–728.
26. Sinobas, L. R., Gil, M., Juana, L., Sanchez, R. (2009). Water distribution in laterals and units of subsurface drip irrigation. II: field evaluation. *Journal of Irrigation and Drainage Engineering*, 135(6):729–738.
27. Van Genuchten, M. T. (1980). A closed-form equation for predicting the hydraulic conductivity of unsaturated soils. *Soil Science Society of America Journal*, 44(5):892–898.
28. Vories, E. D., Tacker, P. L., Lancaster, S. W., Glover, R. E. (2009). Subsurface drip irrigation of corn in the United States Mid-South. *Agricultural Water Management*, 96(6):912–916.

29. Watters, G. Z., Keller, J. (1978). Trickle irrigation tubing hydraulics. Tech. Rep. 78–2015, ASCE, Reston, VA, USA.
30. Yao, W. W., Ma, X. Y., Li, J., Parkes, M. (2011). Simulation of point source wetting pattern of subsurface drip irrigation. *Irrigation Science*, 29(4):331–339.

CHAPTER 7

HYDRAULICS OF DRIP IRRIGATION SYSTEM

BALRAM PANIGRAHI

CONTENTS

7.1 INTRODUCTION

The drip irrigation method has now gained worldwide popularity particularly in water scarce regions, undulating hilly areas and saline soils. It is an appropriate water saving and production augmenting technique for wide spaced orchard and plantation crops. It is also used for commercial crops like cotton, tobacco, sugarcane and also for vegetables. Typically, it has decreased wetted volume thus requiring more frequent irrigations. An irrigation regime with an excessively high irrigation frequency can cause the soil surface to remain wet and the evaporation process persists most of the time, resulting in substantial loss of water. This is one of the disadvantages of the drip irrigation system. The wetted area beneath the emitters particularly in arid and semiarid regions is susceptible to high evaporation. On soils having low infiltration rate, surface application of water by drip irrigation may result in significant surface wetting and ponding, which suffer from high evaporative demand and thereby decreasing the water use efficiency [3]. Advent of subsurface drip irrigation is in fact an approach to curb enhances this efficiency. However, subsurface drip irrigation may suffer from clogging of emitters and microtubes. An alternative approach to the clogging problem is to increase the size of the emitters and microtubes. However, this may increase the discharge of the microtubes and emitters and change the pattern of wetting in the soil thus affecting the water availability to the plants.

Camp [1] made a comprehensive review of published information on subsurface drip irrigation. He concluded that crop yield was superior than that of other irrigation methods including surface drip and required less water in many cases. Lamm et al. [4] studied combinations both preplant surface application and in-season fertigation of nitrogen fertilizer for field corn at three different levels of water application by subsurface drip system. They concluded that nitrogen fertigation through subsurface drip irrigation at a depth of 42 to 45 cm redistributes differently in the soil profile than surface applied preplant nitrogen banded in the furrow. Choi et al. [2] compared the subsurface drip irrigation to sprinkler irrigation of Bermuda grass turf using tertiary treated effluent in Arizona – USA. They concluded that subsurface drip irrigation of turf could be an alternative to sprinkler irrigation system in terms of water saving and also when water was restricted to low quality waste water.

This chapter discusses water front advance under surface and subsurface drip irrigation systems using in-line drippers. Authors also conducted studies to recommend the best drip irrigation system for cultivation of crops in sandy soil.

7.2 MATERIALS AND METHODS

The experimental setup consisted of soil tank model fitted with two fiberglass plates as shown in Fig. 7.1. The length and breadth of the tank were 920 and 915 mm, respectively.

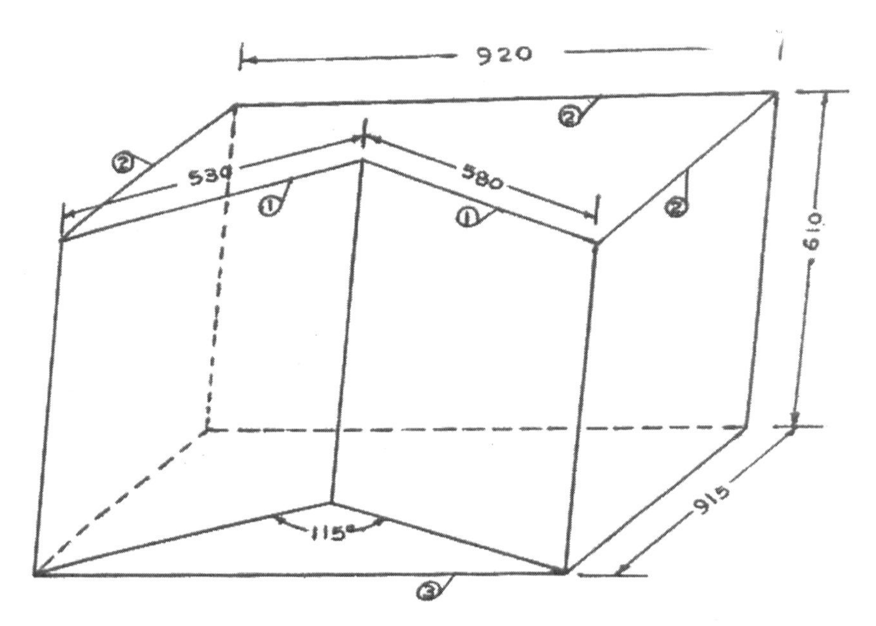

1. **Flexy glass plates fixed on front.**
2. **Mild steel plates fixed on sides.**
3. **Mild steel plate fixed on bottom.**

Note: All dimensions are in mm.

FIGURE 7.1 Isometric view of laboratory model of soil tank.

The bottom and its three vertical sides were made up of metal sheets, and fourth vertical side (front side) was provided with two flexi glass plates at an angle of 115° to observe the water front advance. The tank was filled with soil up to a depth of 50 cm. The soil texture was sandy loam having sand 69%, silt 21.2% and clay 9.8%. A plastic bottle with a capacity of two liters was kept in a particular position on a vertical stand. The bottle was provided with a vertical circular orifice of diameter 6 mm very close to its bottom. A piece of 6 mm diameter microtube was fitted into the orifice and at the end of the microtube a dripper of capacity 4 lph was attached. Then the dripper was calibrated for its design discharge at a constant head. The water was supplied to the overhead bottle continuously from the outside source through siphoning to maintain the constant head. After completion of one experiment, wet soil around the microtube fitted with dripper was carefully taken out of the soil tank. The soil was again dried, pulverized and reused in the subsequent experiment. The microtube was again installed in the soil.

The dripper used in the study had two projections. One projection was of conical shape and the other was of cylindrical shape. Normally water was discharged through the cylindrical projection. At the normal position of the dripper, the conical projection of the dripper was fitted into the microtube coming from the lateral and the cylindrical projection of the same dripper was fitted to the microtube extension through which water was discharged. Twelve observations for each were taken for normal and reverse position of the dripper. Each observation was replicated thrice to calculate the average discharge (Table 7.1).

TABLE 7.1 Average Discharge Rates For Normal and Reverse Positions of the In-Line Microtube

Head	Average discharge, Q Normal position	Average discharge, Q Reverse position	Difference of discharge in normal and reverse positions, ΔQ
m	lph		
0.4	0.597	0.545	0.052
0.5	0.688	0.636	0.052
0.6	0.801	0.750	0.051
0.7	0.910	0.856	0.055
0.8	1.017	0.964	0.052
0.9	1.120	1.069	0.052
1.0	1.216	1.158	0.058
1.1	1.305	1.252	0.053
1.2	1.381	1.326	0.055
1.3	1.447	1.396	0.051
1.4	1.505	1.450	0.055
1.5	1.561	1.508	0.053

After getting discharge data for corresponding hydraulic heads, the graphical relationships were developed between hydraulic head and discharge for normal and reverse positions of the dripper as shown in Fig. 7.2. The model equations were developed for both positions of the dripper.

Tracing paper was fixed on the flexi glass plates of the soil tank model lying in front of the dripper for demarcating the advance of waterfront. The dripper with microtube was placed at a depth of 12 cm from the soil surface in the soil tank in case of sub surface drip irrigation. Water was supplied from the supply bottle under constant head through the dripper and microtube extension into the soil.

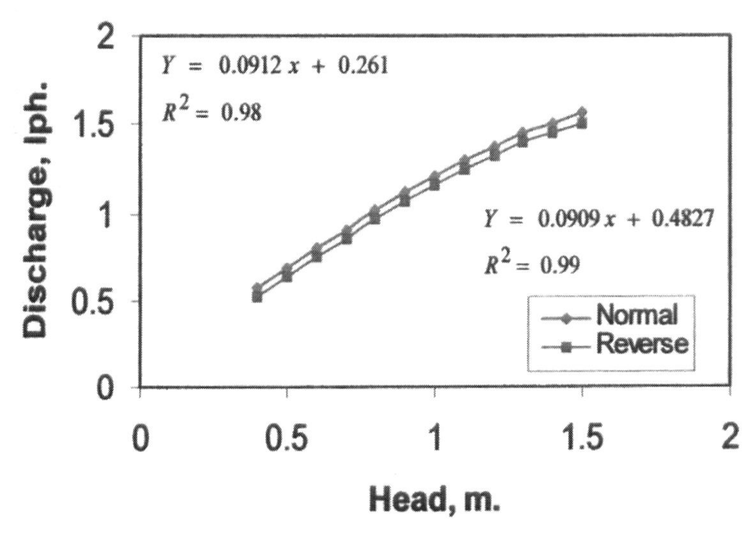

FIGURE 7.2 Head versus discharge relationships for normal and reverse positions of for in-line drip irrigation systems.

Soil in the tank was irrigated at the rate of the capacity of dripper. The maximum waterfront advances both in horizontal and vertical directions were recorded at every 10 min interval. But the first set of reading (horizontal and vertical water front advance) in each case (in-line sub surface and surface) was taken on 20th minute from the start of experiment because the waterfront appeared after 19 min from the start of experiment. The known volume of water supplied during that 20/10 min intervals was also recorded. The water front advances were recorded from 20th minute to 120th minute.

In in-line subsurface drip irrigation, the system was operated until the top soil was wetted at different heads of 0.5 to 1.5 m. The minimum and maximum water front advances were recorded in case of in-line surface and in-line subsurface drip irrigation systems.

7.3 RESULTS AND DISCUSSION

The horizontal and vertical water front advances for in-line subsurface and surface drip irrigation systems were compared at different hydraulic heads and time of application, as shown in Table 7.2 and Figs. 7.3 and 7.4.

TABLE 7.2 Horizontal and Vertical Water Front Advances For In-Line Subsurface and Surface Drip Irrigation Systems at Different Hydraulic Heads and Time of Application

Head	Time of application	In-line sub surface drip irrigation		In-line surface drip irrigation		Difference between subsurface and surface horizontal water front advances	Difference between subsurface and surface vertical water front advances
		Horizontal water front advance	Vertical water front advance	Horizontal water front advance	Vertical water front advance		
m	minutes	cm					
	20	14.0	10.5	11.5	7.0	2.5	3.5
0.5	50	22.0	18.7	19.0	15.2	3.0	3.5
	80	27.0	24.1	24.0	20.0	3.0	4.1
	20	16.0	12.0	13.3	10.0	2.7	2.0
0.6	50	25.0	20.7	22.8	16.5	2.2	4.2
	80	28.0	24.8	26.9	21.0	1.1	3.8
	20	16.5	12.3	14.8	11.0	1.7	1.3
0.7	50	25.5	21.1	23.0	16.8	2.5	4.3
	80	28.5	25.4	27.8	21.3	0.7	4.1
	20	17.0	12.7	15.6	11.5	1.4	1.2
0.8	50	26.0	21.5	23.0	17.0	3.0	4.5
	80	29.0	25.9	27.8	21.5	1.2	4.4
	20	17.8	13.4	16.6	12.0	1.2	1.4
0.9	50	27.0	22.5	23.6	18.2	3.4	4.3
	80	29.8	27.5	28.9	23.5	0.9	4.0
	20	18.5	14.5	17.4	12.6	1.1	1.9
1.0	50	28.0	23.2	24.9	18.6	3.1	4.6
	80	31.0	28.3	29.8	24.2	1.2	4.1
	20	19.0	15.4	17.7	13.0	1.3	2.4
1.1	50	28.3	23.8	25.5	20.2	2.8	3.6
	80	31.4	28.6	30.2	25.0	1.2	3.6

TABLE 7.2 *(Continued)*

Head	Time of application	In-line sub surface drip irrigation		In-line surface drip irrigation		Difference between subsurface and surface horizontal water front advances	Difference between subsurface and surface vertical water front advances
		Horizontal water front advance	Vertical water front advance	Horizontal water front advance	Vertical water front advance		
m	minutes	cm					
1.2	20	19.5	16.3	18.0	13.5	1.5	2.8
	50	28.5	24.9	25.9	22.4	2.6	2.5
	80	32.5	29.8	31.0	27.0	1.5	2.8
1.3	20	19.8	17.2	18.5	14.7	1.3	2.5
	50	29.0	26.3	26.2	23.0	2.8	3.3
	80	Water front advance touched the ground at 76th minute		31.7	28.0	Nil	Nil
1.4	20	20.0	17.8	18.9	15.5	1.1	2.3
	50	29.3	25.8	27.1	23.5	2.2	2.3
	80	Water front advance touched the ground at 70th minute		32.2	29.0	Nil	Nil
1.5	20	20.6	18.1	19.3	16.0	1.3	2.1
	50	29.5	27.4	27.6	24.6	1.9	2.8
	80	Water front advance touched the ground at 65th minute		32.1	29.8	Nil	Nil

It was observed that the horizontal water front advance was always faster than that of vertical water front advance in both cases of drip irrigation systems irrespective of hydraulic head and time of application. However, for horizontal water front advance under both drip irrigation systems, it was observed that the horizontal water front advance of in-line subsurface drip irrigation was always higher than that of in-line surface drip irrigation. Similar trend of vertical water front advance was also noticed in both cases of drip systems. In case of in-line subsurface drip, the horizontal water front advance varied from 5.09 to 25.45% over vertical water front advance under hydraulic head range of 0.5 to 1.5 m, whereas, in case of in-line surface drip, it varied from 2.95 to 30.19%. In general, it was observed that horizontal water front advance of in-line subsurface drip was more than that of corresponding horizontal water front advance of in-line surface drip under all hydraulic heads and time of application.

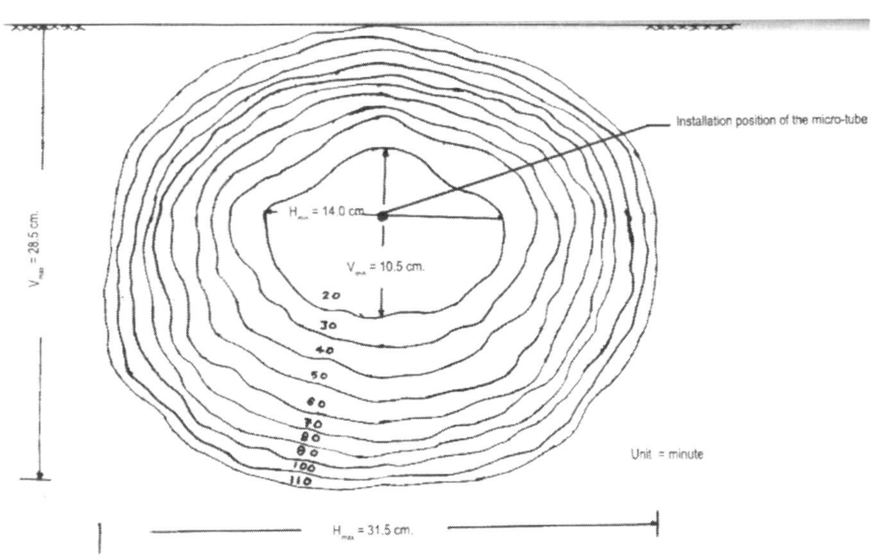

FIGURE 7.3 Water front advance of different durations for in-line subsurface drip irrigation system at 0.5 m hydraulic head.

The maximum horizontal water front advance of in-line sub surface drip was 32.5 cm and the minimum was 14.0 cm, whereas it was 37.3 cm and 11.5 cm, respectively, in case of in-line surface drip. Similar trend was also noticed in case of vertical water front advance. The maximum and minimum vertical water front advances were 30.85 cm and 10.5 cm, respectively, in case of in-line subsurface drip irrigation, whereas it was 36.2 cm and 7.0 cm for in-line surface drip system. The shape of the water front advance curves of in-line subsurface drip was elliptical, whereas it was semielliptical for in-line surface drip system. The in-line sub surface drip system was remained operational for a period of more than six months continuously. It was observed that the system did not show any clogging problem.

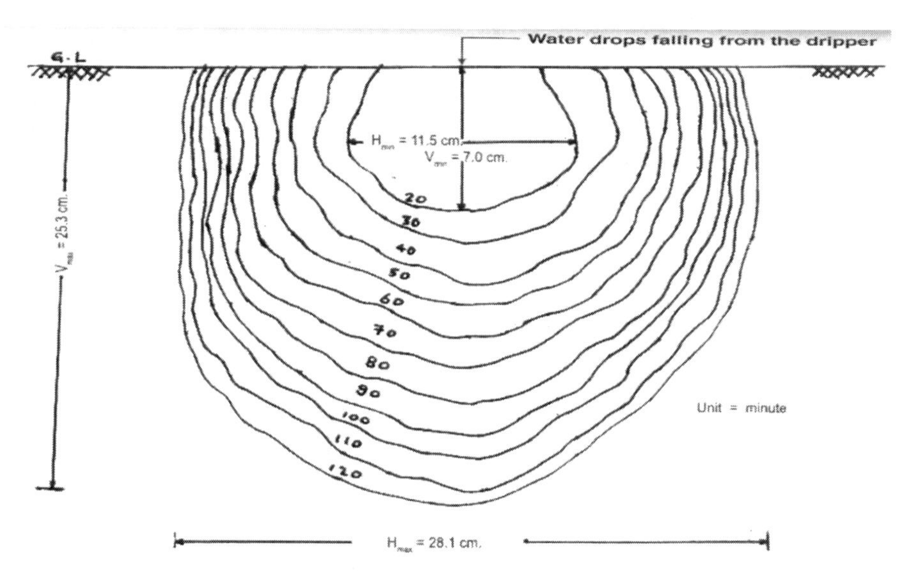

FIGURE 7.4 Water front advance of different durations for in-line surface drip irrigation system at 0.5 m hydraulic head.

7.4 CONCLUSIONS

The surface drip irrigation can always be modified into subsurface drip irrigation without any clogging problems. The loss of water due to evaporation is eliminated in case of in-line subsurface drip irrigation, whereas it is a common phenomenon in case of in-line surface line drip irrigation and hence it increases water use efficiency of in-line subsurface irrigation over in-line surface drip irrigation. Due to considerable amount of water saving in case of in-line subsurface as compared to in-line surface drip irrigation, more area can be brought under irrigation. The horizontal and vertical water front advances of in-line sub surface drip were greater than that of in-line surface drip system.

7.5 SUMMARY

An experiment in the soil tank model with sandy soil was carried out in the Hydraulics Laboratory of College of Engineering and Technology, Bhubaneswar, Odisha, India in 2013. The hydraulics of in-line surface drip irrigation was compared with those of subsurface drip irrigation. Water front movement in both horizontal and vertical directions was also evaluated for both irrigation systems. Under in-line surface and subsurface drip irrigation systems, the horizontal water front movement was faster than the vertical water front movement. However, both horizontal and vertical water front advances were higher under in-line subsurface drip irrigation compared to that of in-line surface drip irrigation. It was further noticed that the water front advance was elliptical in shape under in-line subsurface drip irrigation system whereas the shape was semielliptical under in-line surface drip irrigation. The in-line subsurface drip irrigation was found to be a better option for adoption than in-line surface drip irrigation.

KEYWORDS

- **Bermuda grass**
- **clogging**
- **drip irrigation**
- **dripper**
- **emitter depth**
- **India**
- **irrigation system**
- **nitrogen fertilization**
- **permeable soil**
- **reclaimed water**
- **subsurface drip irrigation**
- **surface drip irrigation**
- **water front advance**
- **water front movement**

REFERENCES

1. Camp, C. R. (1998). Subsurface drip irrigation: A review. *Transaction of the ASAE*, 41(5):1353–1367.
2. Choi, C. Y., Suarez–Rey, E. M. (2004). Subsurface drip irrigation for Bermuda grass with reclaimed water. *Transaction of ASABE*, 47(6):1943–1951.
3. Grimes, S. W., Munk, D. S., Goldhame, D. A. (1990). Drip irrigation emitter depth placement in a slowly permeable soil. In: *Vision of the Future*. Proc. 3rd National Irrigation Symposium, 28 Oct.–1 Nov., Phoenix, Arizona. *St. Joseph, MI–USA, ASAE.*
4. Lamm, F. R., Troien, T. P., Manges, H. L., Sunderman, H. P. (2001). Nitrogen fertilization for subsurface drip irrigated corn. *Transaction of ASAE*, 44(5):533–542.

PART II

PRACTICES IN MICRO IRRIGATION

CHAPTER 8

IMPACTS OF MICRO IRRIGATION SUBSIDIES IN TAMIL NADU FROM WORLD BANK

G. PARTHASARATHI, S. SENTHILNATHAN, and L. SURESH

CONTENTS

This chapter is based on G. Parthasarathi, "Trade and environment impact of micro irrigation subsidies: Experiences from World Bank funded IAMWARM PROJECT: A Study in Palar Sub-basin of Tamil Nadu" (MSc. seminar report, Tamil Nadu Agricultural University, Coimbatore, India, 2012).

8.1 INTRODUCTION

A progressive agriculture serves as a powerful engine of economic growth of any country. In India, agriculture represents a core part of economy and provides food and livelihood activities to much of the Indian population. Agriculture and allied sectors accounted for 13.5% of the GDP in 2012–2013 compared to 51.9% during 1950–1951, thus indicating a steady decline of its share in the GDP. Further, the share of agriculture alone was recorded as low as 11.6% in 2012–2013 compared to 41.8% in 1950–1951 to the GDP. Still agriculture is the largest sector playing a major role in shaping the overall growth trajectories of the Indian economy since independence [2].

Agricultural policy is a set of government decisions and actions relating to domestic agriculture and imports of foreign agricultural products. Governments usually implement agricultural policies with the goal of achieving a specific outcome in the domestic agricultural product markets. Subsidy, the most powerful mechanism, can balance the growth rate of production and trade in various sectors and regions and for an equitable distribution of income for the protection of weaker sections of society [7]. The value of sector wise estimated major subsidies in India were about 231,083,0 million-Rs. for the year 2013–2014. The cost of India's agricultural input subsidies as a share of agriculture output almost doubled from 6.0% in 2003–2004 to 11.6% in 2009–2010, driven mostly by large increases in the subsidies to fertilizer and electricity.

8.2 AGRICULTURAL SUBSIDIES

Agricultural subsidies can play an important role in early phases of agricultural development by addressing market failures and promoting new technologies [1]. The farm subsidies are integral part of the policies supporting farming and farmers. Especially governments of developing countries like India give importance to subsidies to promote agriculture. The government policy of a subsidy is very well for protection of the weaker sections and marginal farmers. The Indian farmers being poor and they were not in a position to buy the expensive inputs. Then the Indian government started the scheme of subsidies on the purchase of various agriculture inputs to facilitate the farmers. Fertilizer, electricity, irrigation water and farm insurance are major items for subsidies given to farmers in India. Government of India pays fertilizer producers directly in exchange for the companies selling fertilizer at lower than market prices. Irrigation and electricity, on the other hand, are supplied directly to farmers by Government of India at prices that are below the cost of production.

India's expenditure on input subsidies has increased sharply in recent years. Fertilizer, electricity and irrigation were the major subsidies to the Indian farmers. In 1993–1994, the value of fertilizer, electricity and irrigation subsidies was 45,620, 24,000 and 58,720 million-Rs., respectively at 1993–1994 prices. During the year 2009–2010, the value of fertilizer subsidy was Rs. 529,800 million-Rs. at 1999–2000 prices. In 2008–2009, the electricity and irrigation subsidies were 274,890 and 236,650 million-Rs., respectively. Drip irrigation has potential to save water [3, 4, 6].

The present study was conducted to find out the impact of micro irrigation subsidies given under Tamil Nadu-IAMWARM project on farm and farmers with the focus on trade and environment. The specific objectives of the study were: (i) to find the investment worthiness of drip irrigation; (ii) to find out the yield difference between the scheme and nonscheme farmers for major crops; and (iii) to find out the water usage in various crops.

8.3 METHODOLOGY

8.3.1 STUDY AREA

The present study was conducted in Udumalpet region of Tiruppur district of Tamil Nadu, where the TN-IAMWARM project implemented successfully. The total area under crop cultivation in Udumalpet block was 29,595.21 ha consisting of 24469.69 hectares under irrigated area and 5125.54 ha under rain-fed area. Therefore, majority of the study area was covered by irrigated conditions. Paddy, maize, onion, tomato, *brinjal* (eggplant), Bengal gram and other pulses were the major crops that were cultivated in this region.

8.3.2 SAMPLING

Based on the objectives and for the purpose of study, both primary and secondary data were collected. The farmers were categorized into TN-IAMWARM scheme (drip) and nonscheme (conventional) farmers. Separate questionnaires were prepared and the data regarding were collected from the farmers by administering pretested interview schedule. Five villages were selected purposively, where the adoption of the scheme is widespread. From each village, six scheme farmers and six nonscheme farmers were identified. Finally the sample size of 60 was considered. The scheme drip farmers were identified through Water Technology Centre, TNAU and the nonscheme farmers were collected randomly in the same region.

8.3.3 THE TAMIL NADU IRRIGATED AGRICULTURE MODERNIZATION AND WATER BODIES RESTORATION AND MANAGEMENT (TN-IAMWARM) PROJECT

The Tamil Nadu Irrigated Agriculture Modernization and Water Bodies Restoration and Management (TN IAMWARM) is a unique World Bank funded project implemented with the prime motive of maximizing the productivity of water leading to improved farm incomes and products. The broader objectives of the project were to achieve sustainable economic growth as well as poverty alleviation through maximizing productivity of water. The IAMWARM project supported the investment in: (i) improving irrigation service delivery including adoption of modern water-saving irrigation technologies and agricultural practices; (ii) agricultural intensification and diversification; (iii) enhancing market access and agri-business opportunities; and (iv) strengthening institutions dealing with water resources.

8.3.4. ANALYSIS

8.3.4.1 INVESTMENT ANALYSIS

To find the worthiness of drip irrigation, the benefit-cost ratio (BCR), net present value (NPV) and internal rate of return (IRR) were calculated for coconut for 12 years.

8.3.4.1.1 BENEFIT COST RATIO

A BCR is an indicator, used in the formal discipline of cost-benefit analysis, which attempts to summarize the overall value for money of a project or proposal.

$$\text{BCR} = \text{Present worth of return/Present worth of cost} \quad (1)$$

8.3.4.1.2 NET PRESENT VALUE

The NPV is a central tool in discounted cash flow (DCF) analysis and is a standard method for using the time value of money to appraise long-term projects.

$$\text{NPV} = [(B_t - C_t)/(1+r)^t] \quad (2)$$

where, B_t denotes the benefits in the year t, C_t is the cost in the year t and t is the time period, and r is the rate of interest.

8.3.4.1.3 INTERNAL RATE OF RETURN

The IRR is used in capital budgeting to measure and compare the profitability of investments.

$$\text{IRR} = \text{LDR} + [\text{Difference between two discount rates (NPPV at LDR} \div \text{Absolute sum of NPV at LDR and HDR)}] \quad (3)$$

8.3.4.2 CONVENTIONAL ANALYSIS

To find out the yield difference between the scheme (drip) and nonscheme (nondrip) farmers for various major crops, the percentages and averages were used.

8.3.5 WATER USE IN EACH CROP

One of the objectives is to find out the water use in various crops between scheme and nonscheme farmers. The water applied for a particular crop in a season is estimated by using the following conversions and express in ha-cm [5]. In order to examine the changes in the water applied for a particular crop in the season, it is estimated as below:.

Water applied for a crop (ha-cm) = {[Area irrigated in ha x number of irrigations/crop x

irrigation duration in hours x average water use in lph]/101,171.26} (4)

8.4 RESULTS AND DISCUSSION

8.4.1 INVESTMENT ANALYSIS

Table 8.1 shows the results of investment analysis for drip-irrigated coconut. Results show the positive impact. The BCR for subsidized farms was 3.22 compared to 1.7 for the nonsubsidized farms. The NPV of scheme farmers was Rs. 261,074 compared to Rs. 123,133 for conventional farmers. The IRR for scheme and nonscheme farmers were 35.62 and 31.76, respectively.

TABLE 8.1 BCR, NPV and IRR

Variables	Scheme farmers	Conventional farmers (Non-scheme)
BCR	3.22	1.70
NPV, Rs.	261,074.8	123,113.0
IRR	35.62	31.76

8.4.2 YIELD DIFFERENCES

The differences in yield were calculated for coconut, tomato, onion and maize. The averages were taken and the results are shown in Table 8.2. The yield of coconut, tomato, onion and maize was increased by 50%, 26%, 43%, and 36%, respectively.

TABLE 8.2 Yield of Study Crops in the Farms of Scheme and Nonscheme Farmers

Item	Coconut (nuts/ year)		Tomato (tons/acre)		Onion (tons/ acre)		Maize (tons/ acre)	
	N S	S	N S	S	N S	S	N S	S
1. Yield	100	150	17.9	22.53	7.03	10.03	2.61	3.54
2. Percentage increasing	50		26		43		36	

NS – Non-Scheme, **S** – Scheme

8.4.3 WATER USE FOR VARIOUS CROPS (HA-CM)

To assess the water usage, the water use for maize, tomato and onion were calculated. Compared to nonscheme farmers, the water usage is low in scheme farmers. The results are shown in Table 8.3. It is depicted that the scheme (drip) farmers were using low quantity of water in a particular season. Comparing nonscheme farmers, the water usage in maize, tomato and onion was reduced up to 25%, 34% and 32%, respectively.

TABLE 8.3 Water Usage in Scheme and Nonscheme Farms

Crop	Water Usage (ha-cm)		
	Scheme farms (drip)	Non-Scheme farms (nondrip)	Percentage saving
Maize	26.33	35.27	25
Onion	25.50	37.58	32
Tomato	20.34	30.92	34

8.5 IMPACTS OF MICRO IRRIGATION SUBSIDY ON PRODUCTION AND TRADE

Apart from saving water in crop cultivation, micro irrigation subsidy has positive impacts in marketing and trade also. Table 8.2 shows the average yield of various crops under drip and conventional irrigation systems. In the study area, mainly the high value crops were growing. If the total area under maize, in the study area comes under drip irrigation, the production will be 15328.51 tons, whereas it will be 11301.53 tons in conventional irrigation. The difference is 4026.98 tons. In case of tomato and onion, the production will be 9787.48 tons and 4908.58 tons for the total area under drip irrigation, respectively compared to 7776.11 tons and 3503.68 tons, respectively under conventional irrigation. The difference in total yield is 1404.9 tons and 2011.37 tons for tomato and onion, respectively. So, apart from conventional irrigation, if the total area comes under drip irrigation we can increase the yield and it will increase the marketable surplus of the farmers and there creates a positive impact on trade also.

Apart from the increase in the yield by drip irrigation, the farmers are able to grow crops in the off-season and get higher prices. The normal market prices (Rs./kg) were 10 for maize, 10 for tomato and 12 for onion. During off-seasons, the price of maize, tomato and onion were approximately 14, 25, and 25 Rs. per **kg**, respectively. By producing maize, tomato and onion in the off-season under drip irrigation, the farmers can get 4, 15. and 13 more Rs. per **kg** of maize, tomato and onion, respectively. This will increase the income for the farmer.

8.6 IMPACTS OF MICRO IRRIGATION SUBSIDY ON NATURAL RESOURCES AND ENVIRONMENT

The effect of micro irrigation subsidy shows positive impact not only in individual farms but also there are positive impacts on the environment also. In conventional irrigation, the water losses are high. Table 8.3 shows the water use for various crops between the drip irrigated and conventional farms. The percentage water saving in maize, tomato and onion was around 25%, 34% and 32%, respectively. The total area under crop cultivation in the study area is 29,595.21 ha that consisted of 4330.09 ha, 434.42 ha and 489.39 ha under maize, tomato and onion, respectively. The tomato and onion were growing only under irrigated conditions. Presently, Tamil Nadu state policy is focusing towards bringing the entire area of garden lands under drip irrigation. If the total area comes under drip irrigation in the study area for tomato and onion, the water usage level is 8836.10 and 12479.44 ha-cm, respectively whereas in conventional irrigation it is 13432.26 and 18391.27 ha-cm, respectively. We can save 4596.16 and 5911.83 ha-cm of water if both tomato and onion comes under drip irrigation in the study area alone. In case of maize, if the total area comes under drip irrigation the water usage level is 114,011.27 ha-cm whereas in conventional irrigation it is 152,722.27 ha-cm. Therefore, we can save 38,711 ha-cm of water in maize cultivation. This will lead to natural resource saving and benefit to the environment by using the saved water for alternate purpose.

8.7 SUMMARY

By subsidizing drip irrigation, the small and marginal farmers also can go for the same. This will lead to increase in production, productivity and saving of natural resources in general. It can be concluded that there are multiple benefits to the farms and farmers because of farm subsidies. The results show that, if the total area comes under drip irrigation, we can save 38711 ha-cm, 4596.16 ha-cm and 5911.83 ha-cm of water in maize, tomato and onion crops, respectively. Apart from water usage level, if the total area comes under drip irrigation, we can increase the production by

4026.98 tons, 1404.9 tons and 2011.37 tons of maize, tomato and onion, respectively. By producing maize, tomato and onion in the off-season (when the availability of water is less for surface irrigation) through drip irrigation, the farmers can get additional price (Rs./kg) of 4 for maize, 15 for tomato and 13 for onion. However, water will be more effective and realistic by incorporating the issues like: free electricity, cropping system dynamics, land use, water harvesting, agriculture/nonagricultural tradeoff, etc. while drafting policies towards subsidizing irrigation.

KEYWORDS

- **agricultural policy**
- **benefit cost ratio**
- **drip irrigation**
- **economic growth**
- **electricity**
- **fertilizer**
- **GDP**
- **Government of India**
- **India**
- **Indian rupee, Rs.**
- **internal rate of return**
- **irrigation**
- **maize**
- **marginal farmers**
- **net present value**
- **onion**
- **small farmers**
- **subsidy**
- **Tamil Nadu**
- **Tamil Nadu Agricultural University**
- **tomato**
- **water harvesting**
- **World Bank**

REFERENCES

1. Fan, Shengyen, Ashok Gulati and Sukhaseo Thorat, (2008). Investment, subsidies and propoor growth in rural India. *Agricultural Economics*, 39(2):163–170.
2. GOI, (2013). *Agricultural Statistics Data for India.* Central Statistics Office.
3. Narayanamoorthy, A. (2010). Can drip method of irrigation be used to achieve the macro objectives of conservation agriculture? *Indian Journal of Agricultural Economics,* 65(3):428–438.
4. Price Gittinger, J. (1988). *Economic analysis of Agricultural Project.* World Bank, Washington D. C.
5. Singh, K, and Y. Katar, (1990). People participation in water resources management. *Workshop Report-8*, Institute of Rural Management, Jaipur, pp. 7–9.
6. Sivanappan, R. K. (1994). Prospects of microirrigation in India. *Irrigation and Drainage Systems,* 8(1):49–58.
7. Suresh Kumar, D., Palanisami, K. (2010). Impact of Drip Irrigation on Farming System: Evidence from Southern India. *Agricultural Economics Research Review,* 23:265–272.

CHAPTER 9

ADOPTION AND ECONOMIC IMPACT MODELS OF MICRO IRRIGATION IN ZAMBIA

SIMEON DIGENNARO and DAVID S. KRAYBILL

CONTENTS

9.1 INTRODUCTION

Agriculture is a vital part of household livelihoods and national economies of developing countries. The role of a country or community's food supply in human survival sets the productivity of the agriculture sector apart from other sectors of production. Historically, throughout the world, the introduction of new technology has been an important factor in increasing the productivity of labor and land and, hence, in improving household livelihoods and food security.

Despite the advances that have been achieved through improved seed varieties and application of chemical fertilizer, hunger and food insecurity still plague many parts of the developing world. The Food and Agriculture Organization (FAO) reports approximately one-quarter of the population in Sub-Saharan Africa is undernourished [11].

Irrigation has the potential to increase agricultural production and improve the livelihoods of small-scale farmers. Adequate water supply is essential to support the production of crops for household consumption or sale. In addition to increasing production levels, irrigation increases the reliability and consistency of production [44]. Reducing vulnerability to shocks and variability of production is extremely important for subsistence farmers. Fewer or less severe shocks mean the household is able to maintain adequate consumption levels and is less likely to deplete savings or productive assets, such as tools and livestock, to cope with shocks. Reduced vulnerability enables poor farmers to maintain their productive assets and avoid indebtedness for consumption [3, 53].

Water is a scarce resource, and the method of irrigation affects the degree of scarcity. Traditional irrigation methods such as furrow irrigation are very inefficient due to evaporation and leaching. Drip, or microirrigation, which uses water much more efficiently, is a potentially important technology in addressing the irrigation needs in areas with limited water resources. In addition to water efficiency, the precision of drip irrigation reduces water logging of the plants, salinization, and leaching of soil amendments, such as fertilizer. In many applications, drip irrigation reduces the cost of production and greatly increase productivity [36].

Despite the benefits of irrigation, adoption of irrigation technology has been very low in Africa. The FAO estimates that only 6% of the cultivated land in Africa is irrigated [16]. In contrast, 35% of the cultivated land in Asia is irrigated. In response to the potential benefits of irrigation and the low adoption rates in rural areas, especially in Africa, many organizations, governmental and nongovernmental, promote small-scale irrigation technology. The irrigation projects implemented in developing countries provide a wide variety of information, services, and financial assistance; however, very little rigorous evaluation has been conducted on the actual impact of these programs on participating households. While the role of irrigation in poverty reduction has been studied more extensively in Asia, relatively little research has been done in Sub-Saharan Africa [21].

The *Prosperity Through Innovation Project* (PTI) is one such program to promote the development of small-scale irrigation in Zambia. The aim of the project is to increase household livelihoods through the use of micro irrigation equipment. Another important goal of the project is to develop supply chains for irrigation technology. The project supplies discount vouchers for micro irrigation equipment to small-scale farmers. The PTI project works jointly with another agriculture development organization, *International Development Enterprises* (IDE), to promote the expansion of irrigation supply chains and provide information and training for its use.

This study evaluates the economic impact of micro irrigation equipment made available through the PTI voucher program. The efficacy of this project is tested using household data collected from farmers participating in the PTI project. The analysis addresses two specific research questions:

1. What are the key factors affecting micro irrigation equipment voucher redemption?
2. What are the economic effects of the project on the livelihoods of participating households?

9.2 LITERATURE REVIEW

Despite major breakthroughs in agricultural technologies such as drip irrigation, agricultural households in many parts of the world still struggle to produce enough to support themselves. Significant portions of the population in Sub-Saharan Africa are food insecure as measured by a variety of nutritional and caloric indicators [11, 45]. With approximately 60% of the labor force employed in agriculture in Sub-Saharan Africa, the level and reliability of agricultural productivity is a critical element in food security. As a result of low priced staple crops coupled with poor yields and erratic rainfall, small-scale agriculture production is both risky and low return [3]. The problems of poor yields and erratic rainfall are expected to worsen in the future as a result of climate change [19, 32]. Small-scale irrigation is frequently cited as an innovation that can bolster rural livelihoods through climate adaptation, food security, and poverty reduction [2, 3, 30, 39, 50]

Polak and Yoder [39] highlight three characteristics of small-scale drip irrigation that enables it to impact the poor and set it apart from other irrigation technology. First, the affordability of small-scale irrigation technology makes it accessible to poor farmers. The cost of a treadle pump for water extraction is approximately one-twentieth the price of a small diesel powered pump. Low-head gravity fed drip systems are much less expensive than conventional high-pressure systems. Second, the divisibility of small-scale irrigation increases the ability of small-scale farmers to adopt the technology. Similar to the scalable nature of the use of improved seed varieties that enabled the Green Revolution to be adopted by both large and small-scale farmers, drip irrigation kits are an investment that can be made in small increments.

Many investments in new technology require large expenditures, such as buying a tractor, because they cannot be divided up into smaller pieces; however, there are low cost drip systems available for as little as US $2.50 for garden sized plots [38]. The third important characteristic of small-scale irrigation technology is its expandability. Small low-cost systems need to be available for entry-level needs, but as a farmer's income increases, a portion of the additional revenues can be reinvested in additional equipment, expanding irrigation capacity. These three characteristics make small-scale irrigation an important tool for increasing the productive capacity of small-scale poor farmers.

Increasing farm output has been shown to have significant impacts on poverty reduction. The poverty reduction elasticity of farm output has been shown to be around −0.35 [20]. That is a 1% increase in farm output leads to a 0.35% reduction in poverty. Both direct and indirect benefits of irrigation contribute to the elasticity of poverty reduction for irrigation. In the long run, indirect benefits can be a major contributor to poverty reduction because the indirect benefits of employment and price effects are realized by the poorest and generally landless class. An estimation by De Janvry and Sadoulet [4] of the direct and indirect effects of agricultural technology change in different regions of the developing world shows that for Africa, direct effects play the biggest role in poverty reduction. Different from Asia and Latin America, Africa has a larger share of rural agrarian households who produce a major portion of their food. Thus, designing agricultural technologies for small-scale rural households and assisting in their diffusion is key to rural poverty reduction [4].

Irrigation in South and South-east Asia has been shown to improve crop productivity, enable households to grow higher valued crops, lead to higher incomes and wage rates for family labor, benefit the poor and landless through increased food availability, and lower food prices [22]. An empirical study by Tesfaye et al. [48] finds access to small scale irrigation leads to increased and stable production, income and consumption in Ethiopia. The adoption of treadle pump irrigation improved the poverty status of households and prevented households from falling into poverty in Malawi [33]. Drip irrigation in India increased production of crops, reduced water consumption and environmental problems such as soil salinization and fertilizer run-off [37]. Analysis of household data from 13 villages in Northern Mali shows increases in total household consumption, agricultural production, caloric and protein intakes, and savings, for households with access to irrigation [8]. In addition to increased production, irrigation reduces the variability of production levels from rainfall shocks. In India, the growth of crop output per year from irrigated areas had a 2.5 times lower standard deviation than growth rates of rain fed areas [31]. Stability in household food production is a major contributor to household food security. Reliable production benefits the rural poor most because of the limited access to coping mechanisms such as credit services, insurance, savings, and other sources of income.

A number of potential barriers to the adoption of a technology such as micro irrigation have been identified [12]. Inadequate information, education, and training are significant barriers to the adoption of new agricultural technology [17]. Uncertainty and risk, also associated with information about the technology, discourages adoption [14]. Lack of access to credit, especially when a significant expenditure is required to purchase equipment, prevents adoption [17]. Absent or unreliable supply of equipment, and insufficient transportation or infrastructure can also be major barriers to adoption of a new technology. A certain level of success has been achieved by programs aimed at addressing these issues; however, the dynamics governing technology adoption vary over time and between localities and groups [12].

While there is literature confirming positive impacts of microirrigation, there is a lack of reliable empirical evaluations of policies and programs, which promote it. Many programs are implemented without any rigorous evaluation of impact. The effectiveness of micro irrigation programs can vary due to the local circumstances of small-scale farmers. There is little empirical work on the impact of micro irrigation in Sub-Saharan Africa. A wider base of research evaluating specific programs will provide needed information about the relative effectiveness of micro irrigation in different settings [46].

9.3 CONCEPTUAL FRAMEWORK

9.3.1 DEMAND FOR IRRIGATION

The neoclassical theory of demand for inputs can be used to provide a framework for how a household uses irrigation water to maximize farm profits. Treating the household as a profit-maximizing firm, it will employ an input if the added benefit provided is greater than the added cost of the input. That is, the level of the input use will occur where the marginal benefit is equal to the marginal cost. Applying this concept to small-scale agricultural households and irrigation water, the household will use irrigation up to the point where the cost of an additional unit of irrigation just equals the added value its use produces. However, in regions where there is no water market, there is no explicit price for irrigation water. In this situation, the cost of irrigation can be represented by the labor costs of irrigating. Furthermore, for rural households, labor markets are often missing or ineffective forcing households to be self-sufficient for their labor supply. The need for labor self-sufficiency is accentuated by the simultaneity of seasonal labor requirement levels, lack of a landless class that provides hired labor services, and similarities in factor endowments [5]. In a context where no well-functioning labor market exists, the shadow wage determines the household's supply and demand of labor rather than a market wage [29, 43]. The shadow wage is how much the household values its labor. It is the implicit wage it needs to receive to engage in an activity. The shadow wage is determined within the household and affected by the household's preferences, technology, other inputs, and market prices [47]. At equilibrium, the shadow wage is equal to the marginal revenue product of labor (marginal product of labor *x* price of output) [24].

Suppose the household labor supply has a tradeoff between irrigating more crops and leisure time. The household will irrigate up to where the marginal cost of irrigation (labor cost) is equal the marginal revenue product of labor (shadow wage). Figure 9.1 illustrates how the adoption of an agricultural technology such as micro irrigation affects the demand and supply of labor for irrigated crop production. The supply curve of labor (S_L) is upward sloping. The household will choose to shift labor away from leisure to irrigation as the implicit wage increases. The labor supply curve also represents the marginal cost of irrigation given that the cost of irrigation is the labor required. The demand for labor for irrigated crop production is represented by the marginal revenue product of labor ($MRP_{L)}$. The MRP_L is downward sloping, because of the law of diminishing marginal returns when assuming capital is fixed.

At equilibrium, the household will supply labor for irrigated crop production up to the point where its shadow wage is equal to the marginal revenue product of labor for irrigation. This is shown in Fig. 9.1 where the labor supply curve (S_L) intersects with the marginal revenue product of labor curve (MRP_L). While a household without micro irrigation equipment may irrigate crops by hand, the level of irrigation will be relatively low as a result of the low productivity of labor. The introduction of a technology such as micro irrigation equipment increases the productivity of labor resulting in an outward shift in the marginal revenue product of labor curve and a higher demand for irrigation labor. The increase in productivity of labor from MRP_L to MRP_L' results in the household supplying more labor for irrigated crop production (from q to q'). The household is able to earn a higher implicit wage with micro irrigation equipment.

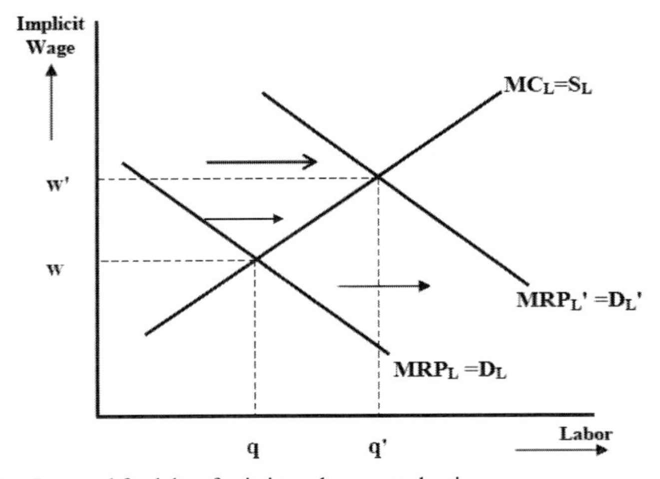

FIGURE 9.1 Demand for labor for irrigated crop production.

9.3.2 DETERMINANTS OF ADOPTION

The literature on agriculture technology adoption identifies a number of factors affecting the decision to adopt a new agricultural technology. [1, 12, 14, 35, 42]. These factors include physical, natural, human, social, community, and financial assets of the agricultural household. The factors affecting the farmer's decision to redeem the equipment voucher and invest in micro irrigation technology are hypothesized below.

9.3.2.1 HUMAN ASSETS

Characteristics of the household, especially those of the household head, affect adoption decisions. The household head's age influences the attitude he or she may have towards a new technology; however, there is not a consensus in the literature with regard to the direction of the influence of age on adoption [25]. It may be that younger farmers are more innovative and open to technological advances and be more willing to adopt a new technology [6]. Therefore, age would have a negative relationship with adoption. Older farmers would be less likely to redeem the equipment vouchers than their younger counterparts. However, it may also be the case that older farmers have gained more agricultural experience that enables them to better apply a new technology [51]. With this expectation, age would have a positive relationship with adoption. Older farmers would be more willing or able to adopt the new technology.

Education and skill level are expected to be positively related to adoption. Farmers with more education have been shown to adopt modern agricultural technologies sooner [12]. In addition to educational attainment, two indicators of the farmer's agricultural ability and propensity to adopt new technologies are included in this study. Farmers' agricultural ability is proxied by the number of correct answers to a short quiz about common agricultural practices. The farmer's ability for adopting new technologies is proxied by an index of modern agricultural practices, such as improved seed varieties, soil conservation techniques, and the use of legumes to fix nitrogen. A farmer with greater innovative ability is better equipped to accurately assess the benefits and risks in the adoption decision and this may enhance the adoption decision [27]. Therefore, the more modern technologies a farmer has adopted, the more likely he or she may be to invest in irrigation equipment.

The labor available to a household for farm work is proxied by household size. The impact of farm labor supply can be different given two scenarios. One argument is that seasonal labor bottlenecks could discourage technology adoption in labor scarce households [12]. The increased production and ability to grow multiple crops per year with irrigation introduce greater labor requirements. Under this assumption, it would be expected that larger households will be more likely to acquire the irrigation equipment because they have access to a larger labor supply. A second argument is that smaller households will be motivated to acquire the equipment because of its labor saving aspect. Micro irrigation technology is much less labor intensive than drawing water and irrigating by hand and therefore may encourage labor scarce households to adopt in order to meet their irrigation needs [42]. It is unclear, then, what affect household size will have on adoption of micro irrigation technology.

Gender is an important determinant in technology adoption [25]. Traditionally, men and women play different roles in small-scale agriculture. Men often control household finances and decisions regarding purchases of agriculture technology and inputs. This social aspect may make men more likely to adopt new agriculture technology. On the other hand, it is often recognized that women are more aware or concerned with the food requirements of the family [49]. Women may therefore be more likely than men to recognize the advantages of irrigation equipment for increasing household food security and be more likely to adopt. The gender of the head of household is hypothesized to impact the adoption decision, but the effect could be either positive or negative.

9.3.2.2 SOCIAL ASSETS

Information plays an important role in the adoption decision [12]. Information regarding the use, productivity, and risk associated with a new technology affects the farmer's choice. The extent of the farmer's information about irrigation technologies is proxied by the number of irrigation meetings, extension or agricultural fair events attended, as well as whether or not a neighbor or friend was using the irrigation equipment before the adoption decision. The amount of information a farmer has is expected to be positively related to adoption. The more information about the benefits, functioning, and cost of the equipment a farmer has, the less uncertainty there is surrounding the new technology and therefore is expected to increase the likelihood of adoption [12].

9.3.2.3 NATURAL ASSETS

The quality and topography of land can also influence the decision whether or not to adopt an agricultural technology [52]. It is expected that the farmer will be more likely to invest in the irrigation equipment if his land is good enough to produce a good return. In order to capture this, a variable classifying the soil quality and slope of the farmer's largest irrigated field is included. Land tenure also influences the adoption decision. The security of land tenure affects the expected returns from investments in the land. Secure land rights reduce the risk of losing the investment in the land and therefore are expected to increase the willingness to adopt an agricultural technology [12].

Water cost and availability are also important determinants of drip irrigation adoption. In the study area of rural Zambia there are no fees or markets for water such as in many areas with larger scale irrigation schemes. With no explicit price for water, access to water is the main issue regarding water for irrigation for these farmers. A farmer with limited access to irrigation water will benefit little from an investment in irrigation equipment. Distance to the nearest water source for irrigation is included as a measure of the farmer's access to irrigation water. The further a farmer has to go to access water, the more costly irrigating will be and is therefore expected to discourage the adoption of drip irrigation equipment.

9.3.2.4 FINANCIAL ASSETS

In order to redeem the PTI equipment voucher, the farmer is required to pay for part of the equipment cost. Therefore, the financial assets of the farmer are expected to impact the adoption decision. Income levels in the adoption year can both determine the adoption decision and be affected by adoption of the equipment. Therefore, a lagged income variable is included to control for varying financial levels that could determine the household's ability to pay for the cost of the equipment. It is expected that the income level of the household will be positively related with adoption.

Access to credit is also an important factor in new technology adoption [12]. However, because of the general lack of financial credit use by farmers in the study sample, it is impossible to test any hypothesis on the impact of credit on technology adoption.

9.3.2.5 COMMUNITY ASSETS

Many of the farmers in the sample are located in rural areas and are separated from markets and suppliers of irrigation equipment by inadequate transportation infrastructure. The distance from markets and the lack of adequate transportation represents a significant cost of time and money in obtaining the equipment. The distance to the nearest market for buying inputs is included and is expected to have a negative impact on adoption.

Characteristics of different localities can affect the adoption decisions [25]. A region variable is included to control for any differences between the two regions from which the study sample was taken. The Table 9.1 is a summary of the variables included in the analysis of the determinants of adoption and their hypothesized sign.

TABLE 9.1 Variables in the Adoption Model and Hypothesized Signs

Variable description		Hypothesized sign
1.	Household head's age (years)	?
2.	Household head's sex (1=male, 0=female)	?
3.	Household head's education level (years of schooling)	+
4.	Household size (number of members)	?
5.	Farmer's agricultural knowledge (an index)	+
6.	Number of technologies already adopted by household	+
7.	Household income in 2007 (natural log)	+
8.	Distance to nearest water source for irrigation	-
9.	Number of irrigation meetings and extension events attended	+
10.	Attended an agricultural fair (1=yes, 0=no)	+
11.	Neighbor or friend using irrigation equipment (1=yes, 0=no)	+
12.	Land tenure status (1=land title or traditional ownership rights, 0=use rights but not ownership rights)	+
13.	Land quality (1=good, 0=fair/poor)	+
14.	Land topography (1=flat, 0=gentle/steep slope)	+
15.	Distance to market for buying inputs (kilometers)	-
16.	Region dummy (1=Kafue, 0=Kabwe)	?

9.3.3 IMPACT MODEL

In order to estimate the impact of irrigation on agriculture productivity, an economic model of farm production is set up. From the microeconomic theory of the firm, output is a function of the relevant factors of production for agriculture including human inputs (both the quality and quantity of labor), land, physical capital, material inputs and the technology of production. Proximity to input markets also influences the productivity of private factor endowments [28]. Economic theory provides a basis for selecting broad categories of explanatory variables for modeling agricultural output. The relevant literature and the data available are used to guide the choice of specific variables.

The change in household assets was chosen as the outcome variable. Because farmers in the study sample have limited access to financial services such as bank accounts, a major portion of the revenue from the sale of crops is invested in household assets. Therefore, changes in household assets are closely linked to farm output and can be used as a proxy. The variables affecting agriculture production, and ultimately household assets, and their hypothesized sign are discussed in the following sections.

9.3.3.1 HUMAN INPUTS

Small-scale rural farmers rely on the household for much of the labor needed for agricultural production. Therefore, the quantity and quality of household labor is an important factor in the output of the farm. The quantity of labor available for farm work is controlled for by including the size of the household. It is unclear what effect household size will have on the outcome variable (change in household assets) for the farmers in this study. According to the neoclassical theory of production, higher levels of production would be expected with a larger available labor supply. However, for rural households with very low income, having more members requires greater expenditures on food, therefore reducing the available funds for expenditures on other household assets. Household size squared is also included to capture the nonlinear effect of household size. The quality of labor is proxied by the age, gender, education level, and agricultural knowledge of the household head. The education level and agricultural knowledge of the household head is expected to have a positive effect on production. While age and gender are expected to impact production, the direction of the impact of these labor characteristics is not clear.

9.3.3.2 LAND

The land endowment of a household is modeled through variables representing total land holdings, irrigated crop area, land quality, and topography. During the survey each farmer was asked to classify his or her field's soil quality as good, fair, or poor and the lay of the land as flat, gently or steeply sloped. Greater land area, better soil quality, and flatter topography are expected to have a positive impact on farm output.

9.3.3.3 MATERIAL INPUTS

The agricultural inputs used by farmers in the study areas include chemical fertilizer, pesticides, animal manure, mulch, and herbicides. Fertilizer, pesticide, and manure use is widespread with almost all farmers using these inputs. The variables for these inputs are binary where a value of one represents the input being used and zero indicates the input not being used. Because a very high percentage of the farmers use these inputs, it is not possible to test for their impact on output. Including these input variables would introduce significant collinearity because each variable would be a column of mostly ones and collinear with the intercept term. Therefore, fertilizer, pesticide, and manure use were not included in the model. For herbicides and mulch, there is much greater variation in the use of these inputs and they are therefore included in the model.

There also are a number of farmers who use irrigation equipment but did not participate in the PTI program by redeeming the voucher. A variable for those households that did not participate in the program but who irrigate with pump or drip is included to control for the effects of non-PTI irrigation on agriculture output. These agricultural inputs are all expected to increase farm output.

9.3.3.4 CROP TYPE

The quantity and market price varies by crop type. When using change in household assets as a proxy for agriculture output, the model has to account for differing values of crop types. In order to control for the effect different crop types have on farm output and income, a variable is included for the two most common crops, tomatoes and rapeseed. It is assumed that the most common crops are the ones that are in highest demand, produce the best, or in some way give the farmer the highest return. A dummy variable is included indicating whether the farmer grew one of these two crops. It is expected that growing these crops will have a positive effect on the outcome variable.

9.3.3.5 VOUCHER REDEMPTION

Farmers participating in the PTI project had a choice of two irrigation equipment types. The recipient could obtain either a pump or drip kit with a discount voucher. To separate the effects of different micro irrigation equipment types on farm output, a variable for voucher redemption is included for each. From theory, it is expected that this irrigation technology will have a positive impact on farm output.

9.3.3.6 DISTANCE TO MARKETS

Transportation in rural Zambia is both time consuming and costly. Therefore, the distance to markets for selling outputs can have a major effect on the profitability of the farm. The distance to the nearest output market is included as a variable to control for the proximity to markets of different households. Greater distance to markets is expected to have a negative impact on farm output.

9.3.3.7 OTHER INCOME SOURCES

When using the change in household assets as an outcome measure, there are other sources of income that affect household asset level that have to be controlled for in order to separate the effects of adoption of irrigation equipment. Household income from sources other than irrigated crops is included to control for this effect. This variable includes wages and salaries from a formal sector job, income from a non-agricultural small business or enterprise owned by the household, as well as income from nonirrigated crops and livestock. Income is expected to have a positive relationship with household asset change. The Table 9.2 is a summary of the variables included in the impact model and their hypothesized signs.

TABLE 9.2 Variables in the Impact Model and Hypothesized Signs

Variable description	Hypothesized sign
1. Obtained pump with voucher (1=obtained pump and it functioned, 0=otherwise)	+
2. Obtained drip kit with voucher (1=obtained drip kit and it functioned, 0=otherwise)	+
3. Average irrigated land area for 2008 and 2009 (hectares)	+
4. Total land area household has rights to (hectares)	+
5. Household head's age (years)	?
6. Household head's sex (1=male, 0=female)	?
7. Household head's educational attainment (years of schooling)	+
8. Size of household (number of members)	+
9. Size of household squared	-
10. Farmer agricultural knowledge index	+
11. Income from nonirrigated crop sources in 2007 (natural log)	+
12. Farmer grows at least one of the two most popular irrigated crop types, tomatoes and rapeseed (1=tomatoes or rapeseed, 0=other)	+
13. Herbicide input use (1=yes, 0=no)	+
14. Mulch use (1=yes, 0=no)	+
15. Land quality (1=good, 0=fair/poor)	+
16. Land topography (1=flat, 0=gentle/steep slope)	+
17. Distance to nearest market for selling outputs	-
18. Region (1=Kafue, 0=Kabwe)	?

9.3.4 DATA

The data used for this evaluation is primary data collected through a household survey conducted in Zambia in August 2009 by the authors.

9.3.4.1 QUESTIONNAIRE

The survey covered two main project sites of PTI. A draft questionnaire was developed designed to fit the requirements of this study and the PTI project. A pretest of the questionnaire was conducted on a small group of farmers participating in the PTI voucher program. With insights from the pretest, the questionnaire was then revised further to more accurately gain the information needed and reduce the time required for interviewing. The final version of the questionnaire consisted of 12 sections and collected data on a variety of household, agricultural, and community characteristics relevant to the evaluation of the project impacts.

9.3.4.2 SAMPLING TECHNIQUE

The PTI project began distributing vouchers in February, 2008. According to project reports, 3,468 vouchers had been distributed by the end of November, 2008 [34]. Distribution of the vouchers was carried out through agricultural fairs, extension events, and farmer groups. The voucher redemption rate for the project was approximately 14% [34]. The survey was carried out in two of the eight project areas. Resource and time constraints of the study did not allow for data to be collected from all eight areas.

A multistage sampling technique was used to obtain the study sample comprised of both farmers who received the voucher but did not redeem it, and farmers who redeemed the voucher. First, two regions were selected. These two regions, Kabwe and Kafue, had the highest participation rates among the project areas. The farmers who had received vouchers were separated into two groups, those who redeemed the vouchers and those who did not. An equal number of farmers from both groups were then randomly drawn. This gave a random sample comprised half of farmers who had received the voucher, redeemed it, and obtained irrigation equipment (treatment) and half of farmers who received the voucher but did not redeem it (control). While this sampling technique does not satisfy the requirement of a true experimental sample, the random sample ensures that each farmer in the "treatment" group has the same chance of being selected. Likewise, each farmer who did not redeem his or her voucher has the same chance of being included in the "control" group. The issue of self-selection bias potentially present in this sample will be discussed later.

9.3.4.3. DATA CHECKS

A total of 101 households from the random sample were surveyed. Analysis of the data revealed a number of observations with extreme values. The influence of extraordinary observations on the model was analyzed using the Cook's distance influence statistic. The Cook's D statistic measures the effect of one observation on all the regression coefficients simultaneously [15]. An influential observation is one with an unusual dependent variable value as well as unusual independent variable value. Only if both parts are unusual will the observation strongly affect the coefficients [26]. Nine observations were found to be influential cases using this method. After close inspection, these nine observations were excluded from the sample. For a more detailed discussion of the excluded observations see [7].

9.3.4.4 DESCRIPTIVE STATISTICS

Table 9.3 gives a summary of selected characteristics of the households sampled. The data shows the majorities are male-headed households that work primarily on the farm and have little off-farm employment. Most of the farmers irrigate in some way using furrow, bucket, drip equipment, or a pump. Most use chemical fertilizer, improved seeds, animal manure and mulch. While many use pesticides, relatively few use herbicides. Access to credit is extremely limited. Only three households in the sample reported having taken a loan in the past year. More farmers used the pump than the drip equipment kit.

TABLE 9.3 Summary Statistics of Study Sample

Characteristics	Number	Percentage
1. Male headed households	92	91.1
2. Households with a member working in a household small business	33	32.7
3. Households with a member working in a formal sector job	9	8.9
4. Households that redeemed voucher for drip kit	21	20.8
5. Households that redeemed voucher for pump	41	40.6
6. Households that irrigate in some way (furrow, bucket, drip, or pump)	99	98.0
7. Households that irrigate with pump or drip	64	63.4
8. Households using fertilizer	99	98.0
9. Households using pesticides	88	87.1
10. Households using herbicides	18	17.8
11. Households using improved seeds	100	99.0
12. Households using animal manure on fields	89	88.1

TABLE 9.3 *(Continued)*

Characteristics	Number	Percentage
13. Households using mulch	75	74.3
14. Households having a neighbor or friend who uses drip or a treadle pump	73	72.3
15. Households taking a loan in previous year	3	3.0
16. Households growing tomatoes or rapeseed, the two most common irrigated crops	80	79.2

9.3.5 EMPIRICAL METHODS

9.3.5.1 OUTCOME VARIABLE

In order to evaluate the impact of the adoption of micro irrigation equipment, an appropriate outcome measure is needed. The following question needs to be answered in evaluating the impact of the project: did participation in the project enable the household to increase production? One way to answer this question would be to measure the output per hectare of the farm before and after the project is implemented. This method is complicated by the fact that the types of crops grown with and without irrigation differ; making comparisons between quantities of two different crops difficult. When considering a program like PTI that enables farmers to adopt irrigation equipment and increase production of vegetables that can be sold as a cash crop, income from irrigated crops may seem to be the first choice for an outcome measure. However, using income level as an outcome measure is problematic when evaluating a voucher program. The voucher provides only a portion of the cost of the equipment, so redeeming the voucher requires the household to supply additional money. At the time of the study, the PTI voucher's value was set at 40% of the total cost of the equipment. The farmer has to cover the remaining 60% of the equipment cost. Therefore, income level can be thought of as a determinant of voucher redemption. On the other hand, adoption of the irrigation equipment also impacts the household's income level through its effect on agriculture output. Therefore, current income level is both a cause and an effect of adoption. Current income is then endogenous with respect to adoption, making it difficult to separate the effects of the irrigation equipment on income level. If baseline data was available from before the program was implemented, previous income levels could be used to estimate the effect of irrigation on current income. However, for this study reliable baseline data on income was not available.

In some studies where baseline data is not available, researchers use recall data in an effort to gain the needed information from a certain point prior to the implementation of the program; however, collecting income data through recall is subject to major errors. Researchers often rely on other outcome variables besides income when measuring changes in household wellbeing [41].

For this study, the outcome variable chosen is change in household assets. This variable is a useful measure because the crops grown with irrigation during the dry season are mainly cash crops such as tomatoes and rapeseed. The farmers in the sample have limited access to financial services such as bank accounts, so a major portion of the profits from the sale of crops is likely invested in household assets. Additional income from increased crop sales allows the household to buy assets such as household items or farm animals.

Using change in household assets as an outcome measure reduces the reporting and recall problems associated with income. Household assets are relatively few in number and are tangible; therefore, households are likely to be able to recall them from previous time periods. For example, it is easier to recall the number of household items such as a mobile phone or sofa than to recall past income amounts. This is particularly the case for agricultural households where sale of crops is often comprised of many small transactions that make recalling year or season totals extremely inaccurate. The use of change in assets also avoids reporting bias. An individual may not be willing to share his or her true income level and simply report an erroneous amount. There may be reporting bias if the interviewee believes there may be some eligibility criteria based on income level that could affect current or future participation in project subsidies. Income may be underestimated if the respondent believes the program specifically targets low-income farm households. Or, income data gathered after an intervention could be overestimated in an effort to not disappoint program sponsors and encourage future assistance. While the change in household assets may not be a perfect proxy for agriculture productivity, it was deemed the best measure available for this study.

9.3.5.2 IMPACT EVALUATION

There is a large literature studying the causal effects of programs and policies. The central question in this literature is what would have been the outcome for a unit (household, village, region, or country) had it not received the particular "treatment" provided by a program or policy [23]. However, a researcher is never able to observe both situations for the same unit. If a household participates in a program or receives treatment, then the alternate outcome for that household, without treatment, is not observed for that particular time period. Therefore, in order to evaluate the effect of participation or treatment, comparison with outcomes of another unit or household receiving a different level of treatment or participation is required. Because this type of comparison involves two different units, a valid comparison group has to be established [23].

In an ideal experimental setup with baseline data and randomized treatment, estimating the counterfactual (that is, what *would* have occurred in the absence of treatment) is straightforward using standard statistical methods. Establishing a credible comparison group can be difficult when the project was planned without an experimental design that includes randomly selected treatment and control groups. The comparison group should be one that would have shown similar outcomes to those of a group that participated in the program if treatment had not occurred. It is the comparison between this control group and the treatment group that enables an estimate of the average impact of the program to be obtained [10]. In this study there is no baseline data available and treatment was not randomized so establishing a valid control group requires more complex techniques.

When observing positive outcomes for farmers participating in a program like PTI, it may be tempting to attribute any positive effects to program involvement; however, this is faulty reasoning. It is not clear what would have been the outcome for these households in the absence of the intervention. The favorable outcome may be due to other causes, and therefore without a valid evaluation, it is inaccurate to attribute the outcome to the presence of the program [9].

Farmers participating in the PTI project self-selected into treatment and non-treatment groups by their choice to redeem the voucher or not. The participation decision may have been driven by systematic differences between those who adopted micro irrigation and those who chose not to. These differences will bias estimates from regression analysis if the factors that determine the selection into the two different groups are not incorporated in the empirical framework [13, 40]. It is often the case that farmers who self-select into a program have favorable characteristics such as initiative, skill, and ability that motivates them to participate. These characteristics could positively influence the outcome estimates and it would be easy to attribute estimated impact to the program when some of the estimated effect may simply be due to more advanced farmers choosing to be in the program. Any difference in these characteristics between the treatment and control groups needs to be controlled for in order to produce a valid estimate of the impact of the program.

Many times there are reasons participants for programs like PTI are not selected randomly. Both ethical and logistical considerations often prevent program managers from conducting randomized experiments especially when dealing with programs addressing poverty, household well-being, and survival. Consequently, many program evaluations find positive impacts of interventions, but without controlling for selection bias, it is impossible to determine if the estimated impact is due to the program or to systematic differences between the treatment and control groups. Credible program evaluations that control for selection bias are relatively rare and therefore directors of development projects and donor agencies lack the evidence needed to choose effective interventions [9]. With the use of statistical techniques to test and control for selection bias, however, reliable results can be achieved [18]. The Heckman two-stage procedure is used in this study to test and control for selection bias. The results of the test for selection bias indicates, that in this sample, the self-selection of farmers into the "treatment group" does not significantly affect the outcome of the impact estimation [7]. Therefore, the impact model can be estimated using an ordinary least squares regression.

9.3.6 ANALYSIS

9.3.6.1 ADOPTION MODEL

In order to answer the question of what key factors affect voucher redemption, an adoption model was evaluated. The dependent variable is equal to one if the household redeemed a voucher for irrigation equipment (either drip equipment or a pump). The explanatory variables were selected based on the concepts discussed earlier. The model was estimated using a probit maximum likelihood regression. The probit model is appropriate when estimating a binomial response variable [26]. The adoption model was defined following multiple linear regression model:

$$D_i = \gamma_0 \pm \gamma_1 * z_{i1} + \gamma_2 * z_{i2} \pm \gamma_3 * z_{i3} + \gamma_4 * z_{i4} \pm \gamma_5 * z_{i5} + \gamma_6 * z_{i6} \pm \gamma_7 * z_{i7} + \gamma_8 * z_{i8}$$

$$+\gamma_9 * z_{i9} + \gamma_{10} * z_{i10} + \gamma_{11} * z_{i11} + \gamma_{12} * z_{i12} + \gamma_{13} * z_{i13} + \gamma_{14} * z_{i14}$$

$$+\gamma_{15} * z_{i15} + \gamma_{16} * z_{i16} + \varepsilon_i \qquad (1)$$

where, D_i = binary adoption variable (1=redeemed voucher for drip or pump equipment); z_{i1} = household head's age (years); z_{i2} = household head's sex (1=male, 0=female); z_{i3} = household head's education level attainment (years of schooling); z_{i4} = household size (number of members); z_{i5} = farmer's agricultural knowledge (an index); z_{i6} = number of technologies already adopted by household; z_{i7} = household income in 2007 (logarithm); z_{i8} = distance to water source for irrigation; z_{i9} = number of irrigation meetings and extension events attended; z_{i10} = attended an agricultural fair (1=yes, 0=no); z_{i11} = neighbor or friend using irrigation equipment (1=yes, 0=no); z_{i12} = land tenure status (1=land title or traditional ownership rights, 0=use rights but not ownership rights); z_{i13} = land quality (1=good, 0=fair/poor); z_{i14} = land topography (1=flat, 0=gentle/steep slope); z_{i15} = distance to market for buying inputs (km); z_{i16} = region dummy (1=Kafue, 0=Kabwe); $\gamma_0, \gamma_1 \ldots \gamma_{16}$ = regression coefficients to be estimated; and $i = 1, 2, 3, \ldots n$ farm households.

The null hypotheses and their alternatives were formalized as follows:

H_0: $\gamma_0, \gamma_1 \ldots \gamma_{16} = 0$

H_a: $\gamma_0, \gamma_1 \ldots \gamma_{16} \neq 0$

9.3.6.2 IMPACT MODEL

An impact model was constructed to answer the second research question: What are the economic effects of the project on the livelihoods of participating households? The impact model was defined as follows using ordinary least squares:

$$Y_i = \beta_0 + \psi_1 * p_i + \psi_2 * d_i + \beta_1 * x_{i1} + \beta_2 * x_{i2} + \beta_3 * x_{i3} + \beta_4 * x_{i4}$$
$$+\beta_5 * x_{i5} + \beta_6 * x_{i6} + \beta_7 * x_{i7} + \beta_8 * x_{i8} + \beta_9 * x_{i9} + \beta_{10} * x_{i10}$$
$$+\beta_{11} * x_{i11} + \beta_{12} * x_{i12} + \beta_{13} * x_{i13} + \beta_{14} * x_{i14} + \beta_{15} * x_{i15} + \beta_{16} * x_{i16} + \beta_{17} * x_{i17} + \beta_{18} * \lambda \quad (2)$$

where: Y_i = change in value of household assets; p_i = obtained pump with voucher (1=obtained pump, 0=otherwise); d_i = obtained drip kit with voucher (1=obtained drip kit, 0=otherwise); x_{i1} = income from nonirrigated crop sources (natural log); x_{i2} = irrigated land area (average 2008 and 2009); x_{i3} = total land area household has rights to (hectares); x_{i4} = household head's age (years); x_{i5} = household head's sex (1=male, 0=female); x_{i6} = household head's educational attainment (years of schooling); x_{i7} = size of household (number of members); x_{i8} = size of household squared; x_{i9} = farmer agricultural knowledge index; x_{i10} = farmer crop type (1=tomatoes or rapeseed, 0=other); x_{i11} = herbicide input use (1=yes, 0=no); x_{i12} = mulch use (1=yes, 0=no); x_{i13} = land quality (1=good, 0=fair/poor); x_{i14} = land topography (1=flat, 0=gentle/steep slope); x_{i15} = farmer irrigates with pump or drip but did not redeem voucher; x_{i16} = distance to nearest market for selling outputs; x_{i17} = region (1=Kafue, 0=Kabwe); u_i = error term; β_0, $\beta_1 \ldots \beta_{18}$ = regression coefficients to be estimated; ψ_1, ψ_2 = PTI participation parameters to be estimated; and $i = 1, 2, 3, \ldots n$ = farm households.

The null hypotheses and their alternatives are formalized as follows:

Hypothesis for the impact of adopting a pump:

$$H_0: \psi_1 = 0 \quad (1)$$

$$\text{vs. } H_a: \psi_1 \neq 0 \quad (2)$$

Hypothesis for the impact of adopting drip irrigation equipment:

$$H_0: \psi_2 = 0 \quad (3)$$

$$\text{vs. } H_a: \psi_2 \neq 0 \quad (4)$$

9.4 RESULTS AND DISCUSSION

9.4.1 ADOPTION MODEL

The results from the estimation of the factors affecting adoption of micro irrigation equipment are presented in Table 9.4. Many, though not all, of the explanatory variables have the expected sign; however, all statistically significant variables show the expected sign for impact on irrigation equipment adoption. Only three of the explanatory variables show a statistically significant impact on adoption.

TABLE 9.4 Factors Influencing Voucher Redemption For Treadle Pump or Drip Kit

Explanatory Variables	Coefficient	Standard error	P-value
Household head's age (years)	0.0282**	0.013	0.033
Household head's sex (1=male, 0=female)	–0.7097	0.596	0.233
Household head's education level (years of schooling)	0.0305	0.050	0.541
Household size (number of members)	0.1761	0.182	0.334
Household size squared	–0.0073	0.008	0.367
Farmer agricultural knowledge (an index)	0.2771*	0.166	0.095
Number of techniques already adopted by household	0.0671	0.107	0.530
Household income in 2007 (natural log)	0.0669	0.106	0.529
Distance to nearest water source for irrigation	–0.0004	0.000	0.336
Number of irrigation meetings and extension events attended	–0.0050	0.026	0.846
Attended an agriculture fair, (1=yes, 0=no)	–0.1778	0.337	0.598
Neighbor or friend using irrigation equipment, (1=yes, 0=no)	0.5811*	0.339	0.087
Land tenure status (1=title or traditional ownership rights, 0=use rights)	–0.1126	0.467	0.810
Land quality (1=good, 0=fair/poor)	–0.0125	0.304	0.967
Land topography (1=flat, 0=gentle/steep slope)	–0.2020	0.320	0.528
Distance to nearest market for buying inputs	0.0162	0.017	0.330
Region dummy (1=Kafue, 0=Kabwe)	–0.3562	0.296	0.229
Constant	–4.4059**	1.965	0.025
Pseudo R-square	0.145		
Chi-square	20.039		
Number of observations	101		

Note: Significance levels marked by * $p<0.10$, ** $p<0.05$, *** $p<0.01$

Household head age is statistically significant and positively related to adoption. That is, older farmers are more likely to redeem the irrigation voucher than younger farmers. This supports the idea that older farmers have gained more agricultural experience that enables them to better apply a new technology [51].

As hypothesized, the farmers' agricultural knowledge of correct agricultural practices positively impacts the adoption of the PTI technology. The third statistically significant explanatory variable is whether or not the farmer had a neighbor or friend who was using drip equipment or a treadle pump before the farmer redeemed his or her voucher. As expected, having connections with a neighbor or friend using micro irrigation equipment increased the likelihood of the farmer choosing to redeem the voucher. This confirms the importance of information in the adoption of new agricultural technology [25]. Social networks and the spillover of information between community members play an important role in the dissemination of information and removing uncertainty about the irrigation equipment.

9.4.2 ECONOMIC IMPACT MODEL

The results from the estimation of impact of participation on household assets are shown in Table 9.5. Both the first null hypothesis were rejected: (i) that the impact of adopting a pump is equal to zero and the second null hypothesis; (ii) that the impact of adopting drip equipment is equal to zero.

TABLE 9.5 Impact of Participation on Household Assets

Explanatory variables	Coefficient	Standard error	P-value
Obtained pump with voucher(1=obtained pump and if functioned, 0=otherwise)	61027.79*	31235.70	0.055
Obtained drip kit with voucher(1=obtained drip kit and if functioned, 0=otherwise)	–86904.46*	46071.95	0.063
Household income from nonirrigated crop sources in 2008 (natural log)	133.53	3888.19	0.973
Average irrigated land area for 08 and 09	–103482.98**	32278.85	0.002
Total land area household has rights to (hectares)	871.39*	498.32	0.085
Household head's age (years)	–610.23	1319.84	0.645
Household head's sex (1=male, 0=female)	2258.54	46242.37	0.961
Household head's education level (years of schooling)	–3832.72	4485.96	0.396
Household size (number of members)	–30200.03*	17920.47	0.096
Household size squared	1821.47**	846.04	0.035
Farmer agricultural knowledge (an index)	–7284.90	16995.47	0.669
Farmer grows tomatoes and/or rapeseed, (1=tomatoes or rapeseed, 0=other)	80356.47	48606.53	0.103

TABLE 9.5 *(Continued)*

Explanatory variables	Coefficient	Standard error	P-value
Herbicide input use, (1=yes, 0=no)	15075.58	41141.44	0.715
Mulch use, (1=yes, 0=no)	40482.60	30917.08	0.195
Land quality (1=good, 0=fair/poor)	–6238.00	30869.03	0.840
Land topography (1=flat, 0=gentle/steep slope)	42089.04	30369.26	0.170
Farmer irrigates with pump or drip but did not redeem voucher (1=yes, 0=no)	–61957.37	53293.30	0.249
Distance to nearest market for selling outputs	22.74	1602.44	0.989
Region dummy (1=Kafue, 0=Kabwe)	–46490.16	32587.33	0.158
Constant	136533.95	140154.90	0.333
R-square	0.330		
Adjusted R-square	0.153		
Number of observations	92		

Note: significance levels marked by * p<0.10, ** p<0.05, *** p<0.01

This indicates a statistically significant impact on the change in household assets with use of micro irrigation equipment. The estimation showed the increase in household assets between the beginning of 2008 and August 2009 was on average 61,028 kwachas (approximately $13 US) plus a $90 increase in farm assets (the pump). Therefore, it can be expected that the economic benefit in subsequent seasons to those households who adopted the pump will be $103 since the pump has been paid for and will continue to generate increased productivity in future growing seasons. The adoption of a drip kit results in a negative change in household assets. On average, household assets of adopters were 86,905 kwachas (approximately $18 US) lower compared to nonadopters. However, adopters have now acquired the drip kit, an asset worth $108 and they have paid it off in the current season. In subsequent years, this drip kit will continue to increase crop productivity and will not need to be purchased again. The net increase in assets as a result of the drip kit is therefore $90 ($108 minus $18).

The economic benefit to households that adopted the micro irrigation equipment is large given that the average value of household assets measured at the beginning of the program was only $160. Therefore as a result of obtaining the micro irrigation equipment, assets held by the adopters increased an average of 60% compared to the assets of the nonadopters.

Besides adopting of micro irrigation equipment, other variables with significant coefficients in the estimation are irrigated land area, total land area, household size and the square of household size. The coefficient on irrigated land area is negative and significant at the 5% level. That is, after controlling for other variables, the model shows farmers with more irrigated land area have a smaller change in household assets. This is the opposite of the effect that production theory predicts. However, total land area has a positive relationship with change in household assets as expected. Every hectare to which a household has property rights adds, on average, 871 kwachas to household assets. Household size and household size squared are jointly significant. Additional household members reduce the change in household assets when compared to those with fewer members.

9.5 SUMMARY

The PTI project in Zambia provided discount vouchers for the purchase of micro irrigation equipment such as treadles pumps and drip irrigation kits. The project's aim was to increase household livelihoods through the development of micro irrigation equipment supply chains and provide information and training on its use. The objectives of this study were to evaluate the factors affecting voucher redemption (adoption of micro irrigation technology) and the effects of the project on livelihoods.

The results of the analysis showed that age of the household head, agricultural knowledge, and the presence of friends or neighbors who already used micro irrigation equipment were significant factors affecting micro irrigation. All three factors exhibited a positive influence on voucher redemption.

Livelihood impacts were measured by the change in household assets between the beginning of the project and the time of the evaluation survey. The economic benefit in the first year-and-a-half of the program to households that adopted a treadle pump was $103 on average. Adopters of the drip kit received an economic benefit of $90. This increase represents a 60% gain in household assets as result of the use of micro irrigation equipment.

The findings of this study provide evidence that micro irrigation can be quite profitable for small-scale farmers. Policies and programs to promote micro irrigation could have a substantial impact on agricultural productivity and rural livelihoods. Virtually none of the farmers in the sample obtained credit for the purchase of the equipment but it seems the provision of credit is warranted given the fact that adoption of this technology resulted in a relatively large increase in farmers' assets. The economic impact of micro irrigation also seems to warrant public or private sector provision of technical assistance to train farmers on the use of micro irrigation equipment.

KEYWORDS

- adoption model
- agriculture technology adoption
- drip irrigation
- economic impact evaluation
- economic impact model
- food security
- Heckman two-stage method
- household assets
- International Food Policy Research Institute, IFPRI
- International Water Management Institute, IWMI
- livelihoods
- micro irrigation
- National Academy of Sciences
- Ohio State University, OSU
- ordinary least squares
- poverty reduction
- productivity of labor
- program evaluation
- Prosperity Through Innovation, PTI
- selection bias
- small scale farmer
- Sri Lanka
- subsidy
- supply chain
- survey
- treadle pump
- voucher
- voucher redemption
- world poverty
- Zambia

REFERENCES

1. Adeoti, Adetola, Boubacar Barry, Regassa Namara, Abdul Kamara, and Atsu Titiati, (2007). *Treadle pump irrigation and poverty in Ghana.* Colombo, Sri Lanka: International Water Management Institute (IWMI), Research Report No. 117.
2. Benin, Samuel, Samuel Mugarura, (2006). *Determinants of change in household-level consumption and poverty in Uganda, 1992/93–1999/00.* International Food Policy Research Institute (IFPRI), Discussion Paper No. 27.
3. Burney, Jennifer, Lennart Woltering, Marshall Burke, Rosamond Naylor, Dov Pasternak, (2010). Solar-powered drip irrigation enhances food security in the Sudano–Sahel. *Proceedings of the National Academy of Sciences of the United States of America,* 107(5):1848–1853.
4. De Janvry, Alain, Elisabeth Sadoulet, (2002). World poverty and the role of agricultural technology: Direct and indirect effects. *Journal of Development Studies,* 38 (4):1–26.
5. De Janvry, Alain, Marcel Fafchamps, Elisabeth Sadoulet, (1991). Peasant household behavior with missing markets: Some paradoxes explained. *Economic Journal,* 101 (409):1400–1417.
6. Diederen, Paul, Hans van Meijl, Arjan Wolters, Katarzyna Bijak, (2003). Innovation adoption in agriculture: Innovators, early adopters and laggards. *Cahiers d'Economie et Sociologie Rurales*, 67:30–50.
7. DiGennaro, Simeon, (2010). *Evaluation of the Livelihood Impacts of a Micro irrigation Project in Zambia.* MSc Thesis. Ohio State University. https://etd.ohiolink.edu/
8. Dillon, Andrew, (2008). *Access to irrigation and the escape from poverty: Evidence from Northern Mali.* International Food Policy Research Institute (IFPRI), Discussion Paper No. 782.
9. Duflo, Esther, Michael Kremer, (2008). Use of randomization in the evaluation of development effectiveness. In: *Reinventing Foreign Aid,* ed. William Easterly, 93–120 pp. Cambridge, MS–USA: MIT Press.
10. Duflo, Esther, Rachel Glennerster, Michael Kremer, (2006). *Using randomization in development economics research: A toolkit.* Massachusetts Institute of Technology, MIT Department of Economics, Working Paper 06–36.
11. FAO, (2014). *The state of food insecurity in the world 2014: Strengthening the enabling environment for food security and nutrition.* Rome, Italy: Food and Agriculture Organization of the United Nations.
12. Feder, Gershon, Richard E. Just, David Zilberman, (1985). Adoption of agricultural innovations in developing countries: A survey. *Economic Development and Cultural Change,* 33(2):255–298.
13. Feder, Gershon, Rinku Murgai, Jaime B. Quizon, (2004). Sending farmers back to school: The impact of farmer field schools in Indonesia. *Review of Agricultural Economics,* 26(1):45–62.
14. Jeremy D. Foltz (2003). The economics of water-conserving technology adoption in Tunisia: An empirical estimation of farmer technology choice. *Economic Development and Cultural Change*, 51(2):359–373.
15. John Fox, (1991). *Regression diagnostics*: *Quantitative applications in the social sciences.* Newbury Park, California–USA: Sage.
16. Karen Frenken (2005). *Irrigation in Africa in figures: AQUASTAT survey–2005.* Rome Italy: Food and Agriculture Organization, FAO Water Report 29.
17. He, Xue-Feng, Huhua Cao, Feng-Min Li, (2007). Econometric analysis of the determinants of adoption of rainwater harvesting and supplementary irrigation technology (RHSIT) in the semiarid loess plateau of China. *Agricultural Water Management*, 89(3):243–250.
18. James J. Heckman (1979). Sample selection bias as a specification error. *Econometrica,* 47(1):153–161.

19. Held, I. M., T. L. Delworth, J. Lu, K. L. Findell, and T. R. Knutson, (2006). Simulation of Sahel drought in the 20th and twenty-first centuries. *Proceedings of the National Academy of Sciences of the United States of America*, 102(50):17891–17896.
20. Intizar Hussain, (2007). Poverty-reducing impacts of irrigation: Evidence and lessons. *Irrigation and Drainage*, 56(2–3):147–164.
21. Intizar Hussain, Munir A. Hanjra, (2004). Irrigation and poverty alleviation: Review of the empirical evidence. *Irrigation and Drainage,* 53(1):1–15.
22. Intizar Hussain, Munir A. Hanjra, (2003). Does irrigation water matter for rural poverty alleviation? Evidence from South and South-East Asia. *Water Policy*, 5(5–6): 429–442.
23. Imbens, Guido W., Jeffrey M. Wooldridge, (2009). Recent developments in the econometrics of program evaluation. *Journal of Economic Literature,* 47(1):5–86.
24. Jacoby, Hanan, (1993). Shadow wages and peasant family labor supply: An econometric application to the Peruvian Sierra. *Review of Economic Studies,* 60(205):903–921.
25. Duncan Knowler, Ben Bradshaw, (2007). Farmers' adoption of conservation agriculture: A review and synthesis of recent research. *Food Policy,* 32(1):25–48.
26. Ulrich Kohler, Frauke Kreuter, (2005). *Data analysis using Stata.* College Station, TX–USA: Stata Press.
27. Koundouri, Phoebe, Céline Nauges, Vangelis Tzouvelekas, (2006). Technology adoption under production uncertainty: Theory and application to irrigation technology. *American Journal of Agricultural Economics,* 88(3):657–670.
28. Kraybill, David, Bernard Bashaasha, (2006). The potential gains from geographical targeting of antipoverty programs in Uganda. *African Journal of Agricultural and Resource Economics,* 1(1):37–48.
29. Le, Kien T. (2009). Shadow wages and shadow income in farmers' labor supply functions. *American Journal of Agricultural Economics,* 91(3):685–696.
30. Lipton, M., J. Litchfield, and J-M Faures, (2003). The effects of irrigation on poverty: A framework for analysis. *Water Policy,* 5(5–6):413–427.
31. Lipton, Michael, (2007). Farm water and rural poverty reduction in developing Asia. *Irrigation and Drainage,* 56(2–3):127–146.
32. Lobell, David, Marshall B. Burke, Claudia Tebaldi, Michael D. Mastrandrea, Walter P. Falcon, Rosamond L. Naylor, (2008). Prioritizing climate change adaptation needs for food security in 2030. *Science,* 319(5863):607–610.
33. Mangisoni, Julius H. (2008). Impact of treadle pump irrigation technology on smallholder poverty and food security in Malawi: A case study of Blantire and Mchinji districts. *International Journal of Agricultural Sustainability,* 6(4):248–266.
34. MEDA, Zambia, (2008). *Project updater November 2008.* Lusaka, Zambia: Mennonite Economic Development Associates (MEDA).
35. Namara, R. E., Nagar, R. K., Upadhyay, B. (2007). Economics, adoption determinants, and impacts of micro irrigation technologies: Empirical results from India. *Irrigation Science,* 25(3):283–297.
36. Narayanamoorthy, A. (2005). *Efficiency of irrigation: A case of drip irrigation.* Mumbai, India: Department of Economic Analysis and Research, National Bank for Agriculture and Rural Development (NABARD), Occasional Paper 45.
37. Narayanamoorthy, A. (2004). Impact assessment of drip irrigation in India: The case of sugarcane. *Development Policy Review,* 22(4):443–462.
38. Polak, P. (2005). Water and the other three revolutions needed to end rural poverty. *Water Science,* 51(8):133–143.
39. Paul Polak, Robert Yoder, (2006). Creating wealth from groundwater for dollar-a-day farmers: Where the silent revolution and the four revolutions to end rural poverty meet. *Hydrogeology Journal,* 14(3):424–432.

40. Puhani, Patrick, (2000). The Heckman correction for sample selection and its critique. *Journal of Economic Surveys,* 14(1):53–68.
41. David E. Sahn, David Stifel, (2003). Exploring alternative measures of welfare in the absence of expenditure data. *The Review of Income and Wealth,* 49(4):463–489.
42. Shrestha, R. B., Gopalakrishnan, C. (1993). Adoption and diffusion of drip irrigation technology: An econometric analysis. *Economic Development and Cultural Change,* 41 (2):407–418.
43. Skoufias, E. (1994). Using shadow wages to estimate labor supply of agricultural households. *American Journal of Agricultural Economics,* 76(2):215–227.
44. Smith, Laurence E. D. (2004). Assessment of the contribution of irrigation to poverty reduction and sustainable livelihoods. *International Journal of Water Resources Development,* 20(2):243–257.
45. Smith, Lisa C., Harold Alderman, Dede Aduayom, (2006). *Food insecurity in sub-Saharan Africa: new estimates from household expenditure surveys.* International Food Policy Research institute, Washington, D. C., Research Report 146.
46. Stewart, Frances, (2005). Evaluating evaluation in a world of multiple goals, interests, and models. In: *Evaluating development effectiveness,* eds. George Keith Pitman, Osvaldo Néstor Feinstein and Gregory K. Ingram, 7:3–32. New Brunswick, NJ–USA: Transaction Publishers.
47. Strauss, John, (1986). The theory and comparative statics of agricultural household models: A general approach. In: *Agricultural household models: Extensions, applications and policy,* eds. Inderjit Singh, Lyn Squire and John Strauss, pp. 71–92. Baltimore, MD–USA: Johns Hopkins University Press.
48. Tesfaye, Abonesh, Ayalneh Bogale, Regassa E. Namara, Dereje Bacha, (2008). The impact of small-scale irrigation on household food security: The case of Filtino and Godino irrigation schemes in Ethiopia. *Irrigation and Drainage Systems,* 22(2):145–158.
49. Thomas, Duncan, (1990). Intra-household resource allocation an inferential approach. *Journal of Human Resources*, 25(4):635–664.
50. White, Howard, (2005). Challenges in evaluating development effectiveness. In: *Evaluating development effectiveness,* eds. George Keith Pitman, Osvaldo Néstor Feinstein and Gregory K. Ingram, 7:33–54. New Brunswick, NJ–USA: Transaction Publishers.
51. Wozniak, Gregory, (1987). Human capital, information, and the early adoption of new technology. *Journal of Human Resources,* 22(1):101–112.
52. Zhou, S. D. (2008). Factors affecting Chinese farmers' decisions to adopt a water-saving technology. *Canadian Journal of Agricultural Economics,* 56(1):51–61.
53. Frederick J. Zimmerman, Michael R. Carter, (2003). Asset smoothing, consumption smoothing and the reproduction of inequality under risk and subsistence constraints. *Journal of Development Economics*, 71(2):233–260.

CHAPTER 10

PERFORMANCE OF MICROSPRINKLER IRRIGATED GROUNDNUT

M. WASEEM and I. KHALEEL

CONTENTS

In this chapter: One US$ = 63.167 Rs. (Indian rupee); One quintal, q = 100 Kg.

Edited and abbreviated version of, "Mohammed Waseem, 2012. Performance evaluation of micro sprinkler irrigation for groundnut (arachis hypogaea l.) under Raichur region. Unpublished M.Tech. thesis, College of Agricultural Engineering, University of Agricultural Sciences, Raichur – Maharashtra, India."

Authors express thanks to Mr. Raj Gopal, progressive farmer in Yeragera, for providing the needed facilities in his field.

10.1 INTRODUCTION

Land and water are the most precious natural resources, the importance of which in human civilization needs no elaboration. The total available land area in the State sets the limits within which the competing human needs have to be met. The needs of agricultural, industrial, domestic and others often result in diversion from one use to the other. Diversion of land from agriculture to nonagriculture uses adversely affects the growth in agriculture sector. Water supports all forms of life on mother earth. It plays a vital role in agriculture. Irrigation is the basic input for enhancing reliability and productivity of agriculture.

Water is one of the most critical inputs for agriculture, which consumes more than 70% of the water resources of the country. Availability of adequate quantity and quality of water are key factors for achieving higher productivity levels. Investments in conservation of water, improved techniques to ensure its timely supply, and improve its efficient use are some of the imperatives, which the country needs to enhance. Poor irrigation efficiency of conventional irrigation system has not only reduced the anticipated outcome of investments made towards water resource development, but has also resulted in environmental problems like water logging and soil salinity thereby affecting crop yields. Thus, are calls for massive investments in adoption of improved methods of irrigation such as drip and sprinkler, including fertigation [15, 16].

Competition for water is increasing rapidly. Water shortage is a worldwide problem for which the only solution is to make efficient use of water in agriculture. Therefore, a better understanding of water requirements and better management of irrigation water will result in large benefits. When irrigation water is insufficient, appropriate scheduling can increase crop yields. A deficit occurring at a certain stage of crop growth may cause a greater reduction in yield than would the same deficit at other growth stages. As the crop water response to water deficits at different periods is not uniform, it is necessary to distribute deficits among intra-seasonal periods optimally for a crop. Several factors are to be considered in irrigation planning, particularly when several crops are grown in the same command area in more than one season in a year. Two distinct decisions to be made are how much water and land should be allocated to each crop at a seasonal level and to each season at inter-seasonal level. The strategy of allocation of land and area at each level is to maximize net income from the project.

The sources of irrigation water are limited and demand for agricultural products is increasing. It has been estimated that the irrigated area in the world is 253 million-ha. The gross irrigated area of India in 2008–2009 increased to 88.42 million-ha from 22.6 million-ha in 1951–1952, thus showing an increase of 250% during the last five decades. Efficient use of water through scientific irrigation management is of utmost importance in providing the best insurance against weather-induced fluctuations in food production. The application of irrigation water by traditional method causes 27 to 42% loss of water through deep percolation depending on the soil type [1]. Due to depletion of water sources and high labor costs, micro irrigation has a significant adaptability all over the world. The worldwide area under micro irrigation has been increasing steadily [16]. Micro irrigation helps to conserve irrigation water and increase water use efficiency by reducing soil evaporation and drainage losses. It also helps to maintain soil moisture conditions that are favorable to crop growth. Thus micro irrigation can help to sustain the productivity of the land.

The total area covered under micro irrigation in India is 4.94 million-ha consisting of 1.90 million-ha under drip irrigation and 3.04 million-ha under sprinkler irrigation. The maximum area under micro irrigation in Maharashtra – India is 0.90 million-ha. Area under micro irrigation in Karnataka is 0.60 million-ha consisting of 0.21 million-ha under drip irrigation and 0.39 million-ha under sprinkler irrigation [6, 45].

Water being a scarce resource, its efficient and economic use is of utmost importance in agriculture. Pressurized irrigation system is quite effective under limited water availability not only in achieving higher productivity but also economizing other inputs such as fertilizers, pesticides, labor, etc., compared to traditional irrigation methods. Micro irrigation is convenient and effective means of supplying water directly to soil and nearer to the plant without much loss of water resulting in higher water productivity. For water to be available to the plant, it is required to wet the root zone only and this can be efficiently achieved through this system. Micro irrigation system avoids unnecessary wetting of soil zones not having roots and minimizes the losses due to surface and deep percolation from such areas [15, 16].

The water users of agriculture have started to realize the importance of water management. To meet the growing demand under domestic and industries, the necessity has arisen for the optimum use of available water. Economic use of water for agriculture is the utmost necessity to bring more area under increased production. Sprinkler irrigation system is one of the water saving technique, which can be adopted for the suitable crops in almost all the soils [32].

Micro sprinkler system is well suited for close growing crops, which require less pressure compared to sprinkler system. For judicious water supply and also to maintain optimum moisture condition during the critical stage, it is assumed that micro sprinkler is more advantageous for getting higher yield. Hence it is necessary to formulate a suitable micro sprinkler irrigation design with simple and efficient scheduling of irrigation of groundnut crop.

Groundnut (*Arachis hypogaea* L.) is native of Brazil (South America). It was introduced in India during first half of the sixteenth century. It is a unique crop, combining the attributes of both oil seed crop and legume crop in the farming system of Indian Agriculture. It is a valuable crop planted in dry areas of Asia, Africa, Central and South America, Australia and Caribbean in view of its economic, food and nutritional value. It is the 13th most important food crop, 4th most important source of edible oil and 3rd most important source of vegetable protein in the world. Groundnut possesses high oil content (44–50%) and protein (25%) and is also a valuable source of vitamins E, K and B. It is a richest plant source of thiamine and niacin, which is low in other cereal crops. The plant, kernels, oil and cake are economically used in one or the other way.

The major groundnut producers in the world are China, India, Nigeria, USA, Indonesia and Sudan. Asia accounts for 54% of the global groundnut area and 67.7% of the production with an annual growth rate of 1.28% for area, 2% for production and 0.71% for productivity [4]. In India, groundnut is grown in an area of 5.95 million-ha with a production of 7.54 million-tons with the yield of 1268 kg.ha^{-1} [5, 6]. Six major groundnut-growing states are Gujarat, Andhra Pradesh, Tamil Nadu, Rajasthan, Karnataka and Maharashtra. These six states contribute 90% of total groundnut area of India.

This chapter presents the performance and evaluation of micro sprinkler irrigation for groundnut crop in Raichur region of India with the following objectives:

1. To determine the water requirements for groundnut crop under micro sprinkler irrigation for Raichur region.
2. To evaluate the effects of micro sprinkler irrigation on the yield of groundnut.
3. To determine the economics of the micro sprinkler irrigation for groundnut crop.

10.2 LITERATURE REVIEW

The scarcity of water is the main limiting factor in getting a good yield from the crops. It is known that surface flooding of water results in extensive runoff, deep percolation and evaporation losses. Micro irrigation system results in enormous saving of water since deep percolation and runoff losses are eliminated.

Larry [27] reported that micro sprinklers can overcome some of the disadvantages of the drip system. The device has larger orifices than emitters, thus reducing the need for filtration to avoid clogging. The device also protects the crop from frost. Under laboratory conditions, Vishnu et al. [73] tested the performance of four spinner emitters and two spray emitters at pressures of 49.03, 98.07 and 147.10 kPa, to study the distribution pattern and uniformity of water application. All tests were conducted with emitters positioned on stakes 0.20 m above the top of the catch cans, which were placed at 0.60 m grid intervals in a matrix. Vishnu observed that spinner emitters had higher distribution uniformity than spray emitters under no wind conditions.

Dilip and Mandakini [12] reported that micro sprinklers incorporate moving parts with greater discharge rate and coverage than dripper and micro jet. They observed that the system is suitable for nursery, lawn, grassland where, infiltration rate is higher. Parikh et al. [38] stated that mini sprinklers are more sensitive to pressure variations than micro sprinklers in terms of discharge rate. The discharge rate of mini sprinkler is higher compared to micro sprinkler within the same pressure range. They also reported that increase in riser height (up to 75 cm) and operating pressure (200 kpa) increases the diameter of water spread of micro sprinkler. At Gujarat Agricultural University – Navsari Campus, Patel et al. [40] carried out field experiment for safflower. They observed that the consumptive use of water was increased with increase in levels of IW/CPE ratios in both methods of irrigation (mini sprinkler and surface irrigation). The mean consumptive use of water recorded under mini sprinkler and surface methods was 221.8 and 279.4 mm, respectively. This was reduced by 20.57% with mini sprinkler compared to surface method.

Singh et al. [63] stated that the spacing of micro sprinkler along the laterals and the spacing of laterals along the submain should be considered equal to the radius of the throw. Less than 50% overlapping leads to poor uniformity coefficient and dry spots are left near the stakes. Dahiwalkar et al. [11] conducted field experiment on groundnut in sandy clay loam soil in summer season at Mahatma Phule Krishi Vidyapeeth, Rahuri, Maharashtra, with four micro irrigation systems viz. micro sprinklers, turbo-key, microtube and in-line drippers. They concluded that the micro sprinkler system with 36 lph discharge is suitable for groundnut crop, due to better yielding ability and judicious water use compared to alternate micro irrigation systems and conventional surface irrigation.

Mukund and Satyendra [33] found that best uniformity coefficient of 89.93% was found with 1.6 kg cm^{-2} pressure. This study also indicates that the micro sprinklers must be operated during morning session in semiarid region. Claudia et al. [10] studied hydraulic characteristics of micro sprinklers after use with wastewater treated by an up-flow anaerobic sludge blanket (UASB anaerobic reactor). The hydraulic characteristics determined for the new and used emitters were manufacturing coefficient of variation, coefficient of use and the pressure-discharge relationship. They observed that there was no appreciable variation in the hydraulic characteristics after using wastewater. They reported that the value of the mean discharge for nominal pressure was reduced by 4.97% in relation to the new, the flow regime did not change and the variation coefficient was increased but continued being of excellent category as per ASAE classification [15, 16].

In the landscaped area of the Tamil Nadu Agricultural University Campus, Suresh and Senthilvel [65] examined the relationships between flow-pressure, profile of distribution of water and Christiansen's Uniformity Coefficient (CUC) related with flow pressure [9]. The value of CUC observed was 94%.

Muralikrishnasamy et al. [35] conducted field experiment during summer and Kharif at Agricultural College and Research Institute, Madurai. They stated that micro sprinkler irrigation provide for higher moisture availability and greater nutrient uptake in onion crop. These mechanisms in turn enhanced plant growth as shown by taller plants, higher dry matter accumulation and more branches per plant and that the conversion of dry matter into economic produce was also much higher as indicated by higher values of harvest index with micro sprinkler irrigation. Sanimer et al. [54] stated that the losses are dependent upon both climatic and operating factors, ranging from 5.49 to 29.93% for the tested mini sprinklers. The evaporation and drift losses indicated that the riser height and nozzle size were the predominant factors affecting the evaporation and wind drift losses.

Shobana and Asokaraja [60] reported that micro sprinkler irrigation once every 2 days numerically recorded higher and constant relative water content throughout the crop period compared to 4 days interval. They observed the decrease in fluctuations in water content compared to surface irrigation. The decrease in relative water content ranged from 80.50 to 49.43% in surface irrigation, due to increase in daily water stress. Suseela and Rangaswami [66] reported that micro sprinkler irrigation system never showed the clogging problem, which is the major problem in the drip irrigation. Based on the experiences in the farmer's field as well as in the fields of the research station, they found that there is no need to install the filter in this irrigation system thus reducing the cost of the system. They concluded that the micro sprinkler system is cheaper than drip system especially in close growing crops.

Kadam et al. [21] found that micro sprinkler may be operated at 1.0 kg. cm^{-2} for spacing of 1.5 m × 1.5 m and 2.0 kg cm^{-2} for spacing of 2.25 m × 2.25 m. They stated that if spacing of the micro sprinkler was increased, these need to be operated at higher operating pressures, thus requiring higher expenditure on energy; and if the spacing was further increased and operating heads were decreased, the uniformity will reduce drastically that may eventually decrease the crop yield. At the Central Farm of Orissa University of Agriculture and Technology, Bhubaneshwar, Sahoo et al. [52] studied uniformity of water distribution at different nozzle pressures and spacing for both plastic and brass type of sprinklers, on a leveled concrete floor. The deviation of uniformity coefficient of plastic sprinklers from that of brass sprinklers ranged from 0 to 2%. They stated that the deviations may be due to human errors, instrumental errors and abrupt changes in wind speed. They found that the plastic sprinklers were working as efficiently as brass sprinklers. The effect of wind velocity on uniformity of water distribution was less for wind velocities below 4 $km.hr^{-1}$. The distribution pattern was distorted at high wind velocities of 15 $km.hr^{-1}$.

Silva et al. [62] reported the water application efficiency for the systems with a 32 lph micro sprinkler per four plants, 60 lph micro sprinkler per four plants and 60 lph micro sprinkler per two plants were 85.01, 79.72 and 89.54%, respectively. Isiguzo [18] observed that the average coefficient of uniformity (CU) and delivery performance ratio (DPR) of the system were 86% and 87%, respectively, indicating satisfactory performance of the sprinkler system. At the experimental farm of the Agricultural and Food Research and Technology Center in Zaragoza-Spain, Sanchez et al. [53] reported that curve of the distribution of irrigation depth along the wetted radius was crucial to characterize the water distribution of a sprinkler. The water distribution closely depends on the shape of the radius. The discharge and the range of a sprinkler are insufficient by themselves to select a sprinkler adequately.

They reported that the shape of the radius is mainly affected by the sprinkler design and that the nozzle diameter and the pressure also modify the shape but to a lesser extent. The characterization of the radius with isolated sprinkler tests requires several precautions because the water distribution will be distorted even at low wind velocities. They concluded that when the tests cannot be performed under indoor conditions, the wind must be observed meticulously to ensure that the vector average of the wind velocity during the tests does not exceed 0.6 m/s, which otherwise will lead to erroneous results.

Neumann and Navashir [36] stated that the wetting front of micro sprinkler irrigation can travel through the soil profile at the rate of 5 to 10 cm.hr^{-1} and reach the depth of 60 cm after 5 to 12 h depending on soil type. During summer in a 6-year-old almond orchard in the Sacramento Valley under micro sprinklers situated mid-way between trees in the tree row, Koumanov et al. [25] observed large evaporation losses during and immediately after the irrigations and the evaporation losses of the wetted area were estimated to be between 2 and 4 mm per irrigation event. Consequently, application efficiencies were only 73–79%, the wetting of the root zone was limited to the 0–30 cm depth interval, the soil profile was depleted of soil water, and daily crop coefficient values at days between irrigation events were 0.6 and 0.8. They recommended that the irrigation must be given in the evening and night hours, thereby largely eliminating the evaporation losses that occur during daytime irrigation hours. At the Irrigation Water Management Laboratory and Instructional Farm of College of Technology and Agricultural Engineering, Udaipur, Rajvir et al. [43] reported that the value of the nozzle exponent of the micro jet was classified as nonpressure compensating and the emission uniformity was more than 90%. They observed that the wetting diameter was increased with increase in operating pressure and stake height. The uniformity coefficient and distribution characteristic were increased with increase in operating pressure and stake height.

Varshney and Raghavaiah [70] concluded that covering the soil surface with paddy husk mulch and irrigating the crop by sprinkler during summer saved water to the extent of about 40.3% and enhanced the crop yield by 36% with returns per hectare when compared to surface irrigation. Kadam et al. [22] suggested that micro sprinkler irrigation system should be scheduled below 20% available moisture depletion level. The frequency and depth of irrigation water can be estimated from pan evaporation and crop coefficient data. There was about 59% water saving in addition to 20% increase in the garlic yield. At the research farm of Dr. Panjabrao Deshmukh Krishi Vidyapeeth, Akola, Arulkar et al. [7] observed that the horizontal spread at 91 lph discharge rate of micro sprinkler and 4 lph of drip irrigation for 1 h and 2 h was 3 m and 0.6 m, respectively. In drip and micro sprinkler with the increase in water application, the soil moisture was increased horizontally and vertically. They also observed that the moisture content was in the range of 76 to 100% and 71 to 100% of field capacity under micro sprinkler and drip irrigation, respectively.

Sakar et al. (2008) reported that in coarse textured soils higher crop yield and optimum water use efficiency (WUE) were achieved through a reduction in drainage loss and there was minimum drainage loss of water if a smaller quantities (2.31–26.5 $L.m^{-2}.week^{-1}$) of irrigation water can be applied with a frequency of once in a week. They concluded that such type of irrigation scheduling can be maintained only under sprinkler system and not in any surface irrigation method. Satyendra et al. [56] conducted the study to compare micro sprinkler, drip and furrow irrigation systems for potato production at Central Institute of Postharvest Engineering and Technology, Abohar, Punjab with 4 irrigation levels (IW: CPE ratio of 1.20, 1.00, 0.80 and 0.60). They observed a better crop performance under micro sprinkler regime. They reported that irrespective of irrigation system, potato tuber yield was increased with increasing irrigation level from 0.60 to 1.20 IW: CPE.

Khade et al. [24] reported that the pod yield in groundnut was increased by 20.76% and less water use by 33%. The WUE was increased from 38.7 to 62.6 kg/ha-cm and the water requirement was decreased from 409 to 380 mm with sprinkler irrigation system compared to check basin method. Chauhan and Srivastava [8] noticed that maximum cabbage yield was under micro sprinkler, followed by drip emitter. Water saving through micro sprinkler was 36.82% over surface irrigation.

In kharif season on sandy clay loam soil on groundnut crop at Mahatma Phule Krishi Vidyapeeth, Rahuri – Maharashtra, Shinde et al. [59] reported that WUE was higher in micro sprinkler method of irrigation (11.9 to 15.5 $kg.ha^{-1}.mm^{-1}$) compared to surface method of irrigation (4.88 to 5.04 $kg.ha^{-1}.mm^{-1}$) and that the water saving was 41.3% in micro sprinkler method of irrigation over surface irrigation method. Jadhav et al. [20] concluded that the maximum yield of chili (8.57 $q.ha^{-1}$) was recorded under micro sprinkler irrigation at five days interval with 0.7 cumulative pan evaporation level and highest WUE was recorded for four days interval with 0.7 cumulative pan evaporation.

Pampattiwar et al. [37] revealed that micro sprinkler spaced at 1.75 m × 1.75 m recorded the maximum coefficient of uniformity (92.13%) influencing the maximum yield of 4.78 $Mg.ha^{-1}$ of garlic bulbs and irrigation WUE of 143.24 $kg.ha^{-1}.cm^{-1}$. Sexton et al. [57] reported that groundnut pod development is sensitive to surface soil (0–5 cm) conditions due to its subterranean fruiting habits. Soil water deficits in the pegging and root zone decreased pod and seed growth rates by approximately 30% and decreased weight per seed from 563 to 428 mg. Peg initiation growth during water stress demonstrated an ability to suspend development during the period of soil water deficit and reinitiate pod development after the water stress was relieved.

Firake and Shinde [13] evaluated effects of drip, microtube, microsprinkler and subsurface drip on summer groundnut in respect of yield attributes, water requirement and WUE. They reported that the microsprinkler irrigation system was superior to the other systems in terms of dry pod (32.53 $q.ha^{-1}$), kernel (22.73 $q.ha^{-1}$), haulm (43.68 $q.ha^{-1}$) and oil yields (11.11 $q.ha^{-1}$). The seasonal water requirement of summer groundnut was 68.52 cm in micro irrigation systems and 95.05 cm in the border irrigation, indicating 27.91% water conservation. The WUE was also highest (4.75 $kg.ha^{-1} mm^{-1}$) with the microsprinkler system. They concluded that the microsprinkler system was considered the most suitable system for obtaining maximum yields of summer groundnut.

Al-Jamal et al. [2] noticed that irrigation WUE under a drip irrigated onion field was low under buried drip irrigation system and by using a sprinkler system can maintain high irrigation efficiency compared to furrow and drip irrigation. They also noticed that the irrigation WUE using the sprinkler system was higher compared to subsurface drip and furrow irrigation methods. In order to conserve water, they advocated to use sprinkler irrigation system.

At GBPUAT – Panthnagar, Manjunatha et al. [29] reported that the saving of irrigation water (36.9%) along with increase in potato yields (25.1%) were recorded in micro sprinkler irrigation compared to furrow irrigation. They observed that various growth and yield parameters registered a significant increase under micro sprinkler irrigation compared to furrow irrigation. Better quality produce was also observed under micro sprinkler irrigation. Manjunatha et al. [28] reported highest yield of brinjal (eggplant) under micro sprinkler (29.33 $tons.ha^{-1}$) followed by drip micro tube (28.74 $tons.ha^{-1}$), drip emitter (26.4 $tons.ha^{-1}$) and surface method (24.2 $tons.ha^{-1}$), respectively. They also observed that the increase in yield over surface irrigation was 7.4, 18.5 and 20.9% for drip emitter, micro tube and micro sprinkler, respectively and water saving was up to 33% in micro sprinkler irrigation over surface irrigation.

At Narmada Irrigation Research Project, GAU, Khanda – Gujarat, Patel et al. [39] reported that the sprinkler irrigation at 0.75 IW/CPE with total water quantity of 380 mm recorded wheat yield of 3833 $kg.ha^{-1}$ compared to 2817 $kg.ha^{-1}$ under surface method with 560 mm of irrigation water. They also noticed that 0.60 IW/CPE under sprinkler registered higher yield with 330 mm of water, which saved 230 mm of water compared to surface method. Sprinklers operated at branching and pod development stages of gram recorded the yield of 1929 $kg.ha^{-1}$ with 100 mm of water, while surface irrigation yielded 1135 $kg.ha^{-1}$ with 240 mm of water.

Shukla et al. [61] conducted field experiment at GBPUAT Pantnagar and reported that total yields of sweet lime obtained after 2 years of planting was 7.04, 6.16, 4.48 and 3.92 $q.ha^{-1}$ for micro sprinkler + drip, micro sprinkler, surface + drip and surface irrigation, respectively, resulting in higher yields of 79.6, 57.1 and 14.3% for micro sprinkler + drip, micro sprinkler and surface + drip, respectively, over surface irrigation. They concluded that the higher WUE of 3.95 $kg.ha^{-1} cm^{-1}$

was recorded under micro sprinkler + drip followed by 3.34 $kg.ha^{-1} cm^{-1}$ with micro sprinkler, 2.28 $kg.ha^{-1} cm^{-1}$ with surface + drip and minimum of 1.96 $kg.ha^{-1} cm^{-1}$ in case of surface irrigation.

Sonune and Palaskar [64] reported that the average groundnut productivity was considerably higher (22.9 $q.ha^{-1}$) with broad bed furrow (BBF) method of planting using sprinkler than the conventional flat-bed planting with sprinkler (17.13 $q.ha^{-1}$). The net irrigation requirement under sprinkler was 770 mm compared to 1050 mm under surface irrigation, resulting 36.4% saving in irrigation water. They concluded that sprinkler irrigation with BBF planting technique was beneficial to increase the crop productivity on shallow medium coarse textured soils with considerable saving in irrigation water. At the Plasticuture Development Center, Hyderabad, Varshney and Raghavaiah [70] observed that the crop with sprinkler irrigation consumed 526 mm of water against 738 mm used with surface irrigation, whereas groundnut yield under sprinkler and surface irrigation were 3042 and 2774 $kg.ha^{-1}$, respectively and with 40.30% saving of water and additional yield of 9.7% under sprinkler over the surface irrigation method.

Krishnamurthi et al. [26] revealed that irrigating the groundnut crop through micro sprinkler at 80% of pan evaporation recorded the highest pod yield, which was 20% increase over surface method of irrigation and 10–12% of water was saved under this method. Vijayalakshmi et al. [71] found that micro sprinkler irrigation scheduled at 100% ET_c recorded total water saving of 28–31%, while micro sprinkler irrigation scheduled at 75% ET_c registered total water saving of 32–38% over surface irrigation. Rank et al. [46] concluded that the lowest and highest pod yield of 1471 and 2550 $kg.ha^{-1}$ were at the IW/CPE ratio of 0.6 and 0.9, respectively, with irrigation water requirement of 523 and 789 mm, respectively under micro sprinkler irrigation.

Satyendra et al. [56] reported that the highest potato yield (31.60 $tons.ha^{-1}$) was obtained with micro sprinkler, followed by drip (29.83 $tons.ha^{-1}$) and furrow (22.6 $tons.ha^{-1}$) irrigation systems when irrigation was scheduled at 1.20 IW: CPE. Irrespective of the irrigation system, potato tuber yield was increased with increasing irrigation level from 0.60 to 1.20 IW: CPE.

Thiyagarajan et al. [68] reported that the yield response factor (k_y) ranged from 0.45 and 0.42 (normal irrigation) to 1.72 and 1.70 (full deficit irrigation) for summer and Rabi seasons, respectively. They observed that the pod formation and flowering stages were more sensitive to moisture stress. Irrigation must be given to ET requirements during these stages for getting maximum yield. At Research Farm of National Research Center for Onion and Garlic, Rajgurunagar (Pune), Tripathi et al. [69] reported that the WUE was higher in drip irrigation (770 kg/ha-cm) than micro sprinkler (344.6 kg/ha-cm), overhead sprinkler (386.5 kg/ha-cm) and surface irrigation (252.5 kg/ha-cm). The highest benefit cost ratio (BCR) was 1.98 in drip irrigation. Sezen et al.[58] reported that effects of the irrigation amounts are significantly important for obtaining higher seed and oil yield of sunflower and not the method of application (sprinkler or drip irrigation). The irrigation levels significantly influenced the sunflower seed and oil yield and yield components.

Raman et al. [44] observed higher net income for summer groundnut under sprinkler irrigation compared to surface irrigation and higher additional income (10,710 Rs/ha) over surface irrigation was obtained under micro sprinkler irrigation at 0.7 IW/CPE ratio. Rajesh and Kamal [42] reported that banana, grape and pomegranate responded better to "Fan-Jet" micro sprayer irrigation than the drip. The cost of the system in these crops was 40% less than the drip system. Manjunatha et al. [30] concluded that the system cost is maximum for drip emitters due to more number of emission devices and laterals and minimum for micro sprinklers. For eggplant, maximum net seasonal income of (80, 649 Rs/ha) was recorded for micro sprinkler irrigation, followed by drip microtube (78, 111 Rs/ha), surface irrigation (63, 321 Rs/ha) and emitters (60, 845 Rs/ha). They also reported that the net profit achieved per cm application of irrigation water was Rs 4,106, 8,920, and 9,520 for furrow irrigation, micro sprinkler with considering system cost, and micro sprinkler without considering system cost, respectively. The benefit – cost ratios of 2.42, 2.79 and 3.17 for furrow irrigation, micro sprinkler irrigation with considering system cost, and micro sprinkler without considering system cost, respectively.

Sahni et al. [51] reported that freedom should be given to the beneficiaries to grow crops in the canal command. Drip irrigation method cannot be adopted directly on canal water with prevailing 14 days rotation period. They also reported that the use of sprinkler irrigation is technically feasible and economically viable in canal command with the proposed irrigation water management strategies.

In Chikamagalur district of Karnataka, Muralidhara et al. [34] reported the microsprinkler had a higher benefit – cost ratio and a greater positive net present value than the corresponding values in conventional sprinkler system. The efficiency gain from the use of the microsprinkler with the sprinkler level of inputs (capital, fertilizer, and labor) was nearly 12%. Thus, the microsprinkler system was considered a viable and economical technology for the irrigation of coffee. The main constraints to the adoption of the system were the clogging of microsprinklers and the need for skilled workers to ensure maintenance.

For close growing crops at the Agronomic Research Station – Chalakudy of Kerala Agricultural University, Visalakshi et al. [72] reported that the investment cost is Rs. 10,000 less than that required for drip irrigation system. Lesser clogging susceptibility, more distribution uniformity, reduced investment cost, etc. were special advantages of the developed micro sprinkler system. Gadge et al. [14] reported that micro irrigation methods are mainly adopted on farms with irrigation from tubewells or dug-wells. However, in view of increasing scarcity of water for irrigation and need to increase food production, it is important to adopt these methods in canal command area. They reported that net returns from the optimal cropping plan were 186 million-Rs. The optimal cropping pattern allocated was 43.17 ha area for papaya, 43.17 ha for sugarcane, 86.35 ha for *Kharif* eggplant, 259.04 ha for cabbage,

85.35 ha for summer onion [55], and 259.04 ha for summer eggplant. It was found that 6% increase in net benefit was achieved through reutilization of the micro irrigation systems among the seasonal crops having similar system requirements.

At the Water Management Research Unit ARS – Chalakudy of Kerala Agricultural University. Suseela and Rangaswami [67] reported that the application depth was found to increase with increase in length of the unit. Minimum Coefficient of variation values in each diameter sprinkler head was obtained for the unit having medium length diameter ratio. They reported that KAU micro sprinklers can be made by a farmer if he would be trained for a day and that it is cost effective (Cost of construction of micro sprinkler head was less than Rs 2/-), farmer friendly and is suitable for small/marginal farmers. Moreover, this system was designed to operate clog free with complete wetting of the basin area (>90%) of the crop due to its rotating action.

10.3 MATERIALS AND METHODS

A field study was conducted from December 2011 to April 2012 in farmer's field at Yeragera village, Raichur. The detailed materials used and methodologies adopted during the field experiment are presented in this chapter.

10.3.1 FIELD LOCATION

The farmers field at Yeragera village, Raichur is located at 16°15' North latitude and 77°20' East longitude and is at an elevation of 389 m above mean sea level (MSL). The climate is semiarid and average annual rainfall is 722 mm.

10.3.2 WEATHER AND CLIMATE

The daily climatologically data during the study was collected from the meteorological observatory at the Main Agricultural Research Station, Raichur. During the period of study, the maximum temperature of 38.9°C was recorded in February, 2012 and the minimum temperature of 8.5°C was recorded in January, 2012. The maximum relative humidity of 98% was recorded in February, 2012 and the minimum relative humidity of 21% in January, 2012. The maximum evaporation of 10 mm/day was recorded in April, 2012 and the minimum evaporation of 2.0 mm/day in December, 2011 and January, 2012. The maximum wind velocity of 6.3 km/h was recorded in March, 2012 and the minimum wind velocity of 1.3 km/h in December, 2011 and January, 2012. The mean monthly maximum temperature, minimum temperature, evaporation, relative humidity and wind velocity are shown in Fig. 10.1.

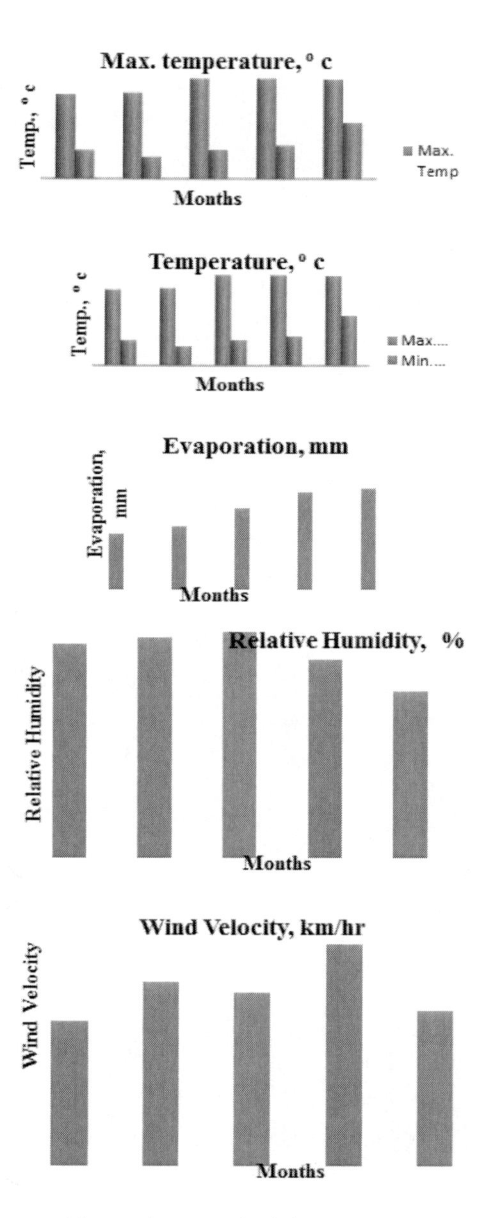

FIGURE 10.1 Mean monthly maximum and minimum temperature, evaporation, relative humidity and wind velocity during study period.

10.3.3 IRRIGATION SOURCE

Irrigation source for the experiment was bore well water. The water was pumped for irrigation with 3 hp submersible pump as and when required.

10.3.4 IRRIGATION WATER QUALITY

The irrigation water was analyzed for its suitability for irrigation. The pH was 7.30 and electrical conductivity (EC) was 1.055 dSm^{-1}.

10.3.5 EXPERIMENTAL SITE

The representative composite soil samples were taken from the experimental site for determination of physical characters (viz., textural composition, field capacity, infiltration rate, Permanent Wilting Point (PWP) and bulk density) and chemical properties (available N, P, K, organic carbon, EC and pH) at 0–30 cm depth and the same were determined using the standard procedures (Table 10.1). Under the textural classification the soil was found to be sandy loam.

TABLE 10.1 Soil Characteristics at the Experimental Field at Raichur

Soil characteristics	Particulars	Value	Method based on
Textural composition	Sand, (%)	68.78	Piper (1966)
	Silt, (%)	18.32	
	Clay, (%)	12.90	
	Soil type	Sandy Loam	
Chemical properties	Available N, Kg ha^{-1}	290.81	Jackson [19]
	Available P_2O_5, Kg ha^{-1}	36.21	
	Available K_2O, Kg ha^{-1}	44.47	
	Organic carbon (%)	0.33	
	pH	7.42	
	EC (dSm^{-1})	0.30	
Physical properties	Bulk density, g cc^{-1}	1.61	Richards [48]
	Field capacity, %	18.0	Piper [41]
	Permanent wilting point, %	7.56	Piper [41]
	Infiltration rate, cm hr^{-1}	2.10	Richards [48]

10.3.6 EXPERIMENTAL SETUP

The experimental set-up consisted of screen filter, mains, sub mains, and micro sprinklers and other accessories required for micro sprinkler irrigation. The plant geometry for micro sprinkler irrigation treatments and layout of micro sprinkler irrigation are shown in Figs. 10.2 and 10.3.

10.3.6.1 PUMPING SOURCE

A 3 hp vertical type submersible pump, with a total head of 50 m and discharge capacity of 11 m^3/h, was used to lift the water from bore well and supply to both micro sprinkler and surface irrigated plots.

10.3.6.2 PIPELINE MANIFOLD

The main and sub main pipelines used for micro sprinkler irrigation were made of PVC of 63 mm and 50 mm diameter, respectively.

10.3.6.3 FILTRATION SYSTEM

A single mesh screen 120 mesh (130 micron) with a maximum capacity of 40 m^3/h and nominal pressure of 2 $kg.cm^{-2}$ was used to filter the irrigation water for micro sprinkler irrigation. The filtration unit was fitted on the main pipeline of micro sprinkler irrigation system.

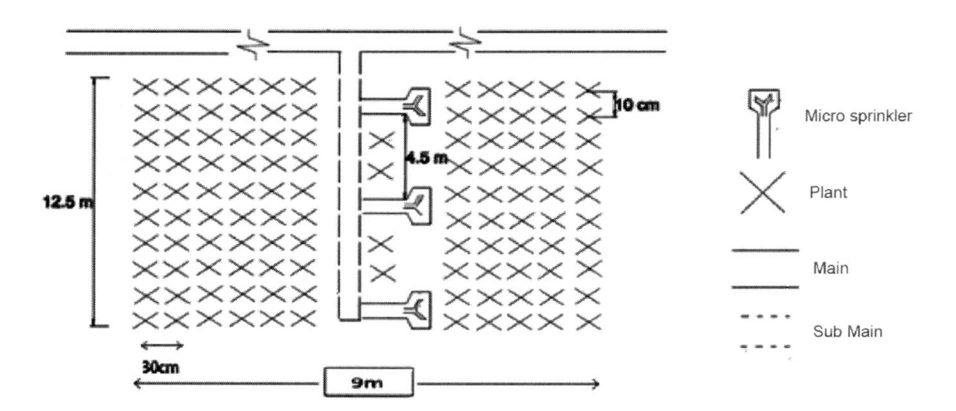

FIGURE 10.2 Experimental layout for microsprinkler irrigation treatments.

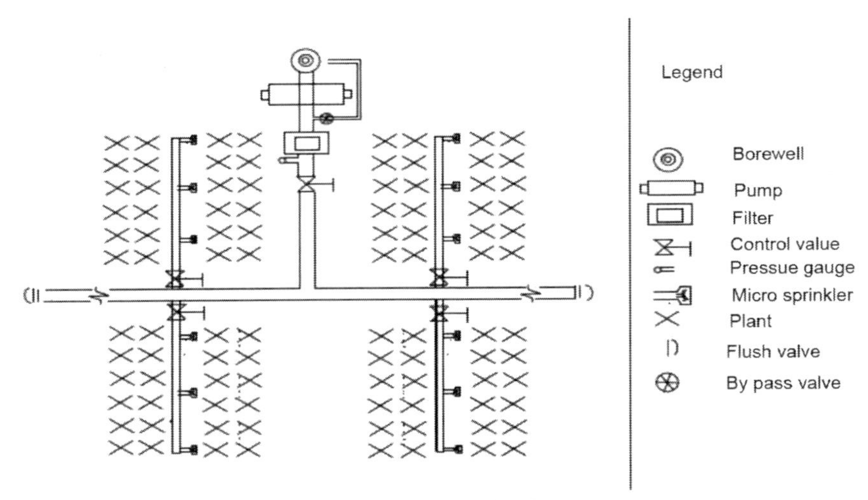

FIGURE 10.3 Field layout of a microsprinkler irrigation system.

10.3.6.4 MICRO SPRINKLERS

Suitability of micro sprinklers in terms of discharge, pressure and wetting area were tested before selection. Micro sprinklers of 360 lph capacity with a throw of 9 m diameter each at the height of 45 cm from the ground surface were selected for micro sprinkler irrigation treatments.

10.3.6.5 UNIFORMITY COEFFICIENT OF MICRO SPRINKLER SYSTEM

Before selection of the over lapping percentage, the single micro sprinkler was taken for finding its distribution pattern and wetting diameter. It was observed during the performance of single micro sprinkler that the less discharge was noticed up to 1.5 m distance from the stake and it varied from 1.2 to 2 mm, whereas the high and uniform discharge was observed from 1.5 m distance from the stake and it varied from 2 mm to 4.2 mm as shown in Fig. 10.4.

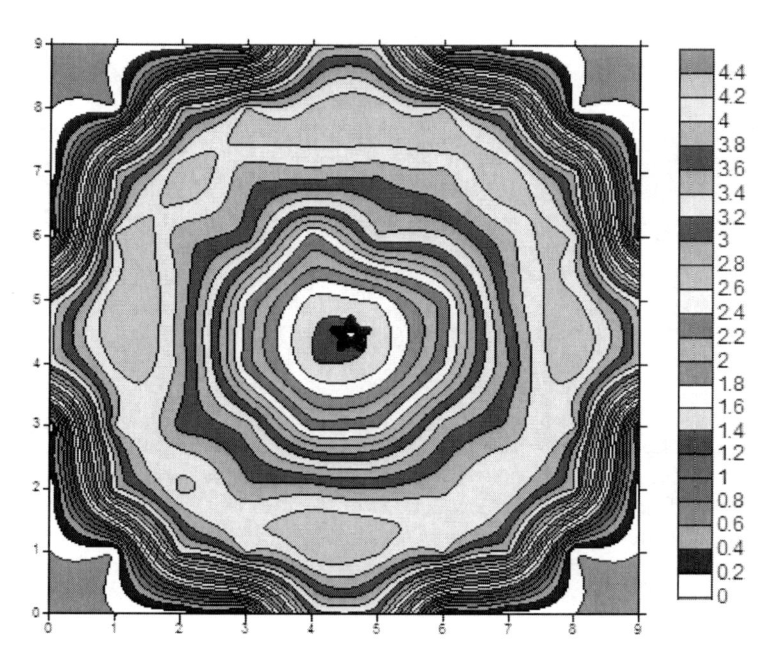

FIGURE 10.4 Uniformity coefficient of single micro sprinkler.

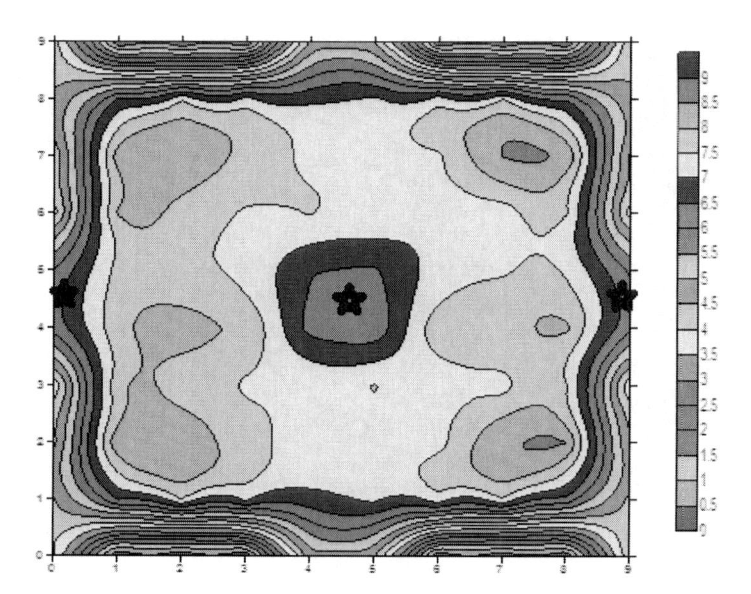

FIGURE 10.5 Distribution pattern of 100% over lapping micro sprinklers.

By observing the pattern of throw and wetting diameter of single micro sprinkler, 100% over lapping was selected to overcome the dry patches and for the uniform distribution. It was observed that the uniform and high discharge was found at 7 m × 7 m spacing with a discharge variation of 6 to 8.5 mm. The distribution pattern of 100% over lapping of micro sprinkler is shown in Fig. 10.5. The uniformity coefficient was calculated by collecting water in the catch cans placed at grid points of the overlapped area and with following Christiansen's equation [9].

$$CU = 100\,(1 - D/M),\ \text{or}\ 100 \times [1 - (A/B)],\ \text{and}$$

$$D = (1/n)\,\Sigma\,|\,Xi - M|$$

$$M = (1/n)\,\Sigma\,Xi \qquad (1)$$

where, CU = Christiansen's Coefficient of Uniformity (%), A is the sum of the absolute value of the deviation of the average catch cup value from each individual catch cup data point, B is the sum of the catch cup observations, D = average absolute deviation from the mean, M = Mean application, Xi = Individual application amount, n = Number of individual application amounts, å = Symbol for summation, and | | = Symbol for absolute value of quantity between the bars. Appendix I indicates list of all symbols in this chapter.

The maximum uniformity coefficient of 89.9% was obtained at 1.4 $kg.cm^{-2}$ pressure for single micro sprinkler at the stake height of 45 cm from the ground surface, whereas the maximum uniformity of 87.7% was obtained at 1.4 $kg.cm^{-2}$ pressure for over lapping micro sprinklers at the stake height of 45 cm from the ground surface.

10.3.6.6 CROP DETAILS

Common name	Groundnut
Scientific name	*Arachis hypogaea* L.
Variety	R-2001–2
Seed rate	125–150 kg/ha
Method of sowing	Dibbling
Seed spacing	30 cm × 10 cm
Effective root zone depth	15 cm
Farm yard manure	7.5 tons/ha
Fertilizers	Recommended dose (kg/ha): N: 25, P: 50, K: 25, $FeSO_4$: 25
Plant protection measures	2 sprays of Qunolphos with dosage of 2 gm l^{-1}, and 2 sprays of Mancozeb with dosage of 2 gm l^{-1}.

10.3.7 EXPERIMENTAL DETAILS

Micro sprinkler irrigation for groundnut crop was designed by considering the design capacity, optimum size of the pipelines, discharge rate of sprinklers, capacity of filter and pump capacity. Groundnut is a close spaced crop having spacing of 30 cm × 10 cm. The operating pressure at the main pipe of the micro sprinkler system was maintained at 1.4 $Kg.cm^{-2}$. This pressure head was sufficient for irrigating the experimental area with micro sprinkler irrigation. From the water source, water was pumped with a 3 hp motor and conveyed to the field by using 63 mm diameter PVC pipe after filtering through the screen filter.

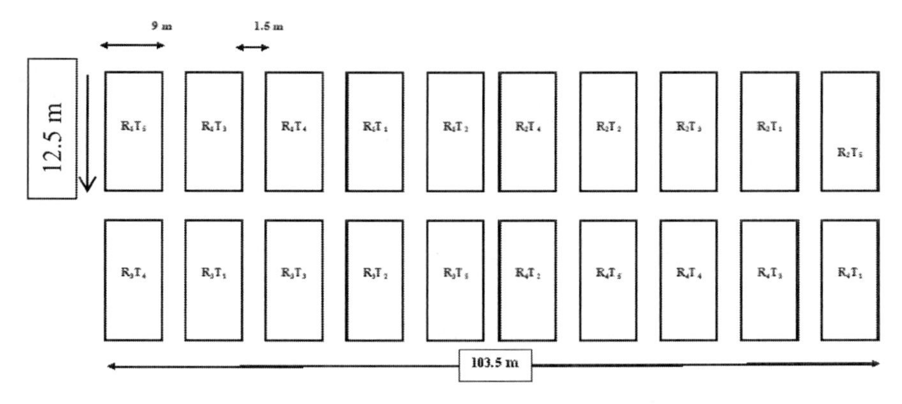

FIGURE 10.6 Statistical layout of the experiment.

The statistical layout of the experimental plot is shown in Fig. 10.6. The experimental plots were laid out in randomized block design (RBD) with five treatments and four replications, viz.

T1 – Water application at 60% of ET using micro sprinkler irrigation,

T2 – Water application at 80% of ET using micro sprinkler irrigation,

T3 – Water application at 100% of ET using micro sprinkler irrigation,

T4 – Water application at 120% of ET using micro sprinkler irrigation,

T5 – Water application at 100% of ET using surface irrigation (control).

Net plot size of 9 m × 9 m was taken and three micro sprinklers were installed in each plot with 100% overlap, to overcome the dry patches around and bottom of the stakes. The flow of water to each plot was controlled by using individual control valves at the start of each sub main line. In order to maintain the effects of a treatment, a buffer of 1.5 m was provided between the plots.

10.3.8 WATER REQUIREMENT OF GROUNDNUT

Amount of irrigation water applied to various treatments were based on daily pan evaporation readings. The irrigation treatments were imposed, once the sowing was done and the total water requirement for groundnut crop was obtained by adding up all the depths of water applied for each treatment. The previous day pan evaporation was used to calculate the water requirement for next day.

10.3.8.1 SURFACE IRRIGATION METHOD

The amount of water to be delivered in check basin method (control) was computed using the following equation:

$$d = [1/100] \times [M_{fc} - M_{bi}] \times [A_s \times d_s] \quad (2)$$

where, d = net amount of water to be applied during irrigation, cm; M_{fc} = moisture content at field capacity, %; M_{bi} = moisture content before irrigation, %; A_s = soil bulk density of soil, g/cm^3, and d_s = effective root zone depth, cm. Quantity of water (L) per plant was estimated as follows:

$$Q = [1/1000] \times [d \times A \times B] \quad (3)$$

where, Q = quantity of water required per plant, liters; d = net amount of water to be applied during an irrigation, cm; A = gross area per plant, cm^2; and B = extent of area covered by foliage, fraction.

10.3.8.2 MICRO SPRINKLER IRRIGATION METHOD

The daily water requirement for micro sprinkler irrigation was computed using pan evaporation data from USDA Class A open pan evaporimeter. The water requirement of groundnut per day under micro sprinkler irrigation was computed as follows [3].

$$Q = [1/E] \times [A \times B \times C] \quad (4)$$

where, Q = quantity of water required mm day^{-1}; A = daily evapotranspiration, mm day^{-1} = pan evaporation × pan coefficient; B = amount of area covered with foliage (canopy factor), fraction; C = crop coefficient, fraction; and E = efficiency of micro sprinkler irrigation system = 80%.

10.3.8.3 IRRIGATION SCHEDULING FOR MICRO SPRINKLER

The daily quantity of water application was computed as explained above. For a known discharge rate of micro sprinkler, the duration of irrigation water application was calculated as follows:

Irrigation duration, hours = [sprinkler discharge, lph] ÷ [area covered by a sprinkler] (5)

Irrigation frequency of once in two days was used in this study according to recommendations by Krishnamurthi et al. [26].

10.3.9 BIOMETRICAL OBSERVATIONS

For periodical field observations, five plants were selected randomly from each treatment and were tagged. Plant height, number of primary branches and leaf area index (LAI) were recorded at 30, 60, 90 and 120 days after sowing (DAS). Plant height (cm) was measured from the ground level up to the base of node on which the first fully opened leaf from the top was borne.

The number of branches emerging directly from main stem was counted and the average of five plants was expressed as number of primary branches per plant. LAI was measured directly by using *Leaf Area Index Ceptometer*. A view of observation of LAI is shown in Fig. 10.7.

FIGURE 10.7 Measurement of leaf area index using Ceptometer.

Fully developed pods were separated from five plants and were counted and the average was taken as the number of pods per plant. The pods from the individual plants were weighed on dry basis and the average weight of five plants was recorded as pod weight per plant in grams. The 100 pods were taken from each plot and their weight was recorded. The same sample was used for recording weight of

100 kernels. Pods from the net plot area were cleaned and pod weight was recorded on the basis of pod yield, kg per plot. The pod yield (quintal per ha) was calculated.

From produce in each net plot, 100 g of cleaned pods were weighed. The kernels were obtained after shelling and were weighed. The shelling percentage was determined as follows:

$$\text{Shelling percentage} = 100 \times [(\text{kernel weight, g})\ x\ (\text{pod weight, g})] \quad (6)$$

10.3.10 QUALITY PARAMETERS

The estimation of oil content (%) was done on dry seed weight basis by using nuclear magnetic resonance spectrophotometer.

10.3.11 IRRIGATION EFFICIENCIES

Appendix II shows an example how to estimate variables for irrigation management in this chapter. The application efficiency of micro sprinkler irrigation and surface irrigation was computed as follows:

$$e_a = [1/100]\ [W_s/W_f] \quad (7)$$

where, e_a = application efficiency (%); W_s = water stored in root zone in liters; and W_f= water delivered to the field in liters.

The WUE for each treatment was computed using the following formula for both micro sprinkler and surface irrigation methods:

$$\text{WUE} = \text{Y}/(\text{WR}) \quad (8)$$

where, WUE = water use efficiency, $kg.m^{-3}$; Y = crop yield, kg; and WR = total amount of water used, m^3.

10.3.12 DUTY AND DELTA

Duty of water is the relationship between the area irrigated and the quantity of water that is used to irrigate it for the purpose of maturing the crop. Delta is the depth of water applied over a base period.

10.3.13 ECONOMICS

Economics of micro sprinkler irrigation and check basin irrigation methods was determined to compute the net returns and benefit-cost ratio (BCR). For this purpose, the life period of polyvinyl chloride (PVC) items was considered as 10 years [49] and that of the submersible pump unit was taken as 15 years [50]. The fixed cost, operation cost and total cost were worked out. Sample calculations are presented in Appendix III.

Fixed cost consisted of interest on initial cost and depreciation on the system. The interest calculated on the capital was @ of 12% per annum. In calculations, apportioned value was taken as fixed cost.

Operating cost is the amount, which is actually paid by the cultivator in cash throughout the crop period for carrying various agricultural operations. Total operational cost of the system is the operating cost plus interest on operational cost @ 12%.

The total cost is a sum of fixed cost and operating cost. The BCR was worked out as follows:

Benefit cost ratio = [Gross returns, Rs./ha]/[Cost of cultivation, Rs./ha] (9)

While calculating gross returns, the prevailing market rate of 4800 Rs./100 kg for groundnut was considered.

10.3.14 STATISTICAL ANALYSIS

The data were analyzed using the standard analysis computer package "DESIGN." The test of significance was carried at 5% level of significance.

$$\text{Degree of freedom} = (T-1) \times (R-1) \quad (10)$$

where, T = number of treatments = 5; and R = number of replications = 4.

10.4 RESULTS AND DISCUSSION

During December 2011 through April 2012, a field experiment was conducted in farmer's field at Yeragera village of Raichur – India, to determine water requirement of groundnut crop under different levels of micro sprinkler irrigation, and to compare with surface irrigation method. The experiment also involved the study of irrigation efficiencies and the economic feasibility of micro sprinkler and surface irrigation methods.

10.4.1 WATER REQUIREMENT OF GROUNDNUT CROP

After sowing the first irrigation was given up to the field capacity to all the plots of different irrigation treatments for better establishment of crop. Then the irrigation treatments were imposed based on ET for both under micro sprinkler and in surface irrigation.

The amount of water delivered to groundnut crop under different levels of micro sprinkler irrigation and surface irrigation are presented in Table 10.2. It was observed that in case of micro sprinkler irrigation at 60% ET, the water application (mm per day) varied from 0.40 to 0.65 in December, 0.59 to 2.86 in January, 1.8 to 3.7 in February, 1.7 to 3.57 in March and 1.76 to 2.84 in April. In case of micro sprinkler irrigation at 80% ET, water application (mm/day) varied from 0.53 to 0.86 in December, 0.78 to 3.82 in January, 2.46 to 4.93 in February, 2.27 to 4.76 in March and 2.34 to 3.78 in April, respectively. Similarly, for micro sprinkler irrigation at 100% ET, water application (mm/day) varied from 0.6 to 1.08 in December, 0.98 to 4.77 in January, 3.08 to 6.16 in February, 2.84 to 5.95 in March, and 2.93 to 4.73 in April. For micro sprinkler irrigation at 120% ET, the water application (mm/day) varied from 0.80 to 1.29 in December, 1.18 to 5.73 in January, 3.70 to 7.39 in February, 3.40 to 7.14 in March and 3.52 to 5.67 in April.

The amount of water application per month for different levels of micro sprinkler irrigation and surface irrigation are presented in Table 10.3. For micro sprinkler irrigation at 60% ET, the mean monthly water requirement (mm/day) varied from 2.73 in December to 39.94 in February; and for 80% ET water requirement varied from 3.63 in December to 53.25 in February. Similarly, for 100% ET water requirement varied from 4.54 in December to 66.56 in February; and for 120% ET the monthly water requirement varied from 5.54 in December to 79.87 in February. For surface irrigation the water requirement varied from 50 in December to 100 in February and March. It is also observed that the monthly water requirement was maximum in February and minimum in December for micro sprinkler irrigation.

TABLE 10.2 Daily Water Application (mm/day) Groundnut Crop Under Different Levels of Micro Sprinkler and Surface Irrigation Methods

Date	Pan evaporation mm/day	Water requirement of groundnut, mm/day				
		T1	T2	T3	T4	Surface irrigation T5
6th Dec *	3	25	25	25	25	25
7th Dec to 20th Dec **	–	–	–	–	–	–
21-Dec	4	0.42	0.56	0.70	0.84	25
23-Dec	3.8	0.40	0.53	0.67	0.80	–
25-Dec	4	0.42	0.56	0.70	0.84	–
27-Dec	4.2	0.44	0.59	0.74	0.88	–
29-Dec	3.8	0.40	0.53	0.67	0.80	–

TABLE 10.2 *(Continued)*

Date	Pan evaporation mm/day	Water requirement of groundnut, mm/day				
		T1	T2	T3	T4	Surface irrigation T5
31-Dec	2.2	0.65	0.86	1.08	1.29	25
02-Jan	4	1.18	1.57	1.96	2.35	–
04-Jan	3.8	1.12	1.49	1.86	2.23	–
06-Jan	4	1.18	1.57	1.96	2.35	–
08-Jan	3.5	1.03	1.37	1.72	2.06	–
10-Jan	3	0.88	1.18	1.47	1.76	25
12-Jan	2	0.59	0.78	0.98	1.18	–
14-Jan	4	1.18	1.57	1.96	2.35	–
16-Jan	4.8	1.41	1.88	2.35	2.82	–
18-Jan	4	1.18	1.57	1.96	2.35	–
20-Jan	4	1.18	1.57	1.96	2.35	25
22-Jan	4.2	1.23	1.65	2.06	2.47	–
24-Jan	4.2	1.94	2.59	3.23	3.88	–
26-Jan	4	1.85	2.46	3.08	3.70	–
28-Jan	6.2	2.86	3.82	4.77	5.73	–
30-Jan	4	1.85	2.46	3.08	3.70	25
01-Feb	5.2	2.40	3.20	4.00	4.80	–
03-Feb	5.2	2.40	3.20	4.00	4.80	–
05-Feb	6	2.77	3.70	4.62	5.54	–
07-Feb	5.2	2.40	3.20	4.00	4.80	25
09-Feb	4	1.85	2.46	3.08	3.70	–
11-Feb	8	3.70	4.93	6.16	7.39	–
13-Feb	5.8	2.68	3.57	4.47	5.36	–
15-Feb	5.8	2.68	3.57	4.47	5.36	25
17-Feb	6	2.68	3.57	4.46	5.36	–
19-Feb	6.3	2.81	3.75	4.69	5.62	–
21-Feb	6	2.68	3.57	4.46	5.36	–
23-Feb	5.2	2.32	3.09	3.87	4.64	25
25-Feb	6	2.68	3.57	4.46	5.36	–
27-Feb	7	3.12	4.17	5.21	6.25	–
29-Feb	6.2	2.77	3.69	4.61	5.53	25

Date	Pan evaporation mm/day	Water requirement of groundnut, mm/day				
		T1	T2	T3	T4	Surface irrigation T5
02-Mar	7.4	3.30	4.40	5.50	6.60	–
04-Mar	6.2	2.77	3.69	4.61	5.53	–
06-Mar	6	2.68	3.57	4.46	5.36	–
08-Mar	8	3.57	4.76	5.95	7.14	25
10-Mar	7.9	3.53	4.70	5.88	7.05	–
12-Mar	8.2	2.32	3.10	3.87	4.65	–
14-Mar	6	1.70	2.27	2.84	3.40	25
16-Mar	6	1.70	2.27	2.84	3.40	–
18-Mar	8.4	2.38	3.18	3.97	4.76	–
20-Mar	8	2.27	3.02	3.78	4.54	25
22-Mar	8	2.27	3.02	3.78	4.54	–
24-Mar	6	1.70	2.27	2.84	3.40	–
26-Mar	8.4	2.38	3.18	3.97	4.76	25
28-Mar	8	2.27	3.02	3.78	4.54	
30-Mar	7.2	2.04	2.72	3.40	4.08	
01-Apr	6.2	1.76	2.34	2.93	3.52	25
03-Apr	8.2	2.32	3.10	3.87	4.65	–
05-Apr	8	2.27	3.02	3.78	4.54	–
07-Apr ***	10	2.84	3.78	4.73	5.67	25
Total		134.37	170.83	207.28	243.74	400

* Soil moisture was brought up to field capacity by applying irrigation.

** No irrigations was applied from 7th Dec to 20th Dec to restrict the vegetative growth.

*** After 7th April 2012 irrigation was stopped.

T1 – Water application at 60% of ET using micro sprinkler irrigation.

T2 – Water application at 80% of ET using micro sprinkler irrigation.

T3 – Water application at 100% of ET using micro sprinkler irrigation.

T4 – Water application at 120% of ET using micro sprinkler irrigation.

T5 – Water application at 100% of ET using surface irrigation.

TABLE 10.3 Monthly Depth of Water Applied to Groundnut Under Different Treatments

Month	Monthly irrigation depth, mm/day				
	Micro sprinkler irrigation				Surface irrigation
	T1	T2	T3	T4	T5
6th December	25	25	25	25	25.00
December	2.73	3.63	4.54	5.45	50.00
January	20.64	27.52	34.41	41.29	75.00
February	39.94	53.25	66.56	79.87	100.00
March	36.88	49.17	61.46	73.76	100.00
April 7th	9.19	12.25	15.31	18.37	50.00
Total	134.37	170.83	207.28	243.74	400.00
% saving water over surface	66.41	57.29	48.18	39.07	–

TABLE 10.4 Irrigation Depth Under Micro Sprinkler and Surface Irrigation Methods, For Each Crop Growth Stage

Growth Stage (days)	Irrigation depth, mm/day				
	Micro sprinkler irrigation				Surface irrigation
	T1	T2	T3	T4	T5
Initial (20 days)	26.24	26.65	27.07	27.48	50.00
Crop Development (35 days)	22.13	29.15	36.88	44.26	100.00
Mid-season (40 days)	59.80	79.74	99.67	119.61	150.00
Late (25 days)	26.2	34.93	43.66	52.39	100.00
Total (120 days)	134.37	170.47	207.28	243.74	400.00

Amount of water applied under micro sprinkler irrigation methods based on crop growth stages are presented in Table 10.4. In Mid-season stage, the highest water application was for irrigation level at 120% ET (119.61 mm), and lowest water quantity was applied for 60% ET (59.80 mm); and surface irrigation was significantly higher (150 mm). The water applied at initial stage for 120% ET (27.48

mm) was highest with lowest value in 60% ET (26.24 mm); and surface irrigation was significantly higher (50 mm) compared to micro sprinkler irrigation treatments.

The water requirement of groundnut crop during growth period showed that the water requirement was maximum in February. This may be attributed to the growth stage of the crop, and higher temperature during this month. The water saving over surface irrigation was maximum for 60% ET treatment (66.41%), followed by 80% ET (57.29%), 100% ET (48.18%) and 120% ET (39.07%) treatments. And there was 9.11% water saving among all the micro sprinkler irrigation treatments compared to each of the individual micro sprinkler irrigation treatments among themselves. From these results, it may be concluded that there is a substantial amount of water saving under micro sprinkler irrigation system as compared to surface irrigation. This may be attributed to the fact that maximum amount of water applied is stored in the root zone in case of micro sprinkler irrigation treatments and the deep percolation losses are eliminated. Further it can be observed that the water loss in surface irrigation is more because of high deep percolation losses. Under surface irrigation, water front does not spread instantaneously over the entire area. Invariably it takes certain time to spread the water and to build up certain water depth. During this time, certain quantity of water might have percolated below root zone. These results are in agreement with the earlier findings of Krishnamurthi et al. [26] and Vijayalakshmi [71].

In this study, the crop water requirements are determined based on soil, plant and climatological factors prevailing during the experimental period at the study site. Therefore, one has to be careful in applying these results for any other location. However, the results may be used as guidelines and not as exact values. The contributing factors to the water requirement namely soil, plant and climatological factors are location-specific and therefore, if similar conditions exist, one can use these results with appropriate allowance so that crop growth and yield are not adversely affected. However, there could be variations in water requirement of groundnut with season due to change in climatological factors.

10.4.2 GROWTH PARAMETERS OF GROUNDNUT UNDER DIFFERENT LEVELS OF IRRIGATION

The effects of different levels of micro sprinkler irrigation on vegetative growth parameters of groundnut were compared with surface irrigation. The plant height, number of primary branches and leaf area index are the most important parameters. The crop under micro sprinkler irrigation had superior growth compared to surface irrigation (Tables 10.5–10.7). Better plant growth led to higher yield in micro sprinkler irrigation treatments as against surface irrigation. The increase in plant height, number of primary branches and leaf area in micro sprinkler irrigation over surface irrigation may be due to frequent application of irrigation water at lower rates, resulting in even distribution of soil moisture in the root zone. Therefore under micro sprinkler irrigated plots, soil moisture was maintained fairly close to the field capacity throughout the crop season, which resulted in high level of plant water use. This shows that adequate supply of soil moisture to groundnut resulted in the development of efficient photosynthetic system and also due to increasing the available nutrient status of the soil due to mineralization and transformation of soil nutrients. The results are in accordance with the findings of Kale et al. [23] and Manjunatha et al. [29].

10.4.2.1 PLANT HEIGHT

The Table 10.5 shows effects of different levels of micro sprinkler irrigation and surface irrigation on plant height at 30, 60, 90 and 120 DAS. The results indicated that at 30 DAS, plants receiving water at 120% ET recorded significantly maximum height (5.50 cm) over surface irrigation (4.53 cm) and 60% ET (4.03 cm), but it was on par with 100% ET (5.05 cm) and 80% ET (5.03 cm). At 60 DAS, plants receiving water at 120% ET recorded maximum height (13.9 cm) that was significantly higher over control treatment (11.68 cm), 100% ET (11.4 cm) and 80% ET (10.78 cm). The minimum height was found in 60% ET (10.6 cm).

TABLE 10.5 Effects of Irrigation Methods and Irrigation Levels on Plant Height (cm) of Groundnut

Treatments	Plant height, cm			
	Days after sowing, DAS			
	30	60	90	120
T1	4.03	10.60	14.22	18.73
T2	5.03	10.78	14.33	18.85
T3	5.05	11.40	15.03	19.75
T4	5.50	13.90	18.65	24.05
T5	4.53	11.68	15.45	20.20
SEM ±	0.34	0.49	0.73	0.93
CD (0.05)	1.04	1.51	2.26	2.86
Mean	4.77	11.67	15.54	20.32

Similar trends were noticed at 90 DAS. The plants receiving water at 120% ET recorded maximum height (18.65 cm) that was significantly higher over control treatment (15.45 cm), 100% ET (15.03 cm) and 80% ET (14.33 cm). The lowest value was found in case of 60% ET (14.22 cm). Finally at 120 days after sowing, plants receiving water at 120% ET recorded the maximum height (24.05 cm) that was significantly higher over control treatment (20.20 cm), 100% ET (19.75 cm) and 80% ET (18.85 cm). The minimum height was found at 60% ET (18.73 cm).

The effect of irrigation methods and irrigation levels on plant height of groundnut at different intervals is shown Fig. 10.8.

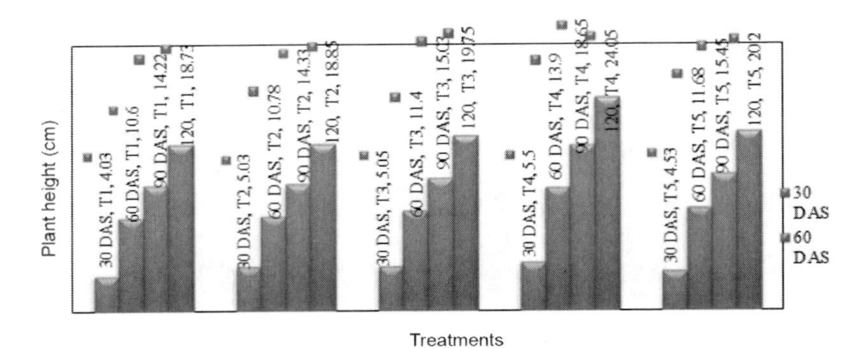

FIGURE 10.8 Effects of irrigation methods and irrigation levels on plant height of groundnut.

TABLE 10.6 Effects of Irrigation Methods and Irrigation Levels on Number of Primary Branches of Groundnut At Different Intervals

Treatments	Number of primary branches of groundnut			
	Days after sowing, DAS			
	30	60	90	120
T1	3.85	4.50	5.50	6.50
T2	4.95	5.75	6.90	7.75
T3	5.40	6.50	7.25	8.75
T4	4.25	5.50	6.65	7.50
T5	4.00	5.25	5.75	7.25
SEM ±	0.05	0.29	0.18	0.25
CD (0.05)	0.17	0.88	0.54	0.76
Mean	4.49	5.50	6.41	7.50

10.4.2.2 NUMBER OF PRIMARY BRANCHES

The Table 10.6 indicates number of primary branches during 30, 60, 90 and 120 DAS. The number of primary branches at different intervals was different significantly among irrigation levels. Among the treatments, plants receiving water at 100% ET recorded significantly maximum number of primary branches (5.4) that was significant over the control treatment (4), followed by 80% ET (4.95), 120% ET (4.25) and it was significantly lower in 60% ET (3.85) at 30 DAS.

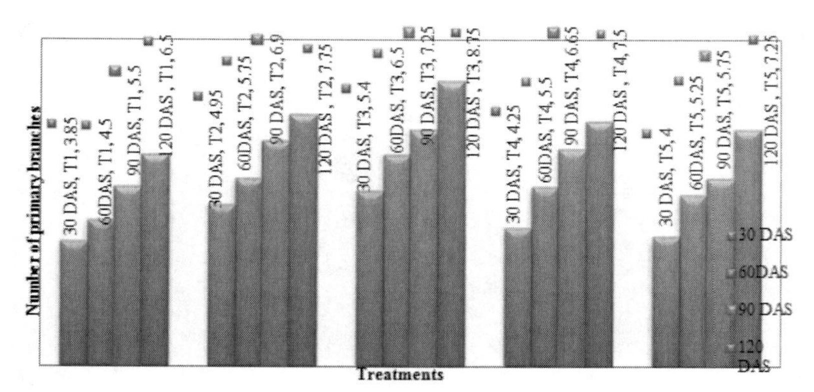

FIGURE 10.9 Effects of irrigation methods and irrigation levels on number of primary branches of groundnut at different intervals.

Similarly at 60 DAS, plants receiving water at 100% ET recorded significantly maximum number of primary branches (6.5) that was significant over control treatment (5.25), followed by 80% ET (5.75), 120% ET (5.5), and it was significantly lower in 60% ET (4.5). Similar trends were noticed at 90 DAS, plants receiving water at 100% ET recorded significantly maximum number of primary branches (7.25) that was significant over control treatment (5.75), followed by 80% ET (6.9), 120% ET (6.65) and it was significantly lower in 60% ET (5.5).

Similar trends were observed at 120 DAS, plants receiving water at 100% ET recorded significantly maximum number of primary branches (8.75) that was significant over control treatment (7.25), followed by 80% ET (7.75), 120% ET (7.5) and it was significantly lower in 60% ET (6.5). The effects of irrigation methods and irrigation levels on number of primary branches of groundnut at different intervals are shown in Fig. 10.9.

10.4.2.3 LEAF AREA INDEX

The Table 10.7 indicates effects of irrigation methods and different levels of irrigation on LAI of groundnut at 30, 60, 90 and 120 DAS. The LAI at different intervals differed significantly due to irrigation levels. The results indicated that at 30 DAS, plants receiving water at 100% ET recorded significantly maximum LAI (0.41) that was significant over control treatment (0.30), followed by 80% ET (0.37), 120% ET (0.33) and it was significantly lower in 60% ET (0.28) which is on par with control treatment. Similarly at 60 DAS, plants receiving water at 100% ET recorded significantly maximum LAI (1.44) that was significant with control treatment (1.26) followed by 80% ET (1.36), 120% ET (1.33) and it was significantly lower in 60% ET (1.23).

TABLE 10.7 Effects of Irrigation Methods and Irrigation Levels on Leaf Area Index of Groundnut At Different Intervals

Treatments	Leaf area index, LAI			
	At days after sowing, DAS			
	30	60	90	120
T1	0.28	1.23	1.59	1.11
T2	0.37	1.36	1.72	1.24
T3	0.41	1.44	1.78	1.36
T4	0.33	1.33	1.66	1.19
T5	0.30	1.26	1.63	1.14
SEM ±	0.01	0.01	0.01	0.01
CD (0.05)	0.03	0.02	0.02	0.02
Mean	0.34	1.32	1.68	1.21

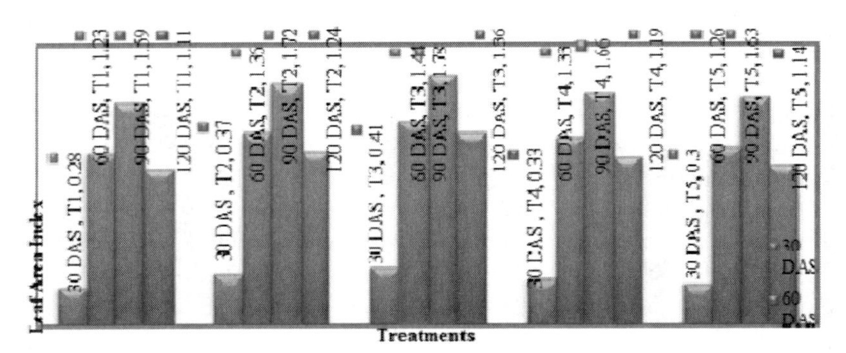

FIGURE 10.10 Effects of irrigation methods and irrigation levels on leaf area index (LAI) of groundnut at different intervals.

Similar trends were noticed at 90 DAS, plants receiving water at 100% ET recorded significantly maximum LAI (1.78) that was significant over control treatment (1.63) followed by 80% ET (1.72), 120% ET (1.66) and it was significantly lower in 60% ET (1.59). Similarly the results indicated at 120 DAS, plants receiving water at 100% ET recorded significantly maximum LAI (1.36) that was significant over control treatment (1.14) followed by 80% ET (1.24), 120% ET (1.19) and it was significantly lower in 60% ET (1.11). The effects of irrigation methods and irrigation levels on leaf area index of groundnut at different intervals are shown in Fig. 10.10. A view of the groundnut crop at different intervals is shown in Fig. 10.11.

FIGURE 10.11 View of the groundnut crop at different growth stages.

10.4.3 PERFORMANCE PARAMETERS OF GROUNDNUT

10.4.3.1 NUMBER OF PODS PER PLANT

The Table 10.8 shows effects of irrigation methods and irrigation levels on number of pods per plant. The maximum number of pods per plant was for 100% ET (28.25) that was significantly higher than the control treatment (22.25), followed by 80% ET (26.50), 120% ET (24.50) and it was significantly lower in 60% ET (20.75). The effects of irrigation methods and irrigation levels on number of pods per plant at different intervals are shown in Fig. 10.12.

TABLE 10.8 Effects of irrigation methods and irrigation levels on yield, number of pods per plant and weight of pods.

Treatments	No. of pods per plant	Weight of pods, gm	Groundnut grain yield	
			kg/plot	q/ha
T1	20.75	12.81	15.50	19.13
T2	26.50	18.02	17.50	21.60
T3	28.25	21.24	19.33	23.86
T4	24.50	16.16	16.28	20.09
T5	22.25	14.53	16.00	19.75
SEM ±	0.46	0.49	0.30	0.37
CD (0.05)	1.41	1.52	0.93	1.15
Mean	24.45	16.55	16.92	20.89

1.00 quintal, q = 100 kg.

10.4.3.2 WEIGHT OF PODS

The Table 10.8 shows the effects of different irrigation treatments on average pod weight. The maximum average pod weight of (21.24 gm) was in 100% ET, which was significantly higher than control treatment (14.53 gm), 80% ET (18.02 gm), 120% ET (16.16 gm) and it was significantly lower in case of 60% ET (12.81 gm) that was on par with control treatment. The effects of irrigation methods and irrigation levels on weight of pods at different intervals are given in Fig. 10.12.

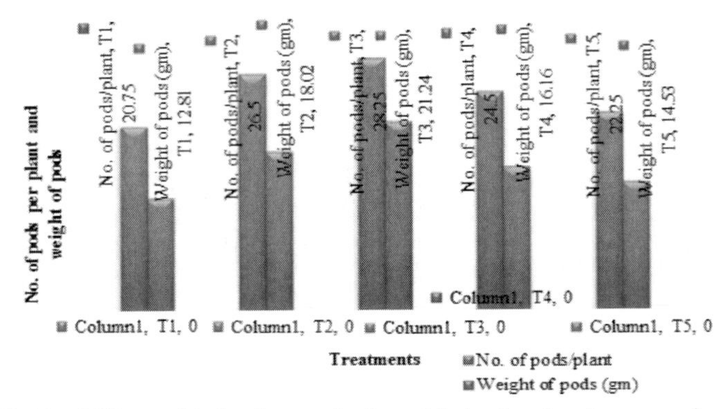

FIGURE 10.12 Effects of irrigation methods and irrigation levels on number of pods per plant and weight of pods at different intervals.

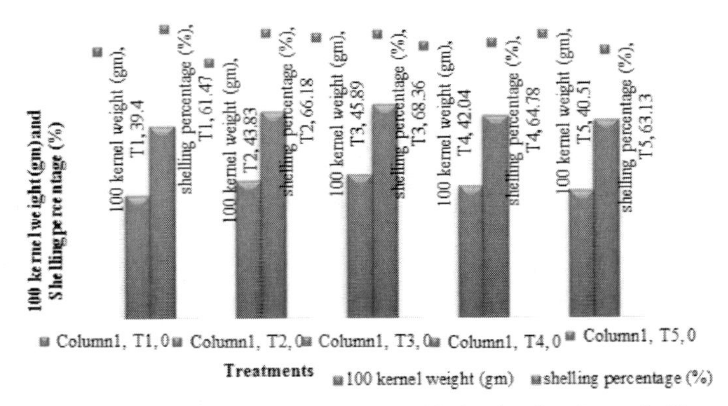

FIGURE 10.13 Effects of irrigation methods and irrigation levels on shelling percentage and weight of 100 kernels.

10.4.3.3 WEIGHT OF 100 KERNELS

The weight of 100 kernels weight (g) is presented in Table 10.9. It is observed that maximum 100- kernel weight was in 100% ET (45.89) that was significantly over control treatment (40.51) followed by 80% ET (43.83), 120% ET (42.04) and was significantly lower in 60% ET (39.40). The effect of irrigation methods and irrigation levels on 100-kernel weight is shown in Fig. 10.13.

10.4.3.4 SHELLING PERCENTAGE

The data on shelling percentage are presented in Table 10.9. The maximum average shelling percentage was in 100% ET (68.36%) that was significantly higher than control treatment (63.13%), 120% ET (64.78%) and 60% ET (61.47%), but it was on par with 80% ET (66.18%). The effects of irrigation methods and irrigation levels on shelling percentage are shown in Fig. 10.13.

10.4.3.5 YIELD PER PLOT

The yield (kg per plot and (100 kg = q)/ha) data under different irrigation treatments are presented in Table 10.8. The plants receiving water at 100% ET recorded maximum yield (19.33 kg) that was significantly maximum over control treatment (16 kg) followed by 80% ET (17.50 kg), 120% ET (16.28 kg) and the lowest yield was noticed in 60% ET (15.50 kg) that was on par with control treatment. A view of groundnut pods is depicted in Fig. 10.14.

FIGURE 10.14 View of groundnut pods under different irrigation treatments.

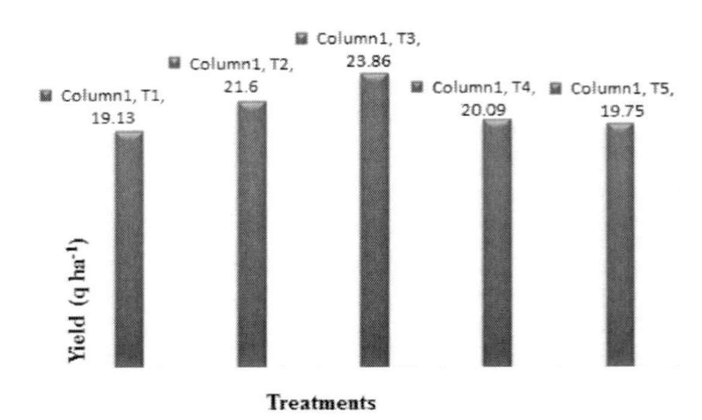

FIGURE 10.15 Effects of irrigation methods and irrigation levels on yield of groundnut.

TABLE 10.9 Effects of Irrigation Methods and Irrigation Levels on 100-Kernel Weight (g) and Shelling Percentage (%).

Treatments	Weight of 100 kernels, g	Shelling percentage (%)
T1	39.40	61.47
T2	43.83	66.18
T3	45.89	68.36
T4	42.04	64.78
T5	40.51	63.13
SEM ±	0.21	0.81
CD (0.05)	0.64	2.50
Mean	42.33	64.78

10.4.3.6 YIELD PER HECTARE

The Table 10.8 shows effects of irrigation methods and levels of micro sprinkler irrigation on total marketable yield (quintals per ha). Significant differences were noticed in yield due to irrigation methods as well as micro sprinkler irrigation levels. The plants receiving water at 100% ET recorded significantly maximum yield (23.86 q.ha^{-1}) over control treatment (19.75 q.ha^{-1}) followed by 80% ET (21.60 q.ha^{-1}), 120% ET (20.09 q.ha^{-1}) and the lowest yield was recorded in case of 60% ET (19.13 q.ha^{-1}) that was on par with control treatment. The effects of irrigation methods and irrigation levels on yield of groundnut are shown in Fig. 10.15.

The groundnut crop performed well in terms of yield and yield contributing factors under micro sprinkler irrigation compared to surface irrigation (Tables 10.7 and 10.8). The performance of crop in terms of number of pods, weight of pods,

100-kernel weight, shelling percentage and yield was superior in 100% ET compared to 60% ET and surface irrigation, which have performed poorly.

The number of pods per plant was higher in 100% ET (Table 10.7). The results are in agreement with the findings of Krishnamurthi [26]. Pod yield of groundnut depends on the number of branches and pods per plant and on LAI that increases production of dry matter [47]. The overall trend for all growth parameters was superior in 100% ET. This may be attributed due to the frequent and consistent application of water, which provided better soil moisture regime in the crop root zone and the better development of kernels at optimum soil moisture conditions, as evidenced by the increase in 100-kernel weight and shelling percentage. The pod weight, 100-kernel weight and shelling percentage were superior in micro sprinkler irrigation treatments than control treatment. The results are in agreement with the findings of Shinde et al. [59] and Varshney [70].

The crop yield decides the superiority of a crop production system. Among all the micro sprinkler irrigation treatments, 100% ET level was superior in yield (Table 10.8). The best yield (kg per plot) was achieved in 100% ET level (19.33 $kg.plot^{-1}$) that was higher when compared to surface irrigation treatment (16 $kg.plot^{-1}$) and 60% ET level (15.50 $kg.plot^{-1}$). Krishnamurthi et al. [26] indicated that the higher yield can be mainly due to high frequency of irrigation, which in turn maintained the soil moisture content in the active root zone at adequate level throughout the crop period. Micro sprinkler irrigation is very important innovation in agricultural production. Plant growth is restricted by water deficits caused by excessive transpiration during hot afternoon. A change of a few degrees leaf temperature can make a major difference in the biological functions of a plant. Small amount of water, intermittently sprinkled, cool the air and plants, reduce the transpiration rate so that a plant, which would wilt on a hot afternoon, can continue to function normally and improves the produce quality and yield. The results confirm the findings of Hubbs [17].

The yield is the ultimate factor, which decides the superiority of any treatment in terms of not only its biological returns but also its economic returns. In the present investigations, highest groundnut yield was recorded in 100% ET level (23.86 $q.ha^{-1}$) followed by 80% ET (21.60 $q.ha^{-1}$) under micro sprinkler irrigation. The yield levels achieved in100% ET were higher compared to surface irrigation (19.75 $q.ha^{-1}$) and 60% ET (19.13 $q.ha^{-1}$). The higher yield was obtained due to the uniform and frequent application of water at right time and right amount; and these are two important factors for higher yields in the plots under micro sprinkler system compared to yields obtained in the plots irrigated through surface method. This advantage is not only in terms of superior yields but also in terms of water saved. These results corroborate the findings of Manjunatha et al. [28].

10.4.4 QUALITY PARAMETERS OF GROUNDNUT CROP

The effects of different levels of micro sprinkler irrigation over surface irrigation on quality parameter of groundnut were analyzed. Any crop production system will be appreciated when it is not only quantitatively superior but also qualitatively promising.

TABLE 10.10 Effects of Irrigation Methods on Oil Content (%)

Treatments	Oil content, %
T1	39.85
T2	43.20
T3	43.38
T4	42.88
T5	42.05
SEM ±	0.40
CD (0.05)	1.23
Mean	42.27

10.4.4.1 OIL CONTENT

The oil content of groundnut as influenced by irrigation methods and different irrigation levels is presented in Table 10.10. The highest oil content was observed in 100% ET (43.38%) which was significantly higher than the control treatment (42.05%) and 60% ET (39.85%). But it was on par with 80% ET (43.20%) and 120% ET (42.88%). The significant increase in oil content in kernels in frequently irrigated treatments may be attributed to better availability and uptake of phosphorus and may be due to increased soil moisture availability. The present results are in accordance with the findings of Mehrotra et al. [31].

10.4.5 IRRIGATION EFFICIENCIES

This section discusses effects of irrigation methods and irrigation levels on application efficiency, distribution efficiency and WUE.

TABLE 10.11 Effects of Irrigation Methods and Different Levels of Irrigation on Irrigation Efficiencies

Treatments	Yield	Water applied	WUE	Application efficiency
	kg/ha	cm	kg/ha-cm	%
T_1	1913	13.43	142.44	82.80
T_2	2160	17.08	126.46	82.05
T_3	2386	20.72	115.15	81.87
T_4	2009	24.37	82.43	80.90
T_5	1975	40.00	49.37	72.70

The irrigation efficiencies are very important factors in deciding the efficiency of micro sprinkler systems and status of availability of water to plants. In the present study the application efficiency was higher in micro sprinkler irrigation treatments than that of surface irrigation (Table 10.11). The higher application efficiency in micro sprinkler irrigation compared to surface irrigation system was due to controlled application of quantity of water to replenish the crop root zone. In micro sprinkler irrigation, water is applied as per plant water requirement over the entire area at a rate less than the infiltration rate. In general, water application efficiency decreases as the amount of water applied in each irrigation increases. This eliminates the deep percolation losses, as water is conveyed through pipes, which results in higher application efficiency. The results are in agreement with the findings of Koumanov et al. (1997).

The lowest WUE in surface irrigation (0.49 $kg.m^{-3}$) may be due to higher irrigation water use with comparatively less yield. The above discussion suggests that higher groundnut yield is possible by adopting micro sprinkler irrigation scheduled at 60 and 80% of ET. The results are in agreement with findings of Krishnamurthi et al. [26].

10.4.5.1 APPLICATION EFFICIENCY

Application efficiency shows: How well the irrigation water is applied to the field?, and the percentage of water applied stored in the crop root zone as required and available for plant use. The application efficiency for different treatments is given in Table 10.11. It is observed that application efficiency ranged from 82.80% in 60% ET to 80.90% in 120% ET for micro sprinkler treatments; and it was 72.70% for surface irrigation treatment. Therefore, the application efficiencies were higher in all the micro sprinkler treatments compared with the surface irrigation treatment.

10.4.5.2 WATER USE EFFICIENCY

The water use efficiency (WUE) for groundnut crop is influenced by irrigation methods and irrigation levels as shown in Table 10.11. The WUE varied from 142.44 in 60% ET to 82.43 kg/ha-cm in 120% ET under micro sprinkler irrigation treatments as compared to 49.37 $Kg.ha^{-1}.cm^{-1}$ in surface irrigation. Among different micro sprinkler irrigation treatments, plant receiving water at 120% ET recorded the lowest WUE (82.43 $Kg.ha^{-1}.cm^{-1}$) that is on increasing trend for 100% ET (115.15 $Kg. ha^{-1}.cm^{-1}$), 80% ET (126.46 $Kg.ha^{-1}.cm^{-1}$) and 60% ET (142.44 $Kg.ha^{-1}.cm^{-1}$). The WUE in 60% ET level was significantly promising over all other treatments.

TABLE 10.12 Irrigation Capacity (Duty) of 1 m^3 of Water and Delta of Water For Different Treatments for the Crop Period

Treatments	Water applied in	Water applied	Duty	Delta
	liters/plot	m^3/ha	ha/m^3	cm
T1	10883.89	1343.69	7.44×10^{-4}	13.44
T2	13808.23	1704.72	5.87×10^{-4}	17.05
T3	16789.28	2072.75	4.82×10^{-4}	20.73
T4	19742.78	2437.38	4.10×10^{-4}	24.37
T5	32400.00	4000.00	2.50×10^{-4}	40.00

10.4.5.3 IRRIGATION CAPACITY (DUTY) AND DELTA

The capacity of unit quantity of water to irrigate a crop is an important factor for any irrigation system. Table 10.12 presents the capacity of one meter^3 of water to irrigate groundnut crop. The irrigation capacity was lowest (2.50×10^{-4} $ha.m^{-3}$) for surface irrigation. The highest irrigation capacity of 7.44×10^{-4} $ha.m^{-3}$ was obtained in 60% ET.

Delta is the depth of irrigation (cm) required during the crop period. Delta of water for different treatments is presented in Table 10.12. The delta was highest (4.94 cm) for surface irrigation, It was lowest (1.66 cm) in 60% ET and was highest (3.01 cm) in 120% ET.

10.4.6 ECONOMICS

For determining the benefit-cost ratio, the fixed cost, operating cost and net returns were calculated for micro sprinkler irrigation and surface irrigation. All costs were expressed in Rs/ha. The sample calculations for cost economics are given in Appendix III.

The performance of micro sprinkler irrigation system can be valued both in terms of biological and economical returns. So far the superiority of micro sprinkler irrigation at 100% ET level in terms of water economy and better crop response have already been discussed in this chapter. However, it is important that a technically feasible proposition should be financially sound for its successful adoption. One of the main constraints under micro sprinkler irrigation is its high initial investment in the form of mains, sub mains, filter, tank and accessories to design the unit. The economic analysis of groundnut crop with micro sprinkler and surface irrigation was determined by considering fixed cost, cost of cultivation, water used and yields obtained.

The initial cost of installing the micro sprinkler irrigation system for field crops are high but over a period of time the cost is recovered and the benefits derived are higher than surface irrigation. Based on Tables 10.13 and 10.14, the micro sprinkler irrigation system at 100% ET and 80% ET showed higher net returns compared to other micro sprinkler irrigation treatments and surface irrigation. The net returns in 100% ET was higher compared with surface irrigation. Similar trend was also exhibited in terms of BCR, which was highest (3.42) in 100% ET and was lowest in 60% ET (2.75). The results agree with the findings of Manjunatha et al. [28].

10.4.6.1 NET RETURNS AND BENEFIT: COST RATIO

The net returns and benefit-cost ratio for irrigation methods and irrigation levels are presented in Table 10.13. It is seen that among all the micro sprinkler irrigation treatments the highest net return of 81,079 Rs.ha^{-1} was obtained in 100% ET, followed by 80% ET (70,231 Rs.ha^{-1}), control treatment (66,244 Rs.ha^{-1}) and 120% ET (62,983 Rs.ha^{-1}). The lowest net return was in 60% ET (58,375 Rs.ha^{-1}).

It was also observed that among all the micro sprinkler irrigation treatments the lowest BCR was 2.75 in 60% ET and the highest BCR was 3.42 in 100% ET, followed by control treatment (3.32), 80% ET (3.10) and 120% ET treatment (2.88).

TABLE 10.13 Economics of Micro Sprinkler and Surface Irrigation Methods in Groundnut

Treatments	**Crop yield**	**Gross returns**	**Total cost of cultivation**	**Net returns**	**B:C ratio**
	q/ha	**Rs//ha**			–
T1	19.13	91,824	33,449	58,375	2.75
T2	21.60	1,03,680	33,449	70,231	3.10
T3	23.86	1,14,528	33,449	81,079	3.42
T4	20.09	96,432	33,449	62,983	2.88
T5	19.75	94,800	28,556	66,244	3.32

One quintal, q = 100 kg.

TABLE 10.14 Projected Additional Return From Saved Water in Micro Sprinkler Irrigation in Groundnut Crop

Treatments	Water saved over surface irrigation	Yield	Net returns	Additional yield with saved water	Total yield = (3+5)	Yield increase over surface irrigation	Increase in net returns with saved water	Projected net returns with micro sprinkler irrigation from water used in surface irrigation = (4+8)
	%	q/ha	Rs /ha	q /ha	q /ha	q /ha	Rs /ha	Rs /ha
(1)	(2)	(3)	(4)	(5)	(6)	(7)	(8)	(9)
T1	66.41	19.13	58,375	37.82	56.95	37.20	1,15,412	1,73,787
T2	57.29	21.60	70,231	28.97	50.57	30.82	94,206	1,64,437
T3	48.18	23.86	81,079	22.18	46.04	26.29	75,384	1,56,463
T4	39.07	20.09	62,983	12.88	32.97	13.22	40,386	1,03,369
T5	-	19.75	66,244	-	-	-	-	66,244

One quintal, q = 100 kg.

10.4.6.2 PROJECTED ADDITIONAL RETURNS FROM SAVED WATER

Availability of water is a main constraint rather than land. Efforts are being made to judiciously use the available water using the concept of deficit irrigation. So, if water can be saved through its judicious use and more efficient irrigation methods, the saved water can be used to irrigate additional area, which will result in additional yield and overall net returns.

Therefore, an attempt was made to calculate additional benefits that can be obtained, if the water needed to raise one ha of groundnut crop through surface irrigation is used through micro sprinkler irrigation. It was assumed that the land is not a constraint and the crop response to other inputs remains constant.

In Table 10.14, it can be observed that by using the saved water, highest additional yield (37.82 q.ha^{-1}) was obtained in 60% ET, closely followed by 80% ET (28.97 q.ha^{-1}). When the water required to irrigate one ha groundnut crop by surface irrigation was completely used by micro sprinkler irrigation, the net returns were 1,73,787 Rs.ha^{-1} in 60% ET and closely followed by 80% ET (1,64,437 Rs.ha^{-1}).

The lowest net returns were observed in 120% ET (1,03,369 Rs.ha^{-1}) and 66,244 Rs.ha^{-1}) for surface irrigation.

Practical applications of the findings are:

1. Water requirements for groundnut under micro sprinkler irrigation were 1343.69 m^3.ha^{-1} at 60% ET, 1704.72 m^3.ha^{-1} in 80% ET and 2072.75 m^3.ha^{-1}100% ET. The maximum was 4000 m^3.ha^{-1} in surface irrigation.
2. Yield in 100% ET (23.86 q.ha^{-1}) was maximum, followed by 80% ET (21.60 q.ha^{-1}).
3. Water use efficiency was higher in 60% ET (142.44 kg.ha^{-1}.cm^{-1}) followed by 80% ET (126.46 kg.ha^{-1}.cm^{-1}).

10.5 CONCLUSIONS

A field experiment was conducted from December 2011 to April 2012 in farmer's field at Yeragera village, Raichur, with a view to work out water requirement of groundnut crop under different micro sprinkler irrigation levels and surface irrigation. The comparison was made in terms of growth, yield and quality parameters between surface and different levels of micro sprinkler irrigation. Further the experiment also aimed to compare various irrigation efficiencies and economics of different levels of micro sprinkler irrigation versus surface irrigation. The irrigation treatments were 60, 80, 100, 120% ET using micro sprinkler and surface irrigation as control. The crop was irrigated once in two days under micro sprinkler irrigation and in surface irrigation the crop was irrigated at 50% depletion of available soil moisture.

The water requirement of groundnut crop under micro sprinkler irrigation was low at initial stage of the crop and it gradually increased in crop development stage, attained peak in mid stage of the crop, and it gradually decreased in late stage of the crop. The net amount of water applied were 134.37 mm for 60% ET, 170.47 mm for 80% ET, 207.28 mm for 100% ET and 243.74 mm for 120% ET, respectively. Under surface irrigation, 400 mm of water was applied. The water saved over surface irrigation was 66.41% in 60% ET, 57.29% in 80% ET, 48.18% in 100% ET and 39.07% in 120% ET.

The plant height was superior in 120% ET. Number primary branches and leaf area index plant receiving water at 100% ET and 80% ET were significantly superior.

Groundnut crop, in all micro sprinkler irrigation treatments (T2, T3, T4) except 60% ET performed very well in terms of number of pods, pod weight, 100-kernel weight and shelling percentage and marketable yield. The highest yield of 2386 kg. ha^{-1} was obtained in100% ET which was closely followed by 80% ET (2160 kg. ha^{-1}). The yield in surface irrigation (1970 kg.ha^{-1}) and 60% ET (1913 kg.ha^{-1}) were significantly lower.

Among the different micro sprinkler irrigation treatments, the treatment 100% ET (43.38%) gave significantly maximum oil content which was closely followed by 80% ET (43.20%). The minimum oil content was in 60% ET (39.85%) and surface irrigation (42.05%).

The application efficiency was higher with micro sprinkler irrigation treatments (T1 to T4) as compared to surface irrigation. Among the micro sprinkler irrigation treatments, water applied at 60% ET was recorded the highest application efficiency followed by 80% ET.

The water use efficiency was highest in 60% ET (142.44 $Kg.ha^{-1}.cm^{-1}$), closely followed by 80% ET (126.46 $Kg.ha^{-1}.cm^{-1}$). The lowest WUE was noticed in surface irrigation (49.37 $Kg.ha^{-1}.cm^{-1}$).

All the micro sprinkler irrigation treatments recorded higher benefit – cost ratio (2.75 to 3.42). In surface irrigation, benefit: cost ratio was 3.32. The highest net return of (81,097 $Rs.ha^{-1}$) was in 100% ET closely followed by 80% ET (70,231 $Rs.ha^{-1}$). The lowest net returns were noticed in surface irrigation (66,244 $Rs.ha^{-1}$) and 60% ET (58,375 $Rs.ha^{-1}$).

The estimation of projected additional returns also showed encouraging trends. The water saved by different micro sprinkler irrigation levels varied from 39.07% for 120% ET to 66.4% in 60% ET. By using the saved water, an additional yield of 37.82 $q.ha^{-1}$ was produced in 60% ET followed by 80% ET (28.97 $q.ha^{-1}$). In conclusion, the projected net return were highest in 60% ET (1,73,787 $Rs.ha^{-1}$).

10.6 SUMMARY

During December 2011 to April 2012, field experiment was conducted at Raichur – India under semiarid climatic conditions. The performance of micro sprinkler irrigation for groundnut at 60, 80, 100 and 120% of ET was compared with surface irrigation. The experiment was laid out with groundnut variety R-2001–2 in a randomized block design with five treatments replicated four times with plot size of 81 m^2. The soil of experimental field was sandy loam. The study revealed that maximum water use efficiency was (142.44 $kg.ha^{-1}.cm^{-1}$) at 60% of ET followed by (126.46 $kg.ha^{-1}.cm^{-1}$) at 80% of ET and minimum of (49.37 $kg\ ha^{-1}\ cm^{-1}$) with surface irrigation.

Irrigating at 100% ET (23.86 100 $kg.ha^{-1}$) recorded the highest yield with the water usage of 207.2. The water saved in micro sprinkler over surface irrigation was 66.41%, 57.29%, 48.18% and 39.07% for 60, 80, 100 and 120% of ET. The uniformity coefficient of the system was 87.69% at 1.4 kg/cm^2 pressure. The maximum application efficiency was for 60% ET (82.80%) and 72% in surface irrigation, respectively. The treatment T_3 (100% of ET) was best in terms of yield and growth parameters in sandy soils under semiarid conditions.

KEYWORDS

- **agricultural statistics**
- **application efficiency**
- **cabbage**
- **ET**
- **FAO**
- **furrow irrigation**
- **Government of India**
- **groundnut**
- **growth parameters**
- **India**
- **Jain irrigation**
- **Maharashtra**
- **micro sprinkler**
- **sandy soil**
- **semi arid**
- **surface irrigation**
- **uniformity coefficient**
- **water management**
- **water use efficiency, WUE**
- **wetting pattern**
- **yield parameters**

REFERENCES

1. Agarwal, M. G., Khanna, S. S. (1983). *Efficient soil and water management.* Haryana Agricultural University, Hisar Bull., 118.
2. Al-Jamal, M. S., Ball, S., Sammis, T. W. (2001). Comparison of sprinkler, trickle and furrow irrigation efficiencies for onion production. *Agricultural Water Management,* 46:253–266.
3. Anonymous, (2008). *Jain irrigation systems manual*, Jalgaon, Maharashtra.
4. Anonymous, (2009). *Agricultural Statistics at a glance*. Agricultural Statistics Division, Directorate of Economics and Statistics. Department of Agriculture and Co-operation, Ministry of Agriculture, Government of India, New Delhi.
5. Anonymous, (2011). *Jain irrigation systems manual*. Jalgaon, Maharashtra.
6. Anonymous, (2011). Ministry of Agriculture, Government of India. *<Indiastat.com>*.
7. Arulkar, K. P., Sarode, S. C., Bhuyar, R. C. (2008). Wetting pattern and salt distribution in drip and micro sprinkler irrigation. *Agric. Sci. Digest,* 28(2):124–126.

8. Chauhan, H. S., Prabhat Srivastava. (1995). Cabbage growth under different irrigation methods. *In Proc. 5th International Micro Irrigation Congress, Orlando–Florida,* April 2–6, pp. 893–897.
9. Christiansen, J. E. (1942). *Irrigation by sprinkling*. University of California, College of Agriculture. Agric. *Expt. Station Bull.,* 670.
10. Claudia, G. da F. Santos, Vera, L. A. de Lima, Jose de A. de Matos, Adrianus C. van Haandel, Carlos, A. V. Azevedo, (2003). Effect of wastewater use on the microsprinkler discharge. *Revista Brasileira de Engenharia Agrícolae Ambiental,* 7(3):577–580.
11. Dahiwalkar, S. D., Andhale, R. P., Powar, A. S., Berad, S. M., (2002). Response of groundnut to various micro irrigation systems. *J. Maharashtra Agric. Univ*., 29:99–100.
12. Dilip, Yewalekar and Mandakini Sonaware, (1996). Micro irrigation systems, design principles and considerations. *In Proc. Modern Irrigation Techniques, June 26–27, Bangalore,* pp. 39–44.
13. Firake, N. N., Shinde. S. H. (2000). Evaluation of different micro irrigation systems for summer groundnut. *Journal of Maharashtra Agricultural Universities,* 25(2):204–205.
14. Gadge, S. B., Kumar Virendra, Kothari Mahesh. (2011). Optimal cropping pattern for adoption of microirrigation methods in canal command area–A case study. *Journal of Agricultural Engineering,* 48:1–11.
15. Megh R. Goyal, (Ed.), (2013). *Management of Drip/Trickle or Micro Irrigation.* Oakville–ON, Canada: Apple Academic Press Inc., pp. 1–408.
16. Megh R. Goyal, (Ed.), (2015). *Research Advances in Sustainable Micro Irrigation, volumes 1–10.* Oakville–ON, Canada: Apple Academic Press Inc.
17. Hubbs, E. H. (1973). Crop cooling with sprinklers. *CAE,* 15(1):6–8.
18. Isiguzo, E. A. (2010). Performance evaluation of portable sprinkler irrigation system in Ilorin, Nigeria. *Indian Journal of Science and Technology,* 3(7):853–857.
19. Jackson, M. L. (1967). *Soil chemical analysis*. Prentice Hall of India Pvt. Ltd., New Delhi.
20. Jadhav S. N., Awari, H. W., Gore, A. K. (1997). Effect of irrigation frequency and irrigation level on yield of chili under micro sprinkler irrigation. *Proc. All India Seminar on Modern Irrigation Techniques, Bangalore, June*, II:105.
21. Kadam, S. A., Kadam, U. S., Gorantiwar S. D., Patil, S. M. (2006). Uniformity in micro sprinkler irrigation system. *Agricultural Engineering today*, 30(3).
22. Kadam, U. S., Kadam, S. A., Shete, G. K. (2006). Effect of different irrigation frequencies on yield of garlic under micro sprinkler irrigation system. *J. Maharashtra Agric. Univ.,* 31(3):295–297.
23. Kale, R. D., Mallesh, B. C., Bano, K., Bagyaraj, D. J. (1992). Influence of vermin-composting application on the available micro nutrients and selected microbial populations in paddy field. *Soil biology and Biochemical,* 24:1317–1320.
24. Khade, V. N., Patil. B. P., Khanvilkar, S. A., Dongale, J. H., Thorat, S. T. (1989). Evaluation of sprinkler method of irrigation for green gram and groundnut. *Journal of Maharashtra Agricultural Universities,* 14(1):117–118.
25. Koumanov, K. S., Hopmans, J. W., Schwankl, L. J., Andreu, L., Tuli, A. (1997). Application efficiency of microsprinkler irrigation of almond trees. *Agricultural Water Management,* 34:247–263.
26. Krishnamurthi, V. V., Manickasundaram, P., Vaiyapuri, K and Gnanamurthy, P. (2003). Micro sprinkler–A boon for groundnut crop. *Madras Agric. J.,* 90(1–3):57–59.
27. Larry, G. J. (1988). Principles of farm irrigation system design. *John Wiley & Sons, New York.*
28. Manjunatha, M, V., Shukla, K. N., Singh, K. K., Singh, P. K and Chauhan, H. S. (2001). *Response of micro irrigation on various vegetable crops: Micro irrigation.* Central board of Irrigation and Power Publication 282, pp. 398–402.

29. Manjunatha, M. V., Shukla, K. N., Chauhan, H. S. (2001). *Effect of micro sprinkler and surface irrigation methods on yield and quality of potato: Micro irrigation.* Central Board of Irrigation and Power Publication 282, pp. 403–406.
30. Manjunatha, M. V., Shukla, K. N., Chauhan, H. S., Singh, P. K., Rameshwar Singh, (2001). Economic feasibility of micro irrigation systems for various vegetables, Micro irrigation. Central Board of Irrigation and Power Publication 282, pp. 357–362.
31. Mehrotra, O. N., Garg, R. C., Ali, S. A. (1973). Influence of soil moisture on growth, yield and quality of groundnut. *Indian Journal Plant Physiology,* 11:158–163.
32. Michael, A. M. (1989). *Irrigation theory and practices.* Vikas publication Pvt. Ltd., New Delhi. pp. 801.
33. Mukund, N., Satyendra K. (2002). Optimization of system pressure for maximizing the uniformity of micro sprinkler spray in semiarid condition. *Journal of Soil and Water Conservation,* I(2 & 3)
34. Muralidhara, H. R., Hariyappa, N., Manjunatha, A. N., Raghuramulu, Y. (2006). Micro sprinkler system considered a viable and economical technology for irrigation of robusta. *Journal of Coffee Research (India),* 30(1):40–46.
35. Muralikrishnasamy, S., Veerabadaran, V., Krishnasamy, S., Kumar, V., Sakthivela, S. (2004). Research report. Dept of Agronomy, Agricultural College and Research Institute, *Tamil Nadu Agricultural University, Madurai* 625104, India.
36. Neumann, Nava Shir, (1995). Fertigation is a physiological advantage. In: *Proceeding of Dahlia Greidinger International Symposium on Fertigation, Israel*, pp. 259–261.
37. Pampattiwar, P. S., Gorantiwar, S. D., Deshmukh, M. M., Pawar, C. Y. (1997). Yield response of garlic crop to micro sprinkler irrigation method coupled with solar photovoltaic (SPV) pumping system. *J. Water Management,* 5(1&2):31–35.
38. Parikh, M. M., Savani, N. G., Raman, S. (1996). Micro and mini sprinkler performance evaluation based on field observation. *In: Proc. all India Seminar on Modern Irrigation Techniques, June 26–27, Bangalore,* pp. 102–105.
39. Patel, D. B., Patel, C. L., Raman, S. (2001). Effect of sprinkler irrigation on performance of wheat and gram in narmada command area, Micro irrigation. *Central Board of Irrigation and Power Publication 282*, pp. 689- 692.
40. Patel, P. G., Patel, Z. G., Lad, A. N., Raman, S. (1997). Comparative study of mini sprinkler versus surface method of irrigation under different IW/CPE ratios in safflower (*carthamus tinctorius* L.) grown on vertisol of south Gujarat. *J. Water Management,* 5(1&2):79–80.
41. Piper, C. S. (1996). *Soil Analyzer*. Academics Press, New York, pp. 152–158.
42. Rajesh, Raut and Kamal Kacholia, (2001). Fan-Jet micro sprayer irrigation–A proven water management technology in Indian condition, Micro irrigation. *Central Board of Irrigation and Power Publication 282*, pp. 149–154.
43. Rajvir, S., Kale, M. U., Avdesh Chandra, (2001). Performance evaluation of micro jets, Micro irrigation. *Central Board of Irrigation and Power Publication 282*, pp. 155–162.
44. Raman, S., A. N. Lad., A. G. Patel., Z. C. Patel., S. N. Patel and G. K. Timbadia. (1996). Economic feasibility of micro sprinkler in some pulses and oil seed crops. *In: Proc. All India Seminar on Modern Irrigation Techniques, Bangalore,* pp. 286–292.
45. Rane, N. B. (2011). Development, scope and future potential of fertigation in India. *Proceedings of National Seminar on Advances in Micro Irrigation*, NCPAH, Ministry of Agriculture, GOI, pp. 44–54.
46. Rank, H. D., Gontia, N. K., Venkariya, P. B and Ghaghada, R. H. (2005). Crop performance and economics of micro sprinkler irrigation systems under deficit water application for summer groundnut. *In: Natural resources engineering and management and agro-environmental engineering, Anamaya publishers, New Delhi*, pp. 249–254.
47. Reddy, S. R. (2006). *Agronomy of field crops*. Kalyani publishers, New Delhi.

48. Richards, L. A. (1954). *Diagnosis and improvement of saline alkali soils*. Agriculture Hand Book 60, US Department of Agriculture, Washington.
49. Safanotas, J. E and Dipoala, J. C. (1985). Drip irrigation of maize. *In Proceedings of the 3rd International Drip/Trickle Irrigation Congress,* California, USA, 2:275–278.
50. Sahay, J. (1986). *Elements of pumps and tubewells*, Agro-Book Agency, Jakkanpur, pp. 39.
51. Sahni, B. M., Holsambre, D. G., Pandav, R. M. (2001). Techno economic feasibility of drip and sprinkler irrigation in canal command, Micro irrigation. *Central Board of Irrigation and Power Publication 282*, pp. 745–751.
52. Sahoo, N., Pradhan, P. L., Anumala, N. K., Ghosal, M. K. (2008). Uniform water distribution from low pressure rotating sprinklers. *Agricultural Engineering International: the CIGRE journal*. Manuscript LW 08 014, X:1–10.
53. Sanchez, I., Faci, J. M., Zapata, N. (2011). The effects of pressure, nozzle diameter and meteorological conditions on the performance of agricultural impact sprinklers. *Agricultural Water Management.*
54. Sanimer, K., Gulati, H. S., Kaushal, M. P. (2005). Development of production functions using line- source sprinkler technique. *J Res Punjab Agric Univ,* 42(2):202–207.
55. Sarkar, S., Goswami, S. B., Mallick, S., Nanda, M. K. (2008). Different indices to characterize water use pattern of microsprinkler irrigated onion (*Allium cepa* L.). *Agricultural Water Management,* 95(5):625–632.
56. Satyendra, K., Ram Asrey, Mandal, G., Rajbir Singh. (2009). Micro sprinkler, drip and furrow irrigation for potato (*Solanum tuberosum*) cultivation in a semiarid environment. *The Indian Journal of Agricultural Sciences,* 79(3).
57. Sexton, P. J., Benet, J. M., Boote, K. J. (1997). The effect of dry pegging zone soil on pod formation of for tunner peanut. *Peanut Sciences,* 24:19–24.
58. Sezen, S. M., Yazar, A., Kapur, B., Tekin, S. (2011). Comparison of drip and sprinkler irrigation strategies on sunflower seed and oil yield and quality under Mediterranean climatic conditions. *Agricultural Water Management, 98*:1153–1161.
59. Shinde, S. H., Kabra, S. H., Firake, N. N. (1995). Performance of planting technique and micro irrigation system layout in kharif groundnut. *J. Water Management,* 3(1&2): 102–104.
60. Shobana, R., Asokaraja, N., (2005). Effect of micro sprinkler irrigation on physiological characters in raddish. *Madras Agric. J.,* 92(7–9):549–554.
61. Shukla, K. N., Manjunatha, M. V., Chauhan, H. S. (2001). Studies on effect of combined micro irrigation systems on growth and yield of sweet lime, Micro irrigation. *Central Board of Irrigation and Power Publication 282*, pp. 326–336.
62. Silva, A. J. P. da., Coelho, E. F., Miranda, J. H. de, (2008). Water application efficiency of a microsprinkler irrigation system in banana crop. *XXXVII Brazilian Congress of Agricultural Engineering, International Livestock Environment Symposium–ILES VIII*, Lguassu Falls City, Brazil.
63. Singh, A. L., Chaudhari, V., Koradia, V. G., Zala, P. V. (1998). Effect of excess irrigation and iron and sulfur fertilizers on the Chlorosis, dry matter production, yield and nutrients uptake by the groundnut in calcareous soil. *Agrochimica,* 39(4):184–198.
64. Sonune, S. P., Palaskar, M. S. (2001). Performance of rabi-cum-hot weather groundnut under sprinkler irrigation with broad bed furrow, Micro irrigation. *Central Board of Irrigation and Power Publication 282*, pp. 698–700.
65. Suresh, S., Senthivel, S. (2003). Hydraulic design and performance evaluation of landscape.
66. Suseela, P., Rangaswamy, M. V. (2005). KAU micro sprinklers–a promising irrigation technique. *Madras Agric. J.,* 92(7–9):599–602.
67. Suseela, P., Rangaswamy, M. V. (2012). Discharge and distribution characteristics of low cost KAU micro sprinklers. *Karnataka J. Agric. Sci.*, 25(1):96–99.

68. Thiyagarajan, G., Ranghaswamy, M. V., Rajakumar, D., Kumareperumal, R. (2010). Deficit irrigation on groundnut (*Arachis hypogeae* L.) with micro sprinklers. *Madras Agric. J.*, 97(1–3):40–42.
69. Tripathi, P. C., Sankar, V., Lawande, K. E. (2010). Influence of micro irrigation methods on growth, yield and storage of rabi onion. *Indian J Hort.,* 67(1):61–65.
70. Varshney, K. G., Raghvaiah, R. (2001). Effect of sprinkler irrigation on performance of summer groundnut variety ICGS-11, Micro irrigation. *Central Board of Irrigation and Power Publication 282*, pp. 701–704.
71. Vijayalakshmi, R., Veerabadran, V., Shanmugasundram, K., Kumar, V. (2004). Micro sprinkler irrigation and fustigation and land configuration as a best management technology package for groundnut. www.google.com.
72. Visalakshi, K. P., Reena Mathew, Suseela, P., Bridjit, T. K. (2007). Development and performance evaluation of KAU micro sprinklers. *Journal of Indian Water Resources Society,* 27(3–4).
73. Vishnu, B., Xavier, Jacobs, L., Santhana Bosu, (1995). Performance evaluation of spinner and spray type micro irrigation emitters. *J. Water management,* 3(1&2):1–3.

APPENDIX I – LIST OF ABBREVIATIONS

%	Percent
@	At the rate of
ASAE	American Society of Agricultural Engineers
BBF	Broad bed furrow
CD	Critical difference
cm	Centimeter
cm hr^{-1}	Centimeter per hour
cm^{-2}	Square centimeter
CPE	Cumulative pan evaporation
CU	Coefficient of uniformity
CUC	Christiansen's uniformity coefficient
CV	Coefficient of variation
DAS	Days after Sowing
DPR	Delivery performance ratio
dS/m	Deci Siemens per meter
e_a	Application efficiency
EC	Electrical conductivity
e_d	Distribution efficiency
ET	Evapotranspiration
FAO	Food and Agriculture Organization
FYM	Farm yard manure
gm cc^{-1}	gram per cubic centimeter
gm l^{-1}	gram per liter
ha	hectare
ha cm	hectare centimeter

hp	Horse power
hr	Hour
IW	Irrigation water
IWUE	Irrigation water use efficiency
k_c	Crop coefficient
kg	Kilogram
kg cm^{-1}	Kilogram per centimeter
kg ha^{-1}	Kilogram per hectare
kg ha^{-1} cm^{-1}	Kilogram per hectare per centimeter
kg ha^{-1} mm^{-1}	Kilogram per hectare per millimeter
kg m^{-3}	Kilogram per cubic meter
kgf cm^{-2}	Kilogram force per square centimeter
km hr^{-1}	Kilometer per hectare
K_p	Pan coefficient
Kpa	Kilo Pascal
LAI	Leaf area index
LLDPE	Linear low density polyethylene
lpd	liters per day
lph	liters per hour
m	Meter
M ha	Million hectare
m^2	Square meter
Mg ha^{-1}	Million gram per hectare
MIS	Micro irrigation system
mm	millimeter
MSI	Micro sprinkler irrigation
MSL	Mean sea level
Mt	Metric ton
NPK	Nitrogen, Phosphorus and Potassium
PVC	Poly vinyl chloride
q ha^{-1}	quintals per hectare
RBD	Randomized block design
RPM	Revolutions per minute
Rs	Rupees
Rs ha^{-1}	Rupees per hectare
SEM	Standard error mean
SMC	Soil moisture content
t ha^{-1}	Tons per hectare
TSS	Total soluble salts
WUE	Water use efficiency

APPENDIX II – SAMPLE CALCULATION OF WATER REQUIREMENT OF GROUNDNUT

a. The daily water requirement of groundnut crop for micro sprinkler irrigation was calculated by using the following equation:
$Q = [A \times B \times C]/[E] = [(6 \times 0.7) \times 0.85 \times 1]\ [0.8] = 4.46$ mm/day
where, Q = quantity of water required mm/day, A = daily evapotranspiration, mm/day = pan evaporation *x* pan coefficient, B = amount of area covered with foliage (canopy factor), fraction, C = crop coefficient, fraction, E = efficiency of micro sprinkler irrigation system, % (80%)

b. The amount of irrigation water to be delivered through surface method of irrigation for groundnut crop was calculated by using the following equation:
$d = AWC \times A_s \times d_s = [(18 - 7.56)/100] \times 1.61 \times 15 = 2.52$ cm = 25.2 mm or 25 mm

APPENDIX III – ECONOMICS OF GROUNDNUT CROP UNDER MICRO SPRINKLER AND SURFACE IRRIGATION

A. Establishment Cost of Micro Sprinkler Irrigation System, Rs ha^{-1}

S. No.	Particulars	Cost Rs.	Apportioned value (Rs yr^{-1})
1	Submersible pump	17,400	1,160
2	Screen filter (m^3 hr^{-1})	3,500	350
3	Main line PVC 63 ϕ mm	23,435	2,343.5
4	Sub main PVC 32 ϕ mm	51,150	5,115
5	Micro sprinklers	3,705	370.5
6	Control valves 63 ϕ mm	1,260	126
7	Control valves 32 ϕ mm	8,928	892.8
8	PVC Tee 32 ϕ mm	1,815	181.5
9	Coupler PVC 32 ϕ mm	300	30
10	Elbow PVC 32 ϕ mm	913	91.3
11	Tee four way PVC 63 ϕ mm	3,100	310
12	PVC fitting and accessories	2,500	250
13	Installation charges	1,200	120
	Sub Total	119,206	11,340.6
	Interest on fixed investment @ 12%	14,304.72	1,360.87
Total		133,510.72	12,701.47

Fixed cost for submersible pump = = 17,400 /15 =1,160 Rs/Year
Fixed cost for PVC pipe = = 3,500/10 =350 Rs/Year

B. Cost of Micro Sprinkler Irrigation

S. No.	Particulars	Cost Rs.
1	Ploughing (Rs ha^{-1})	350.00
2	Harrowing (Rs ha^{-1})	350.00
3	Manure spreading (Rs ha^{-1})	480.00
4	Bed preparation (Rs ha^{-1})	480.00
5	Weeding (Rs ha^{-1})	1600.00
6	Spraying (Rs ha^{-1})	240.00
7	Harvesting (Rs ha^{-1})	1000.00
8	Electricity charges (Lump sum)	800.00
9	Seed (Rs ha^{-1})	9,600
10	FYM (Rs ha^{-1})	3,125
11	Fertilizers (Rs ha^{-1})	9073.30
12	Variable cost (Rs ha^{-1})	27,098.3
13	Fixed cost (Rs ha^{-1})	6,350.74
14	Total cost (Rs ha^{-1})	33,449.04

Fixed cost for micro sprinkler irrigation is the apportioned value of the establishment cost of micro sprinkler irrigation system. Apportioned value is the life of the material.

- Sub Total of the apportioned value (fixed investment) = Rs 11,340.60
- Interest on fixed investment @ 12% = Rs 1,360.87
- **Fixed cost = Rs 12,701.47**

The system can be used for two seasons in a year. Therefore, the fixed cost for one season will be Rs. 12,701.47/2 = Rs. 6,350.74 per season.

C. Cost of Surface Irrigation

S. No.	Particulars	Cost Rs.
1	Ploughing (Rs ha^{-1})	350.00
2	Harrowing (Rs ha^{-1})	350.00
3	Manure spreading (Rs ha^{-1})	480.00
4	Weeding (Rs ha^{-1})	2,200.00
5	Spraying (Rs ha^{-1})	640.00
6	Harvesting (Rs ha^{-1})	1,000.00
7	Electricity charges (Lump sum)	800.00
8	Seed (Rs ha^{-1})	9,600
9	FYM (Rs ha^{-1})	3,125
10	Fertilizers (Rs ha^{-1})	9,073.30
11	Variable cost (Rs ha^{-1})	27,618.30
12	Fixed cost (Rs ha^{-1})	938.28
13	Total cost (Rs ha^{-1})	28,556.58

Fixed cost of the surface irrigation is calculated by considering apportioned value of

- Submersible pump = Rs 1,160.00
- Pipeline = Rs 365.50
- Fittings and accessories = Rs 100.00
- Installation charges = Rs 50.00

Sub Total = **Rs 1,675.50**

- Interest on fixed investment @ 12% = Rs 201.06

Grand Total = **Rs 1, 876.56 for two seasons**

= Rs 938.28 for one season.

CHAPTER 11

IRRIGATION SCHEDULING OF WHEAT UNDER MICRO-SPRINKLER IRRIGATION

J. P. VISHWANATH and M. M. DESHMUKH

CONTENTS

Adapted from Jadhao Pandit Vishwanath, "Effect of open pan evaporation based irrigation scheduling on growth and yield of wheat" (M. Tech thesis, Post Graduate Institute at Dr. Panjabrao Deshmukh Krishi Vidyapeeth, Akola, India, 2013).

11.1 INTRODUCTION

Agriculture contributes about 35% of national income in India. Since agricultural production is mainly affected by inputs provided, one of the most important input is irrigation. The average annual rainfall in the Maharashtra State is about 1300 mm of which 88% occurs during the south-west monsoon. The net irrigated area of Maharashtra State in 2006–07 was 3.246 million-ha with gross irrigated area of 3.958 million-ha. The percentage of gross irrigated area to gross cropped area in 2006–07 was 17.5 [2].

Water is a key input for all recommended agronomic practices and therefore efficient utilization of irrigation water is essential. The objective of irrigation is to maintain the soil moisture at optimum levels in the plant root zone, so that root zone will have a constant supply of moisture with adequate aeration.

Wheat is the leading source of protein in human food, having higher protein content than either maize (corn) or rice and the other major cereals. In terms of total production used for food, it is currently second to rice, as the main human food in India. Wheat grain is a staple food used to make flour for leavened, flat and steamed breads, biscuits, cookies, cakes, breakfast cereal, pasta, noodles, couscous and for fermentation to make beer, other alcoholic beverages or biofuel. Wheat is one of the most important cereal crops of the world on account of its wide adaptability to different agro climatic and soil conditions. Global wheat production has declined substantially in past few years. In 2007–08, the global wheat production was 604 million-tons.

Wheat is the most important cereal crop in India also. It is sown in 27.5 million-ha with total production of 80.58 million-tons (2008–2009). The average yield of wheat is 2700 kg/ha. Area and production of wheat in Maharashtra was 1.022 million-ha and 14.83 million-tons, respectively during 2008–2009 with the productivity of 1452 kg/ha [10]. Average productivity is much lower than the national average because of high temperature and low humidity. In cultivation of high yielding wheat varieties, irrigation assumes greater importance because during growing season of crop (October to April), weather remains relatively dry.

In Wheat, different growth stages are crown root initiation, tillering, jointing, boot flowering, milk and dough that are well delineated. Experiments conducted to study the important stages critical in their demand for water have clearly indicated that some stages can tolerate moisture stress to a certain extent. Most of the researchers have observed that in case of dwarf varieties of wheat, irrigation at the crown root initiation stage (20–25 days after sowing) resulted in maximum production per unit of water applied and therefore this stage is considered as the most critical stage for irrigation.

Crop water requirement is the quantity of water needed for normal growth, development and yield and may be supplied by precipitation or by irrigation or by both. Water is needed mainly to meet the demands of evaporation (E), transpiration (T) and metabolic needs of the plants. The water requirement of any crop is dependent upon: variety, growth stage, duration, plant population and growing season; and soil factors: texture, structure, depth, and topography; climatic factors: temperature, relative humidity and wind speed; and crop management practices: tillage, fertilization, weeding, etc.

Irrigation aims to restore soil water in the root zone to a level at which crop can fully meet its evapotranspiration (ET) requirement. The amount of water to be applied at each irrigation and how often a soil should be irrigated depends on several factors: the degree of soil water deficit before irrigation, soil type, crops and climatic conditions.

Over irrigation wastes water, energy and labor, leaches expensive nutrients below the root zone of the plants and reduces the soil aeration and thus crop yields. Whereas under irrigation stresses the plant and causes yield reduction. Therefore, adequate irrigation scheduling requires a sound basis for making irrigation decisions. The level of sophistication for decision making is based on personal experience to expensive computer aided instrument and using soil water and atmospheric parameters.

Irrigation scheduling is the systematic method by which producer can decide: when to irrigate and how much water to apply. The goal of effective scheduling program is to supply the plants with sufficient water while minimizing losses to deep percolation or runoff. Irrigation scheduling depends on soil, crop, atmosphere, irrigation systems and operational factors. Several approaches for scheduling irrigation have been used by scientists and farmers.

Irrigation scheduling techniques can be based on soil water depletion approach, plant indices, climatic approaches, critical growth stage approaches and plant water status itself. In soil water depletion approaches, the available soil moisture in the root is a good criterion for scheduling irrigation. When the soil moisture in a specified root zone depth is depleted to a particular level (which is different for different crops), it is replenished by irrigation. In plant indices, as the plant is the user of water, it can be taken as a guide for scheduling irrigation. The deficit of water will be reflected by plants itself such as dropping, curling or rolling of leaves and change in foliage color as indication for irrigation scheduling. In critical growth approaches, irrigation scheduling at growth stage of crop at which moisture stress level reaches to irrevocable yield loss. These stages are known as critical period or moisture sensitive period. In plant water status approaches, water content in plant itself is considered for scheduling irrigation however it is yet common use for wants of standard low cost techniques. Whereas in climatologically approach, the amount of water lost by evapotranspiration is estimated from climatic data. When ET reaches to a particular level, irrigation is scheduled. The amount of irrigation given is either equal to ET or fraction of ET. Different methods of climatic approaches are IW/CPE ratio method and pan evaporation method. In IW/CPE approach, known amount of irrigation water is applied when cumulative pan evaporation reaches predetermined level. For practical purpose, irrigation should be started when allowable depletion of available moisture in the root zone reaches. The available water is soil moisture, which lies between field capacity and permanent wilting point.

Thus scheduling provides information to the managers to develop irrigation strategies for each plot of field. Keeping these points in view, this study was conducted:

1. To assess optimum water requirement of wheat crop using different levels of IW/CPE approach of irrigation scheduling.
2. To study the soil moisture depletion and water use efficiency for different irrigation scheduling.
3. To study the effects of different irrigation scheduling on growth and yield of wheat crop.

11.2 REVIEW OF LITERATURE

11.2.1 IRRIGATION SCHEDULING BASED ON CRITICAL STAGES WHEAT

Choudhary and Kumar [10] compared mild, moderate and severe stress treatments with a no stress control. At all stages of moderate and severe water stress decreased plant height, leaf area, ear number, 1000 grain weight, grain yield and water use efficiency (WUE) of wheat. In stage 3, the effect of water stress on straw yield was not marked. Wheat was most sensitive to water stress during stage 1, when the reduction in grain yield was caused by a reduction in number of ears and grains per ear. In stage 2, grain yield reduction was due to fewer grains per ear and a lower 1000-grain weight. On rewatering, mild stressed plants showed recovery of plant height, tiller number and yield. Results indicated that wheat crop should be irrigated at a IW/CPE ratio of 0.75. Under limited water resources with an unlimited water supply, the ratio may be increased to 1.2 in stage 2 to maximize the yield.

Hassan et al. [18] identified wheat growth stages that are most sensitive to soil moisture stress. Treatments included three seeding rates (75, 100 and 150 per ha), seven drought treatments affected by missing one or two consecutive irrigations at different growth stages. With holding irrigation at any growth stage prior to anthesis had detrimental effects on most of the growth characters and stress during crown root initiation (CRI) and reduced most of the yield characters. Missing two consecutive irrigations at any given growth stage reduced grain yield more than did missing one irrigation. The greatest reductions in grain yield were 13 and 65% when one irrigation was missed at CRI and two consecutive irrigations were missed at jointly (the most sensitive growth stage), respectively of grains, spike number, spike length, leaf area and number of tillers. A seeding rate of 100 kg/ha produced best wheat performance.

Chavan and Pawar [9] carried out investigations on wheat (*Triticum aestivum*) to find out crop coefficient (Kc) values for various stages using CU of water by wheat and ET crop (mm) computed by modified Penman equation. The results indicated that the Kc values were 1.783, 0.878, 0.919, 0.869, 1.081 and 1.591 for sowing to crown root initiation (21 days), CRI to tillering (19 days), tillering to jointing (15 days), jointing to flowering (20 days), flowering to milk (15 days) and milk to maturity (21 days) stages, respectively. Vijaykumar et al. (1987) stated that the number of effective tillers was significantly influenced by irrigation frequencies. Maximum number of grains per spike was observed up to three irrigations, further increases with increase in number of irrigations was not significant. Ear-head length increased with the increase in frequencies of irrigation up to 4000 grain weight followed the same trend as followed by the grain yield of the wheat.

Chauhan [8] concluded that 6 ds/m EC of water with 4 irrigations (at cross root initiation, tillering, flowering and milk stage) and EC 12 ds/m (high salt concentration) with 2 irrigations (cross root and milk stages) were effective to grow wheat.

Rathore and Patel [36] found that five irrigations (at cross root initiation, tillering, jointing, flowering and dough stages) in late sown wheat were optimum during dry crop season. Under the normal crop season (with rainfall 108.9 mm), only three irrigations (at cross root initiation, jointly and milk stage) were needed. Three irrigations were optimum in late sown wheat in clay loam soil, irrespective of seasonal variations. Application of 120 kg N/ha gave significantly highest grain yield, and water use efficiency was higher at lower levels of irrigation. Maximum net returns were recorded with five irrigations and 120 kg N/ha of fertilizer.

Deshmukh et al. [10] observed that most critical stages were CRI and flowering (F) stages based on moisture stress. Significant reduction in productivity was noticed due to deleting irrigation at the CRI (33%) and flowering (25%) stages compared to no stress. Maximum tillering, late jointing and milk stages were less sensitive (critical) stages in Vidarbha region of Maharashtra State. Bhoi et al. [5] concluded that wheat can be grown successfully with 1 to 4 irrigations applied at most critical growth stages, depending on the number of irrigations available. Satisfactory wheat yield may be obtained under lower investment of fertilizers when NPK doses are decided on the basis of soil test crop response.

Maliwal et al. [25] indicated that 4 irrigations at presowing, cross root initiation, tillering and flowering under restricted water supply gave consistently higher grain yield of wheat during all years. The WUE was also higher. The quality of wheat grain (Lysine and tryphoton) was not affected due to reduction in number of irrigations under constraint of irrigation water than normal recommended irrigation.

Thakur et al. [43] conducted an experiment on sandy-loam soil at Ranchi to estimate the energetics of wheat (*Triticum aestivum*) under different levels of irrigation, seed rate and fertilizer. It was observed that wheat cultivation required 92.9% more indirect sources of energy than direct sources of energy input. The crop with 4 irrigations at CRI, maximum tillering, best and milk stress produced 4071 Kg of grain/ha and consequently gave 33.8 and 32.4% higher energy through grain and total biomass respectively along with higher grain (3.08), total biomass (6.55) energy use efficiency and required less specific energy (4,769 MJ/ton) than 3 irrigations at maximum tillering, root and milk stores (44,717 MJ/ha grains and 96,092 MJ/ha biomass energy output). Thakur et al. [43] also found that wheat crop receiving 4 irrigations at crown root initiation, maximum boot and milk stages gave maximum grain (2,707 kg/ha) and straw (4,164 kg/ha) yields, with net return of Rs. 11,855/ha and benefit-cost ratio (BCR) of 1.31. This treatment showed significant edge over all over irrigation treatments except the crop receiving 3 irrigations at maximum tillering, boot and milk stages (2,523 kg/ha grain yield) and Rs. 10474/ha of net return and 1.23 BCR). At 2 irrigations, wheat irrigated at maximum tillering and milk stages gave the maximum grain (112,011 kg/ha) and straw (2,823 kg/ha) yields, with net return of Rs. 7,328/ha and BCR of 0.92/ha.

11.2.2 EFFECTS OF IRRIGATION SCHEDULING BASED ON IW: CPE RATIOS ON PERFORMANCE OF WHEAT

Prihar et al. [33] suggested that most of the scheduling methods are not sufficiently simple to be adopted by farmers. Recently a more practicable approach is based on the ratio of fixed amount of irrigation water (IW) to pan evaporation. IW/EPAN of 0.75 irrespective of growth stage produced as much grain yield as gravity irrigation. However, the farmer used 12 cm less irrigation depth. There was no grain in the yield by combining the IW/EPAN with growth stage. These results indicated that irrigating wheat, sown after a presowing irrigation, on the basis of IW/EPAN irrespective of growth stages, offered practical approach to economic irrigation water without reduction in yield.

Chaudhary and Bhatnagar [7] concluded that the wheat crop developed a more extensive root system when the first irrigation was applied after 26 days than after 40 and 54 days with the first irrigation on the 26th day. The crop, receiving subsequent irrigations less frequently but at a heavier rate, developed a deeper root system than the crop receiving frequent and light irrigations. The water extraction pattern was corresponded with the root distribution pattern. A relatively small difference in root density in the deeper layers caused a greater difference in soil water content than in the upper layers. Light and frequent irrigations produced maximum grain yields. However, for developing on extensive root system and enhancing water utilization in the subsoil, an early light irrigation seems desirable with subsequent irrigations applied less frequently at a relatively heavier rate.

Reddy et al. [37] evaluated that irrigation scheduling of 60 mm depth of irrigation when CPE was 60 mm (IW/CPE = 1.00) was optimum for wheat, while 60 mm depth of irrigation when CPE was 50 mm (IW/CPE = 1.2) was optimum for maize. The ratios IW/CPE were able to increase with the advance of the season. The ratios were higher in maize than wheat. The CCE during the different stages of crop growth varied from 60.6 and 69.0 mm in wheat and from 51.5 to 94.0 mm in maize. Paradkar et al. [27] used IW/CPE ratio of 0.8, 1.0 and 1.2 in wheat. Four irrigation depths (30, 50, 70 and 90 mm) were used as sub treatments. The highest grain yield was obtained at IW/CPE ratio of 1.2 with 30 to 50 mm water per irrigation. The preirrigation soil water potential was highest in plots under irrigation based on IW/CPE of 1.2 and it decreased as the number of irrigations was reduced and quantity of irrigation water was increased. Based on two years crop performance, it was concluded that the irrigation scheduling in moderately sodic clay soil should be based on the IW/CPE of 1.2 with 50 mm depth of water per irrigation.

Gill and Lenvain [14] found that irrigation interval between 50 to 75 mm COPE and irrigation depth based on $R_{0.75}$ to $R_{0.90}$ were satisfactory. Lower wheat yields indicated that physical condition at the soils contributed towards wheat response to irrigation. They recommended evaluation of open pan based irrigation scheduling for other climate conditions and consideration of the physical conditions of a soil to predict crop response to irrigation. Sharma [38] found that an increase in IW/CPE ratio resulted in greater relative growth rate, yield attributes and yield of wheat. The seasonal ET of crop with 1.0 IW/CPE ratio was 332.4 and 338.5 mm of which 75.7 and 69.0 mm was extracted from 0–0.90 m soil profile during 1986–1987 and 1987–1988, respectively. Most of the seasonal ET of crop took place from 0–0.15 m soil layer and increased from lower layers with an increase in IW/CPE ratio. However, the total among the seasonal ET from all the layers was greater under higher IW/CPE rate. Maximum WUE of 6.49 and 6.55 kg of grain/mm of ET was recorded, when irrigation was given at 1.0 IW/CPE ratio than at the other rates.

Mishra et al. [26] concluded that irrigation of the crop at 0.75 IW/CPE ratio gave the grain yield and yield component similar to that at five critical growth stages of wheat. However, at 0.75 IW/CPE ratio, 32% irrigation water was saved. The wheat crop exhausted more moisture at 0–60 cm depth that at deeper layers. The response of grain yield and yield component to N was significant up to 90 kg/ha. Singh and Mohan [39] used 6 cm of irrigation depth on the basis of IW/CPE ratio of 0.6,0.8, 1.0 and 1.2 in sugarcane field. CPE values were computed after accounting for rainfall in each year individually. The yield and yield attributes were highest. At IW/CPE ratio of 1.0, WUE was decreased with increase in irrigation efficiency. Nevertheless, relatively more water was extracted from the upper layer in 1.2 IW/CPE ratio as compared to other treatments. Soil moisture extraction from deeper layer was comparatively higher under lower ratios (0.6 to 0.8) than 1.0 and 1.2 ratios.

Bandyopadhyay [3] found that the actual evapotranspiration was 239.08 mm and water uptake was maximum (56.5%) from the 0–15 cm depth and it gradually changed with soil depth. Also higher rainfall and its good distribution resulted in sizeable deep drainage and nonsignificant yield response to irrigation regimes.

Imtiyaz et al. [20] found that the higher mean marketable yield of cabbage (71.65 tons/ha), spinach (33.63 tons/ha), rapeseed (73.22 tons/ha), carrot (56.66 tons/ha), tomato (46.81 tons/ha) and onion (56.05 tons/ha) were recorded for irrigation scheduling at CPE of 22, 11, 22, 22, 22 and 11 mm, respectively. The irrigation at CPE of 22 mm resulted in higher irrigation production efficiency of cabbage (11.32 kg/m^3), spinach (3.35 kg/m^3), carrot (9.83 and 6.66 kg/m^3), tomato (5.9 kg/m^3) and onion (6.26 kg/m^3) but rapeseed (12.03 kg/m^3) gave higher irrigation production efficiency at IW/CPE of 33.55 mm. Irrigation scheduling at 22 mm of CPE resulted in a higher net return of 4713100, 10439800, 7669100 and 9319200 kg/ha for cabbage, rapeseed, carrot and tomato, respectively. The spinach and onion gave a higher net return of 27086 and 839434 P/ha at 11 mm of CPE (1 US $ = 4.55 P). Irrigation scheduling at 22 mm of CPE gave a higher B:C ratio of 2.92, 1.94, 5.40, 4.98, 4.91 and 4.82 for cabbage, spinach, rapeseed, carrot, tomato and onion, respectively. Seasonal water applied and marketable yield of vegetative crops exhibited quadratic relationship (R^2 = 0.85–0.99). The fitted regression models attained the maximum yield, net return and B:C ratio at CPE of 16–18 mm. The results revealed that rapeseed is the most remunerative crop followed by tomato, onion, carrot, cabbage and spinach.

Mohamod (23) conducted the experiment on irrigation scheduling on the basis of cumulative pan evaporation (CPE). Wheat at seed rates of 100, 125 and 150 kg/ha was irrigated using IW:CPE ratio of 0.70, 0.90, 1.10 and 1.3. Data on agronomic traits like plant height, tillers/m, grains/spike, 100-grain weight, grain yield, straw yield, harvest index and WUE were collected. Statistical analysis suggested that wheat was quite responsible to increase in irrigation as well as seed rate. Highest irrigation and seed rate produced greater plant population/m^2 and thus, the competition among plants was increased. There was reduction in important yield attributes like grains/spike and 1000 grains weight, which ultimately produced less grain yield. It as inferred that wheat should be sown at the rate of 125 kg/ha and should be irrigated at IW:CPE ratio of 0.9, to obtain maximum production.

Singh et al. [40] found that narrow row spacing of 15 cm using 205 kg of seed/ha resulted in significant higher grain yield (2990 kg/ha) compared with 2790 kg/ha under normally used spacing of 22 cm @ 140 kg seed/ha. Significant increase in yield was recorded upto 150 kg of N/ha and it did not differ significantly with that obtained with 180 kg of N/ha. Irrigations at IW:CPE of 0.8, with each irrigation of 6.0 cm, recorded significantly higher grain yield (2950 kg/ha) than with irrigation at IW:CPE 0.6. Increase in irrigation levels decreased WUE. The highest WUE of 76.4 and 66.9 kg grain/ha-cm during 1998–99 and 1999–2000, respectively was recorded with narrow spacing. Narrow row spacing recorded higher net return (Rs. 11,729 and 10156/ha) and benefit-cost ratio. The highest net returns and benefit-cost ratio were obtained with 6 cm irrigation depth at IW:CPE ratio of 1.0 and IW:CPE ratio of 0.8 during 1998–99 and 1999–2000, respectively.

Parihar and Tiwari [28] concluded that irrigation at 1.2 IW:CPE ratio gave significantly higher grain yield than 0.6 and 0.9 ratios. Application of 120 kg N/ha gave higher yield and yield attributes than 80 kg N/ha. Total nutrient uptake was positively influenced by irrigation and N-levels. Total water use was lowest (29.60 cm), while WUE was highest with 0.6 IW:CPE ratio. The WUE was decreased with increase in frequency of irrigation. The moisture depletion from upper soil layers (0–60 cm) was higher than lower layers (60–90 cm). However, total profile moisture contribution was highest with limited water supply (0.6 IW:CPE) and decreased with the increase in number of irrigations. Canopy temperature was comparatively lower than ambient temperature under higher moisture (1.2 or 0.9 IW:CPE) than lower moisture regime (0.6 IW:CPE).

Channagoudar and Janawade [6] studied performance of onion under four irrigation scheduling's based on 0.1, 1.1, 1.3 and 1.5 IW/CPE ratios four sulfur levels @ 0, 20, 40 and 60 kg of S/ha. The results indicated significantly higher bulb yield (18929 kg/ha) and yield components like bulb length, bulb diameter and weight of 20 bulb at 1.5 IW/CPE ratio irrigation scheduling compared to 0.9, 1.1 and 1.3 IW/CPE ratios. The growth components (plant height, number of leaves, leaf area, leaf area index, leaf area duration and total dry matter production per plant) were also higher at 1.5 IW/CPE ratio. Application of 40 kg S/ha recorded significantly higher bulb yield (17060 kg/ha). Similar trend was observed in yield components and sulfur uptake. Significantly higher TSS (12.26%) and pyruvic acid (301 μ mol/g) content in onion bulbs were recorded with 60 kg of S/ha compared to 20 kg S/ha and no sulfur application but was on par with 40 kg S/ha.

Kumar et al. [22] studied the feasibility of using microsprinkler, drip irrigation system for vegetable production in a canal command area. These systems were compared with the existing flood irrigation method for onion production with four irrigation levels based on 0.60, 0.80, 1.00 and 1.20 IW/CPE ratios. Micro-sprinkler and drip irrigation systems were also compared with fertigation rate of 100 (50:25:25 NPK), 150 (75:37.5:37.7 NPK) and 200 (100:50:50 NPK) kg/ha. Micro irrigation systems resulted in higher onion yield and greater profitability than surface irrigation at each irrigation level. However, microsprinkler irrigation indicated higher economic benefits than drip irrigation. Micro-sprinkler, drip and surface irrigation system based on 1.20 IW/CPE ratio produced maximum crop yields of 34.34, 33.10 and 22.57 tons/ha, respectively. Increased crop yield with microsprinkler and drip irrigation resulted in higher profitability than existing surface irrigation. Reduction in nutrient application by 25% in fertigation from the standard dose in flood irrigation did reduce yield and net returns significantly. However, net return was significantly higher in microsprinkler (67334 Rs./ha) than drip fertigation ((59930 Rs./ha)). The overall results of the present study favored microsprinkler over existing irrigation methods for onion production in a canal command area with higher profit under limited available surface water.

Patel et al. [29] studied the effect of irrigation levels based on IW:CPE ratio (0.6, 0.8 and 1.0) and six levels of time of nitrogen application (100% based, 50% as basal + 50% N at branching, 50% N as basal + 25% N at branching and 25% N at flowering, 50% as basal + 25% N at branching + 25% N at podding, 50% N as basal + 25% N at flowering + 25% N at podding, 25% N as basal + 25% N at branching + 25% N at flowering + 25% at podding shade) on seed yield, straw yield and economics of French bean. Irrigation scheduling at 1,0 IW:CPE ratio recorded significantly higher seed and straw yield, net returns and benefit-cost ratio. Among the time of nitrogen application, application of half nitrogen as basal and remaining half nitrogen at branching stage was most effective mode of nitrogen application by recording the highest seed yield, straw yield, net returns and BCR.

11.2.3 EFFECTS OF IRRIGATION SCHEDULING ON WATER USE EFFICIENCY

Doorenbos and Pruitt [12] stated that the management allowable depletion (MAD) is the degree to which volume of water in the soil is allowed to be depleted before the next irrigation is applied. MAD is considered as the ratio of readily available water and available water. The desired MAD value of maize crop was 0.65.

Batra and Rai [4] conducted a field studies on root crop production using six main plot treatments comparing three irrigation intensities (ID/CPE 0.4, 0.8 and 1.2) and on seed crop using three irrigation intensities (ID/CPE 0.2, 0.4 and 0.6). Plant height, fresh foliage, weight /plant and root weight/ plant increased significantly under irrigation scheduling of root crop at 0.8 and 1.2 over 0.4. Length and diameter of roots were also improved. Irrigation scheduling at 0.8 and 1.2 ID/CPE ratios resulted in higher yield than under 0.4 ID/CPE. Dry matter content of both root and leaf increased significantly under 0.8 ID/CPE than 0.4 and 1.2. Care diameter appearance of premature bolters and number of split roots increased with the increase in irrigation intensity. Fertility was significantly lowered under ID/CPE 0.8 than 0.4 and 1.2. Carotene content declined with increase in irrigation intensity. Reducing sugar increased but nonreducing sugar decreased with increased in irrigation intensity. Plant height and number of primary/ secondary branches per plant were significantly higher under ID/CPE 0.4 and 0.6 than 0.2, in seed crops. Incidence of lodging was increased with increased irrigation frequency. Seed yield/plant and per hectare increased significantly under ID/CPE 0.4 and 0.6 than 0.2. Consumptive water use increased with the increase in the irrigation frequency, but higher WUE was observed in the driest moisture regime (ID/CPE 0.4). Leaf water potential was decreased with aging of crop, irrespective of different irrigation treatments. However, higher leaf water potential was observed in carrot roots as well as seed crop irrigated at higher available soil moisture status. Alam [1] observed similar results.

Prasad et al. [32] concluded that optimum irrigation scheduling was at IW/CPE ratio of 1.0 receiving 3 to 4 irrigations of 6 cm each. Response to N ranged from 60 to 120 kg/ha due to significant interaction between irrigation and N. Water use efficiency was higher in the drier regime. Phogat et al. [31] studied sugarcane performance under with three irrigations levels of 0.9, 0.7 and 0.5 IW/CPE ratios. Irrigation applied at 88 mm open pan evaporation (IW/CPE = 0.9) produced maximum yield of 85.2 tons/ha. Gajri et al. [13] stated that a small early irrigation and/or deep tillage reduced soil strength and stimulated crown root development and increased the rate of root extension down the profile. The better root growth in tilled /early irrigated crop increased profile water depletion, dry matter production and grain yield. Singh and Uttam [41] found that water use for wheat was higher with 3 irrigations.

Solanki and Patel [42] found that Irrigation at IW/CPE ratio of 1.25 produced significantly higher dry fodder yield (9920 kg/ha) of Lucerne. Kibe and Singh [21] evaluated effects of irrigation, nitrogen, zinc and their interactions on yield and moisture use by wheat. Progressive increases, in irrigations from 0 to 4 and N levels from 0 to 100 kg/ha, increased wheat yield attributes to grain yield significantly over the control treatments (I_0 and N_0). The average seasonal consumptive water use (Cu) by wheat was increased with every additional irrigation level to a maximum of 328.4 mm and 307.7 mm in the first and second season, respectively. On an average WUE was the highest (1.38 kg grain/m^3) with I_2. However, the straw yield (680.85 kg/ha) and grain yield (3962.5 kg/ha) were highest with I_3. The average moisture use rate was increased with increase in irrigation to a maximum of 2.63 mm/day. Maximum moisture extraction of 59.4 to 65.8% was from the 0–30 cm and minimum (7.10 to 5.32%) from 90–120 cm soil depth.

Wajid et al. [44] studied the effects of cultivars and irrigation regimes on water use efficiencies that were calculated from total water applied and crop evapotranspiration. Four irrigations (I_0 = control, I_1 = Irrigation up to stem elongation, I_2 = Irrigation from stem elongation to maturity and I_3 = full irrigation treatments based on soil moisture deficit) were applied to each cultivar. Irrigation treatments were managed to induce a range of drought from full irrigation to no irrigation from the emergence to physiological maturity. They concluded that for each mm of crop ET, 3.27 g/m^2 of total dry matter (TDM). The yield was increased by 10.7%. Fully irrigated crop produced 93.18%, higher yield over control treatment. In treatments with drought before or later anthesis, the primary cause of reduced efficiencies was a decrease in intercepted light.

Hosmani and Janawade [19] found that the groundnut pod yield was higher in I_1S_6 (2230 kg ha^{-1}) and I_1S_3 (2140 kg ha^{-1}) over no sand application I_1S_7 (1665 kg ha^{-1}). The harvest index was also significantly higher in I_1S_6 (0.40) and I_1S_3 (0.42) over no sand application I_1S_7 (0.38). Highest consumptive use of water was recorded in I_3S_7 (control) 522 mm and it was reduced to 494 mm in I_3S_6. The highest water use efficiency was observed in I_2S_3 (5.35 kg ha^{-1} mm^{-1}) while minimum was 2.80 kg ha^{-1} mm in I_3S_1. The available nutrients in soil after harvest of groundnut was also higher in I_1S_6 P (25) and K (690 kg ha^{-1}) compared to no sand application P (16.5) and K (510 kg ha^{-1}). The soil moisture extraction was higher from surface layer (0–15 cm) 37.53 in I_1 to 42.23% in I_3. Sand application marginally increased the soil moisture extraction from surface soil layer 0–15 cm, 38.27% in S_3 and 37.09% in S_6 compared to S_7 (no sand application) 36.40%.

Maheshwari et al. [24] evaluated aerobic rice (*Oryza sativa* L.) productivity under four irrigation regimes (at IW/CPE ratio of 0.8, 1.0, 1.2), micro sprinkler irrigation once in three days, four N levels (100, 125, 150 and 175 kg/ha). Irrigation at 1.2 IW/CPE ratio recorded significantly higher crop growth rate and yield with no moisture stress, minimal praline accumulation and sterility coefficient. N levels with 150 and 175 kg N/ha produced higher yield. Hence, irrigation at 1.2 IW/CPE ratio with 150 kg N/ha was optimum to realize the maximum productivity under aerobic rice cultivation.

Patel et al. [30] found that summer groundnut irrigated at 40 mm cumulative pan evaporation (CPE and 17 irrigations) gave higher values of consumptive use (795.8 mm), WUE (4.76 kg/ha/mm) and pod yield (3.79 tons/ha) compared to irrigation scheduling at 50 and 60 mm of CPE. The soil moisture extracted from surface (0–30 cm) layers was maximum (49.88%) under 40 mm CPE, whereas from 45–60 cm layer maximum (20.18%) was noted with 50 mm CPE. From deeper layer (45–75 cm), highest moisture extraction was noticed with 60 mm CPE.

Hallikeri et al. [17] found that early sown cotton produced significantly higher yield (2227 kg/ha) than late sowing in July (1809 kg/ha) and August (1004 kg/ha). Irrigation at 0.8 IW/CPE ratio produced significantly more cotton yield (1807 kg of seed /ha) compared to 0.4 or 0.6 IW/CPE ratios but was at par with 0.6 IW/CPE ratio (1731 kg/ha). Productivity was affected to a greater extent by date of sowing than irrigation levels. Loss in yield due to late sowing was not compensated with increase in irrigation level. Different moisture regimes did not affect the fiber length, fiber strength, fiber fineness and uniformity percentage. Fiber maturity was significantly increased with 0.6 and 0.8 IW/CPE ratios. Cotton with best fiber quality was produced with June sowing at 0.6 IW/CPE ratio, which can produce highest fiber length, fiber strength, uniformity percentage and maturity ratio under transitional tract of Dharwad.

Ramamoorthy et al. [34] found that hybrid sunflower yield and yield attributes varied significantly during both summer and *kharif* seasons of 2004 and 2005. Irrigation at 0.80 and 0.60 IW/CPE ratios under alternate furrows and paired row system of irrigation (80/40 × 30 cm) provided better performance than other treatment combinations with regard to WUE, gross return, net return and B-C ratio.

Alan et al. (2010) found that a significant increase in average dry bean seed yield (15% in 2006 and 46% in 2007) and in WUE (30% in 2005 and 50% in 2007) was found in more frequently irrigated treatments (0–30 m root zone, split 0–30/0–60 cm root zone, and 12-mm water depth) compared to less frequently irrigated treatments (0–60 cm root zone and 50 mm water depth). Dry bean seed yield and WUE may be maximized by keeping the majority of roots moist.

Rathod and Vadodaria [35] concluded that to obtain higher grain yield, the crop should be irrigated either at critical crop growth stages (Crown root initiation, tillering late jointing, root, flowering, milking and drought stage) or at 0.9 IW/CPE ratio by irrigating the crop with 8 irrigations each of 50 mm depth including one common irrigation of 50 cm after sowing for getting uniform plant stand in wheat. Significantly highest WUE (20.28 and 24.07 kg/ha-mm) was recorded under 0.7 IW/CPE. However, post emergence application of 1.0 kg isoproturon/ha significantly recorded higher WUE (17.59 and 19.23 kg/ha-mm) than the other treatments during 1997–1998 and 1998–1999, respectively.

11.3 MATERIALS AND METHODS

During winter (Rabi season) of 2010–2011, a field experiment was conducted to study effects of open pan evaporation based irrigation scheduling on growth and yield of wheat crop at the Wheat Ram Research Unit of Dr. Panjabrao Deshmukh Krishi Vidyapeeth, Akola – India. The topography of the field was fairly uniform and leveled. The climate of Akola is subtropical semiarid characterized by three distinct seasons namely: summer becoming hot and dry from March to May, the warm and rainy monsoon from June to October; and winter with mild cold from November to February. Akola is situated in the subtropical zone at the latitude of 20° 42' North and longitude of 77° 02' East. Altitude of the place is 307.41 m above the mean sea level. Average annual precipitation is 750 mm and the major amount is received during the period from June to September. Winter rains are few and uncertain. The normal mean monthly maximum temperature during the hottest month (May) is 42 °C while the normal mean monthly minimum temperature in the coldest month (December) is 10.7 °C. The mean daily evaporation reaches as high as 19.0 mm in the month of May and as low as 3.00 mm in the month of August.

To determine important physical/ chemical properties of the soil, samples were drawn from 0–60 cm depth at randomly selected spots spread over the experimental area before sowing of the crop. A composite sample was prepared and analyzed. Physical analysis was done as per the standard procedure. Similarly standard procedure was used to perform chemical analysis of soil (Table 11.1).

TABLE 11.1 Physical and Chemical Properties of Soil

Soil property	Value	Analytical method
Mechanical composition		
Sand (%)	14.78	Bouycous hydrometer method (Piper, 1966)
Silt (%)	33.69	
Clay (%)	51.53	
Soil texture	Clay	
Chemical composition		
Soil organic carbon (%)	0.63	Walkyey and Black method (Piper, 1966)
Available nitrogen (kg/ha)	238.36	Modified Kjeldahl Method (Piper, 1966)
Available phosphorus (kg/ha)	24.52	Olsen's Method (Jackson, 1967)
Available potassium (kg/ha)	266.85	Flame photometric (Jackson, 1967)
pH (1:2.5 soil water ratio)	7.84	Beckman's Glass electrode pH meter (Jackson, 1967)
EC (ds/m)	0.77	Conductivity bridge from 1:2:5 soil water ratio paste.

The soil moisture constant in terms of field capacity and permanent wilting point was determined using the standard procedure (Table 11.2).

TABLE 11.2 Soil Moisture Constants

Constant	Value	Analytical method
Bulk density (g/cm^3) 0–30 cm layer	1.18	Core sampler method (Piper 1966)
Field Capacity (%) 0–60 cm	38.25	Pressure plate and pressure membrane apparatus (Dastane, 1972)
Permanents wilting point (0–60 cm)	17.21	Pressure plate and pressure membrane apparatus Dastane 1972)

The source of water was open well in wheat Research Unit. The water was conveyed to the field through pipe line. Before starting experiment, water was analyzed to the check its suitability for irrigation (Table 11.3).

TABLE 11.3 Chemical Analysis of Irrigation Water

Item	Value
pH	7.24
Ec (ds/m)	0.33
CO_3	2.00
HCO_3	8.00
Cl (meq/lit)	2.40
Ca (meq/lit)	2.00
[Ca + mg] (meq/lit)	2.40
Na (meq/lit)	3.20
SAR	2.63
RSC	0.40

11.3.1 *EXPERIMENTAL DETAILS*

The field experiment was laid out in randomized block design, with four replications and five treatments. In four treatments out of five, irrigation was scheduled on the basis of various IW/CPE ratios and in one control treatment irrigation was scheduled at critical growth stages of wheat. The layout of experiment is shown in the Fig. 11.1. The pipe line with water meter is shown in Fig. 11.2. Treatments and experimental details were:

I_1	IW/CPE = 0.6
I_2	IW/CPE = 0.8
I_3	IW/CPE = 1.0
I_4	IW/CPE = 1.2
I_5	Control with six irrigations at Crown Root Initiation (CRI), Maximum Tillering, Late Jointing, Flowering, Milking Stage and Dough Stage.

Crop details:	
Crop	**Wheat**
Scientific name	*Triticum aestivum* L
Variety	AKAW – 4627
Experimental Design	Randomized block design
Number of replications	4
Number of treatments	5
Number of plots	20
Interspace between replication/plot	2 m
Season	Winter 2010
Crop spacing	18 cm (row to row)
Seed rate	140 kg/ha
Recommended fertilizer dose	80:40:40(N:P:K)
Duration of crop	95 days
Date of sowing	13 Dec. 2010
Date of harvesting	30 Mar. 2011

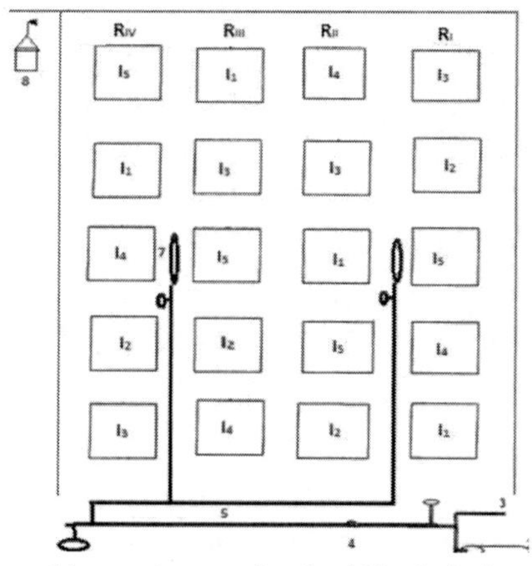

FIGURE 11.1 Experimental layout in a randomized block design.

FIGURE 11.2 Irrigation scheduling with pipeline and water meter.

11.3.2 IRRIGATION SCHEDULING

For the purpose of irrigation scheduling the irrigation in various treatments, predetermined soil moisture constants were used. Total available water (TAW) is the amount of water that is available for plant use. It is actually the difference of soil moistures between field capacity and permanent wilting point. The total available water was calculated using following formulae.

$$TAW = \left[\frac{\theta_{FC} - \theta_{PWP}}{100}\right] \times \gamma \times Z_r \times 1000 \tag{1}$$

where, TAW = total available water, (mm), Q_{Fc} = field capacity (%); Q_{pwp} = permanent wilting point, γ = bulk density (g/cc), and Z_r = root zone depth of 60 cm soil layer (cm).

Using soil moisture constants, firstly total available water was determined for the experimental soil. For this purpose, depth of effective root zone was taken 60 cm for wheat crop. After determining TAW, depth of irrigation (IW in mm) was determined as follows considering the maximum allowable depletion of 50%:

$$IW = 0.50 \times TAW \qquad (2)$$

Different moisture regimes were created by different irrigation scheduling based on IW/CPE ratio. For this purpose, CPE for respective treatments of IW/CPE ratios were determined using predetermined IW and values of ratios by using following equation:

$$CPE = IW \div ratio \qquad (3)$$

Pan evaporation data were recorded daily and cumulative figures were calculated subtracting the rainfall. In IW/CPE approach, a known amount of irrigation water was applied when CPE reached a predetermined level, determined as per equation (3).

In the control treatment, six irrigations were scheduled at six critical growth stages of wheat crop: crown root initiation (CRI), maximum tillering, late jointing, flowering, milking stage and dough stage. In this treatment, depth of irrigation was determined by observing actual soil moisture before each irrigation, as follows:

$$TAW = \left[\frac{\theta_{FC} - \theta_{BI}}{100}\right] \times \gamma \times Z_r \times 1000 \qquad (4)$$

where, IW = irrigation water, (mm); θ_{FC} = moisture content at field capacity, (%); θ_{BI} = Moisture content before irrigation, (%); γ = Bulk density, (gm/cm^3); and Z_r = Effective root zone depth, (m).

First common irrigation was given to all treatments just after sowing to bring the experimental plots to field capacity. For this purpose soil moisture content was determined before sowing to calculate the depth of irrigation of first common irrigation. In all plots, water was conveyed through pipeline and measured quantity of water was applied using water meter (See Fig. 11.2).

11.3.3 AGRONOMIC PRACTICES

Wheat variety AKW-4627 is sown late and matures in 95 to 105 days. The seed was obtained from the Wheat Research Unit at Dr. Panjabrao Deshmukh Krishi Vidyapeeth, Akola. The details of cultural operations carried out in the experimental plot during the crop season are presented below:

Preparatory tillage:		
Harvesting with disc harrow	1	18–11–2010
Harrowing with a blade harrow	1	19–11–2010

Collection of crop residues	1	21–11–2010
Leveling	1	27–11–2010
Layout and opening of irrigation channels	1	10–12–2010
Fertilizer application:		
Recommended half dose of N at sowing	1	13–12–2010
Recommended full dose of P_2O_5 and K_2O at sowing	1	13–12–2010
Recommended half dose of N top dressing	1	12–1–2011
Sowing:		
Marking of row by hand marker	1	13–12–2010
Sowing	1	13–12–2010
Other post sowing operations:		
First hand weeding	1	5–1–2011
Second hand weeding	1	28–1–2011
Irrigation – as per treatment:		
Harvesting and threshing:		
Harvesting	1	30–3–2011
Threshing	1	30–3–2011

On 13th December 2010, sowing was done using seed rate of 140 kg/ha. After sowing, irrigation was given to each plot for seed germination and it was observed till 6th day after sowing (DAS). The recommended dose of fertilizer for wheat under late sown condition (80 kg N + 40 kg P_2O_5 + 40 kg K_2O /ha) was applied to the crop. Nitrogen was applied in two split doses. Half dose of nitrogen (40 kg/ha) and full dose of P_2O_5 (40 kg/ha) and K_2O (40 kg/ha) was applied at the time of sowing. Remaining half dose of nitrogen was applied at CRI stages. No incidence of pest and diseases was noticed crop growth was normal. The crop from net area of plot was harvested at maturity. The harvested produce was tied in separate bundles, labeled, weighed and kept for sun drying. As preproduce was completely dried, it was threshed with wooden batons. The grains were winnowed, cleaned and weighed separately from each net area of plot.

11.3.4 SOIL MOISTURE DEPLETION PATTERN AND MOISTURE AVAILABILITY INDEX

Evaluation of soil moisture depletion patterns was necessary to determine amount of irrigation water that was infiltrated into and percolated through the soil mass around the root spread of the crop. To evaluate the soil moisture depletion pattern before sowing, the soil samples were collected by screw auger and immediately kept in the oven for 24 h. Then, soil moisture content were determined by gravimetric method on dry weight basis, and depth of irrigation and volume of water were calculated.

An each before and after irrigation, the moisture was measured by a micro gopher profiler. The axis tube was fixed to each plot of field by using screw auger. One plot moisture content was calibrated with a moisture meter attached with sensor at bottom of meter. It can automatically measure moisture content at a particular depth and record it. Daily reference evapotranspiration (ET_o) for the growing period of wheat was calculated using standard Penman-Monteith equation and daily climatic data, as follows:

$$ET_o = \frac{0.408\Delta(R_n - G) + \gamma\left(\frac{900}{T+273}\right)u_2(e_s - e_a)}{\Delta + \gamma(1 + 0.34u_2)} \quad (5)$$

where, ET_o = reference evapotranspiration (mm day^{-1}), D = slope of saturation vapor pressure curve (kPa °C^{-1}), T = mean air temperature (°C), g = psychrometric constant (kPa °C^{-1}), R_n = net radiation at the crop surface (MJ m^{-2} day^{-1}), G = soil heat flux density (MJ m^{-2} day^{-1}), u_2 = wind speed at 2.0 m height (ms^{-1}), e_a = actual vapor pressure (kPa), e_s = saturation vapor pressure (kPa), and (e_s–e_a) = saturation vapor pressure deficit (kPa).

Total reference evapotranspiration (ET_o) during crop season was determined and it was divided by total water requirement to get moisture availability index.

11.3.5 BIOMETRIC OBSERVATIONS

The various biometric observations were recorded at periodical intervals is given as shown below:

Pre-harvest observations:		
Number of effective tillers per m^2	1	At harvest
Plant Height (cm)	3	30 days, 60 days, At harvest
Post harvest observations:		
Length of earhead	1	At harvest
Length of peduncle (cm)	1	At harvest
Number of grains per earhead	1	At harvest
Test weight (g)	1	At harvest
Yield:		
Grain yield (kg/ha)	1	At harvest
Straw yield (kg/ha)	1	At harvest

Within each net area of plot, the representative row of one meter or length were selected randomly and then marked by fixing pegs at the two extreme ends of each unit. Five observation plants were labeled for recording various observations at fixed intervals. Throughout the growth period, the *plant height* was measured at 30 days interval on five randomly selected plants. The height of main shoot was measured in cm from the ground level, up to the auricle of the fully open top leaf. After earhead emergence, the plant height was measured up to the base of the earhead.

Total number of effective tillers inclusive of main shoot per meter of row length were counted and recorded, just before harvesting. *Length of peduncle* was measured from base of plant to first inter node of plant from top in centimeter. *Length of earhead* was measured (cm) from the ring at the basic of the ear to its tip and mean earhead length was calculated. About 10 earheads were randomly selected from each plot and were threshed separately. The grains obtained from earheads were counted and the average number of grains per earhead was calculated.

From the produce in each plot, representative samples of wheat grains were collected. The 1000 grains were counted, and weighed. The produce from each plot was threshed separately. The grains were cleaned, dried and weighed (grain yield per plot). **Straw yield per plot** was calculated by deducting the grain yield from the weight of total produce per plot.

11.3.6 WATER USE EFFICIENCY

Water use efficiency (WUE) of wheat for different irrigation treatments was calculated using grain yield, straw yield and irrigation water for the entire season [15, 16].

$$\text{WUE} = [\text{Y}]/[\text{mm}] \tag{6}$$

where, WUE = water use efficiency (kg/ha-mm), Y = economic yield (kg/ha), and ET = total evapotranspiration (mm).

11.3.7 STATISTICAL ANALYSIS

The statistical analysis was using analysis of variance. The F – test revealed significant effects. The critical difference (CD) was determined at P = 0.05% to compare the treatments.

11.4 RESULTS AND DISCUSSION

11.4.1 CROP WATER REQUIREMENT

The estimation of the crop water requirement is one of the basic needs for crop planning on the farm. Unless this is given proper consideration, it results in poor crop growth. Crop water requirement varies with kind of crop, degree of maturity, water availability and climatic conditions (humidity, wind velocity, sunshine hours and temperature). For this purpose, the authors used the climatological data recorded at Meteorological Observatory of Department of Agronomy at Dr. PDKV, Akola – India during the period 13th December 2010 to 30th March 2011.

TAW was determined using soil moisture constants of the soil. Depth of IW per irrigation was calculated considering 50% maximum allowable depletion. Then, CPE at predetermined IW and at different IW/CPE ratios was calculated. Accordingly, irrigation scheduling were calculated as listed below:

Total available water (TAW), mm		149.0
Depth of irrigation (IW), mm		75.0
Cumulative pan evaporation at which irrigation was scheduled (CPE), mm	I_1 (IW/CPE = 0.6)	125.0
	I_2 (IW/CPE = 0.8)	93.8
	I_3 (IW/CPE = 1.0)	75.0
	I_4 (IW/CPE = 1.2)	62.5

Thus, irrigation was scheduled at 125 mm CPE in treatment I_1 (IW/CPE = 0.6), at 93.8 mm CPE in treatment I_2 (IW/CPE = 0.8), at 75 mm CPE in treatment I_3 (IW/CPE = 1.0), and at 62.5 mm CPE in treatment I_4 (IW/CPE = 1.2). However, in control treatment I_5, irrigation was scheduled at 14, 28, 36, 57, 75 and 82 days after sowing at six critical growth stages of wheat.

Accordingly the evaporation data, cumulative pan evaporation and dates of irrigation are shown in Table 11.4. The underlined and bold cumulative pan evaporation shows that irrigation was given on that respective day.

Details of irrigation water applied are given in Table 11.5. Irrigation water is conveyed through pipe and water meter was used to apply the measured amount of water at each irrigation. First common irrigation was applied in each treatment to bring the soil to field capacity. Depth of irrigation of this first common irrigation was determined on the basis of actual soil moisture content available before sowing. The depth of irrigation water to be applied in control treatment was determined by observing the soil moisture before irrigation.

TABLE 11.4 Evaporation Data During Growing Period of Wheat

Date	**Evaporation (mm)**	**Treatments**				
		Cumulative Evaporation, mm				**I_5 (Control)**
		I_1 (125)	**I_2 (93.8)**	**I_3 (75)**	**I_4 (62.5)**	
13-Dec-2010	Date of sowing	Common irrigation after sowing				

14-Dec-2010	2.9	2.9	2.9	2.9	2.9	
15-Dec-2010	3.4	6.3	6.3	6.3	6.3	
16-Dec-2010	3.6	9.9	9.9	9.9	9.9	
17-Dec-2010	3.8	13.7	13.7	13.7	13.7	
18-Dec-2010	3.9	17.6	17.6	17.6	17.6	
19-Dec-2010	3.4	21	21	21	21	
20-Dec-2010	4.6	25.6	25.6	25.6	25.6	
21-Dec-2010	3.3	28.9	28.9	28.9	28.9	
22-Dec-2010	3.2	32.1	32.1	32.1	32.1	
23-Dec-2010	3.2	35.3	35.3	35.3	35.3	
24-Dec-2010	3.4	38.7	38.7	38.7	38.7	
25-Dec-2010	3.8	42.5	42.5	42.5	42.5	
26-Dec-2010	3.4	45.9	45.9	45.9	45.9	
27-Dec-2010	6.2	52.1	52.1	52.1	52.1	First irrigation
28-Dec-2010	4.8	56.9	56.9	56.9	56.9	
29-Dec-2010	4.1	61	61	61	61	
30-Dec-2010	6.1	67.1	67.1	67.1	67.1	
31-Dec-2010	3.2	70.3	70.3	70.3	3.2	
1-Jan-2011	5.6	75.9	75.9	75.9	8.8	
2-Jan-2011	3	78.9	78.9	3	11.8	
3-Jan-2011	3.1	82	82	6.1	14.9	
4-Jan-2011	3.5	85.5	85.5	9.6	18.4	
5-Jan-2011	3.2	88.7	88.7	12.8	21.6	
6-Jan-2011	3.8	92.5	92.5	16.6	25.4	
7-Jan-2011	3.6	96.1	96.1	20.2	29	
8-Jan-2011	3.8	99.9	3.8	24	32.8	
9-Jan-2011	3.6	103.5	7.4	27.6	36.4	
10-Jan-2011	4.8	108.3	12.2	32.4	41.2	Second irrigation
11-Jan-2011	3.2	111.5	15.4	35.6	44.4	
12-Jan-2011	3.2	114.7	18.6	38.8	47.6	
13-Jan-2011	3.8	118.5	22.4	42.6	51.4	
14-Jan-2011	4.4	122.9	26.8	47	55.8	
15-Jan-2011	5	127.9	31.8	52	60.8	

16-Jan-2011	5.2	5.2	37	57.2	66	
17-Jan-2011	4.3	9.5	41.3	61.5	4.3	
18-Jan-2011	4.3	13.8	45.6	65.8	8.6	Third irrigation
19-Jan-2011	5.8	19.6	51.4	71.6	14.4	
20-Jan-2011	5.6	25.2	57	77.2	20	
21-Jan-2011	3.4	28.6	60.4	3.4	23.4	
22-Jan-2011	3.6	32.2	64	7	27	
23-Jan-2011	4.4	36.6	68.4	11.4	31.4	
24-Jan-2011	4.4	41	72.8	15.8	35.8	
25-Jan-2011	4.2	45.2	77	20	40	
26-Jan-2011	5.4	50.6	82.4	25.4	45.4	
27-Jan-2011	4	54.6	86.4	29.4	49.4	
28-Jan-2011	4.4	59	90.8	33.8	53.8	
29-Jan-2011	6.4	65.4	97.2	40.2	60.2	
30-Jan-2011	6	71.4	6	46.2	66.2	
31-Jan-2011	3.8	75.2	9.8	50	3.8	
1-Feb-2011	4.7	79.9	14.5	54.7	8.5	
2-Feb-2011	5.2	85.1	19.7	59.9	13.7	
3-Feb-2011	6	91.1	25.7	65.9	19.7	
4-Feb-2011	6	97.1	31.7	71.9	25.7	
5-Feb-2011	5	102.1	36.7	76.9	30.7	
6-Feb-2011	7.8	109.9	44.5	7.8	38.5	
7-Feb-2011	6.8	116.7	51.3	14.6	45.3	
8-Feb-2011	5.2	121.9	56.5	19.8	50.5	Forth irrigation
9-Feb-2011	4.6	126.5	61.1	24.4	55.1	
10-Feb-2011	5.9	5.9	67	30.3	61	
11-Feb-2011	7.6	13.5	74.6	37.9	68.6	
12-Feb-2011	6.6	20.1	81.2	44.5	6.6	
13-Feb-2011	6	26.1	87.2	50.5	12.6	
14-Feb-2011	6.4	32.5	93.6	56.9	19	
15-Feb-2011	7.4	39.9	101	64.3	26.4	
16-Feb-2011	7.8	47.7	7.8	72.1	34.2	
17-Feb-2011	6.8	54.5	14.6	78.9	41	

18-Feb-2011	6.4	60.9	21.4	6.4	47.4	
19-Feb-2011	6.2	67.1	27.2	12.6	53.6	
20-Feb-2011	7.6	74.7	34.8	20.2	61.2	
21-Feb-2011	4.9	79.6	39.7	25.1	66.1	
22-Feb-2011	5.2	84.8	44.9	30.3	5.2	
23-Feb-2011	7.8	92.6	52.7	38.1	13	
24-Feb-2011	6.4	99	59.1	44.5	19.4	
25-Feb-2011	5.7	104.7	64.8	50.2	25.1	
26-Feb-2011	6	110.7	70.8	56.2	31.1	Fifth irrigation
27-Feb-2011	6.6	117.3	77.4	62.8	37.7	
28-Feb-2011	5.4	122.7	82.8	68.2	43.1	
1-Mar-2011	7.6	130.3	90.4	75.8	50.7	
2-Mar-2011	7.2	7.2	97.6	7.2	57.9	
3-Mar-2011	6.2	13.4	6.2	13.4	64.1	
4-Mar-2011	7.6	21	13.8	21	7.6	
5-Mar-2011	5	26	18.8	26	12.6	Sixth irrigation
6-Mar-2011	6.8	32.8	25.6	32.8	19.4	
7-Mar-2011	6.4	39.2	32	39.2	25.8	
8-Mar-2011	8.2	47.4	40.2	47.4	34	
9-Mar-2011	10	57.4	50.2	57.4	44	
10-Mar-2011	9.8	67.2	60	67.2	53.8	
11-Mar-2011	10.6	77.8	70.6	77.8	64.4	
12-Mar-2011	6.8	84.6	77.4	6.8	6.8	
13-Mar-2011	9.4	94	86.8		16.2	
14-Mar-2011	7.2	101.20	94		23.4	

TABLE 11.5 Details of Irrigation Water Applied

Irrigation	Irrigation treatments									
	I1 (IW/CPE=0.6)		I2 (IW/CPE=0.8)		I3 (IW/CPE=1.0)		I4 (IW/CPE=1.2)		I5 (Control)	
	Irrigation water (IW) applied, mm									
	Date	IW	Date	IW	Date	IW	Date	IW	Date	IW
Post-sowing	75 mm of common irrigation after Sowing on 13–12–2010									
First	15–1-2011	75	7–1-2011	75	1–1-2011	75	30–12–2011	75	27–12–2010	67
Second	9–2-2011	75	29–1-2011	75	20–1-2011	75	16–1-2011	75	10–1-2011	83.2
Third	1–3-2011	71.5	15–2-2011	75	5–2-2011	75	30–1-2011	75	18–1-2011	61.7
Fourth	–	–	2–3-2011	71.5	17–2-2011	75	11–2-2011	75	8–2-2011	100.8
Fifth	–	–	14–3-2011	72	1–3-2011	71.5	21–2-2011	75	26–2-2011	80.7
Sixth	–	–	–	–	11–3-2011	72	3–3-2011	71.5	5–3-2011	71.5
Seventh	–	–	–	–	–	–	11–3-2011	72	–	–
Total		296.5		443.5		518.5		593.5		539.9
Rainfall received during the crop season = 6.50 mm.										

It is seen from Table 11.5 that highest number of irrigations (seven) were applied in treatment I_4 (IW/CPE = 1.2), whereas six irrigations were applied in treatments I_3 (IW/CPE=1.0) and I_5 (Control), followed by treatment I_2 (IW/CPE=0.8) with five irrigations. However, in treatment I_1 (IW/CPE=0.6), only three irrigations were applied. During the crop season, 6.5 mm rainfall was received. It is observed that the total amount of irrigation water applied during crop season was highest in treatment I_4 (593.5 mm), followed by I_5 (539.9 mm), I_3 (518.5 mm), I_2 (443.5 mm) and I_1 (296.5 mm). Table 11.6 indicates crop water requirement for each treatment and at different growth stage.

TABLE 11.6 Crop Water Requirements

Crop growth stage	DAS, days	Water requirement, mm				
		I_1	I_2	I_3	I_4	I_5
Post sowing	0	75	75	75	75	75
Crown root initiation	14	-	75	75	75	67
Maximum tillering	28	75	-	-	75	83.2
Late jointing	36	-	75	150	75	61.7
Flowering	57	75	75	75	150	100.8
Milking stage	75	75	75	75	75	84.2
Dough stage	82	3	75	75	75	74.5
Seasonal water requirement		303	450	525	600	546.5

TABLE 11.7 Total Water Requirement of Wheat

Treatment	No. of Irrigations	Irrigation water applied	Rainfall	Total water requirement	Saving of water over control
	–	mm			%
I_1 (IW/ CPE=0.6)	3	296.5	6.5	303	45
I_2 (IW/ CPE=0.8)	5	443.5	6.5	450	18
I_3 (IW/ CPE=1.0)	6	518.5	6.5	525	4
I_4 (IW/ CPE=1.2)	7	593.5	6.5	600	(-) 10
I_5 (Control)	6	539.9	6.5	546.4	–

It is seen from Table 11.6 that irrigations were scheduled for all growth stages in I_4 and I_5, whereas irrigation was not scheduled during maximum tillering stage in I_2 and I_3. Similarly in I_1, irrigation was not scheduled during three growth stages: crown root initiation, late jointing and dough stage.

Total water requirement and saving of water for different treatments is presented in Table 11.7. Number of irrigations applied were highest in I_4 (7), followed by I_3 and I_5 (6), I_2 (5), I_1 (3). Irrigation water applied under different irrigation scheduling varied from 593.5 to 296.5 mm. Total water requirement of wheat was highest

(600 mm) under irrigation scheduling at I_4 (IW/CPE = 1.2), followed by I_5 (Control: 546.4 mm), I_3 (IW/CPE = 1.0, 525 mm) and I_2 (IW/CPE = 0.8, 450 mm). It was lowest (303 mm) under irrigation scheduling at I_1 (IW/CPE = 0.6). Hence highest saving of water over control treatment was achieved in I_1 (45%), followed by I_2 (18%) and I_3 (4%), whereas more water was required in I_4 (10%) as compared to control treatment.

11.4.2 SOIL MOISTURE DEPLETION PATTERN

The moisture depletion in soil under different irrigation treatments are presented in Table 11.8. The soil moisture depletion curves are presented in Fig. 11.3.

TABLE 11.8 Soil Moisture Data Obtained Under Different Irrigation Treatments

Days after sowing	Soil moisture at depth, %			Average soil moisture content, %	Depletion before next irrigation, %
	0–15 cm	15–30 cm	30–45 cm		
		IW/CPE ratio = 0.6, I_1 treatment			
2	37.7	38.35	39.05	38.37	-
33	21.05	25.29	28.04	24.79	63.97
35	36.04	37.80	37.24	37.02	-
57	20.12	24.34	27.20	23.88	68.30
60	35.40	35.47	36.85	35.91	-
77	19.24	23.33	26.29	22.95	72.72
81	33.42	35.05	35.60	34.69	-
		IW/CPE ratio = 0.8, I_2 treatment			
2	37.70	38.35	38.53	38.19	-
24	25.47	28.23	29.60	27.77	49.81
27	37.23	38.14	38.85	38.07	-
42	24.25	27.00	29.04	26.76	54.61
46	36.54	37.48	38.14	37.38	-

60	22.85	26.65	27.84	25.78	59.27
63	35.78	36.40	37.40	36.53	-
IW/CPE ratio = 1.0, I_3 treatment					
2	37.70	38.35	39.05	38.37	-
19	25.51	29.47	30.37	28.45	46.58
22	37.60	38.55	39.17	38.44	-
37	26.35	28.22	31.40	28.66	45.58
40	37.27	37.89	38.95	38.03	-
53	26.05	29.47	30.93	28.82	44.82
57	37.22	38.28	38.99	38.16	-
65	26.69	29.07	31.40	29.05	43.73
67	36.87	37.40	38.55	37.61	-
77	26.88	29.40	31.14	29.14	43.30
81	36.45	38.00	38.83	37.76	-
88	27.35	30.54	31.83	29.91	39.64
91	36.60	37.49	38.47	37.52	-
IW/CPE ratio = 1.2, I_4 treatment					
2	37.70	38.35	39.05	38.37	-
17	25.90	29.23	30.33	28.48	46.43
19	36.95	38.07	39.79	38.27	-
33	25.69	30.14	30.95	28.92	44.34
37	38.00	38.82	39.76	38.86	-
48	26.05	29.47	31.68	29.07	43.63
50	37.82	38.08	39.00	38.30	-
60	26.94	29.19	31.90	29.34	42.35

64	38.35	38.15	39.85	38.78	-
70	26.45	31.93	32.30	30.23	38.12
73	38.02	38.03	39.70	38.58	-
79	26.79	31.92	33.60	30.77	35.55
82	37.82	37.12	39.95	38.30	-
88	28.52	32.19	33.19	31.30	33.03
91	37.70	36.73	38.63	37.69	-
		I_4 control treatment			
2	37.70	38.30	38.55	38.18	-
14	26.55	29.23	30.58	28.78	45.01
17	37.87	37.52	38.35	37.91	-
28	23.68	25.83	30.00	26.50	55.85
31	38.18	39.13	39.70	39.01	-
36	25.60	30.27	32.73	29.53	41.44
39	36.37	38.27	39.07	37.90	-
57	21.67	23.27	27.10	24.01	67.68
60	36.92	37.48	39.29	37.89	-
75	22.94	28.00	29.63	26.85	54.18
77	36.10	37.37	38.59	37.35	-
82	25.67	28.49	30.30	28.15	48.00
85	36.25	37.42	38.20	37.29	-

FC = 38.25%, PWP = 17.21%, and MAD = 27.73%.

It can be observed in Table 11.8 that soil moisture depletion before each irrigation in treatment I_1 was more than 50%, which shows the inadequacy of moisture available to crop. Figure 11.3a shows that soil moisture depletion was inadequate in treatment I_1, as it was depleted below maximum allowable depletion level before each irrigation. Therefore sufficient amount of soil moisture was not maintained throughout of growing period of wheat crop.

In treatment I_2, Table 11.8 reveals that soil moisture depletions before four irrigations were depleted slightly below maximum allowable depletion, which shows the inadequacy of soil moisture in later stages of crop. Figure 11.3b indicates that soil moistures were depleted below allowable depletion limit before three irrigations. The favorable soil moisture was not maintained throughout growing period, as it was depleted slightly below allowable depletion limit.

In treatment I_3, Table 11.8 reveals that soil moistures before irrigation were within the allowable depletion limit of 50%, which shows favorable soil moisture was maintained throughout the growing season of crop. Figure 11.3c indicates that favorable soil moisture was maintained, as the depletion of moisture was within allowable limit.

In treatment I_4, Table 11.8 shows that soil moisture depletions before each irrigation were always more than maximum allowable depletion that shows the adequacy of soil moisture throughout the growing season of crop. Figure 11.3d indicates that soil moisture was sufficient throughout the growing period of wheat crop

In the control treatment I_5, Table 11.8 shows that soil moisture was depleted below maximum allowable limit of 50%, before three irrigations only. It may be due to the interval of those stages of crop were enough to deplete the soil moisture below allowable limit. Figure 11.3e shows that in control treatment I_5, soil moisture depleted below allowable depletion before three irrigations only.

11.4.3 MOISTURE AVAILABILITY INDEX

Daily reference evapotranspiration rate for the growing period of wheat was calculated using standard Penman-Monteith model (FAO-56) using daily climatic (temperatures, relative humidity, sunshine hours and wind speed). Total water requirement of wheat was used to determine the moisture availability index (MAI), as shown in Table 11.9.

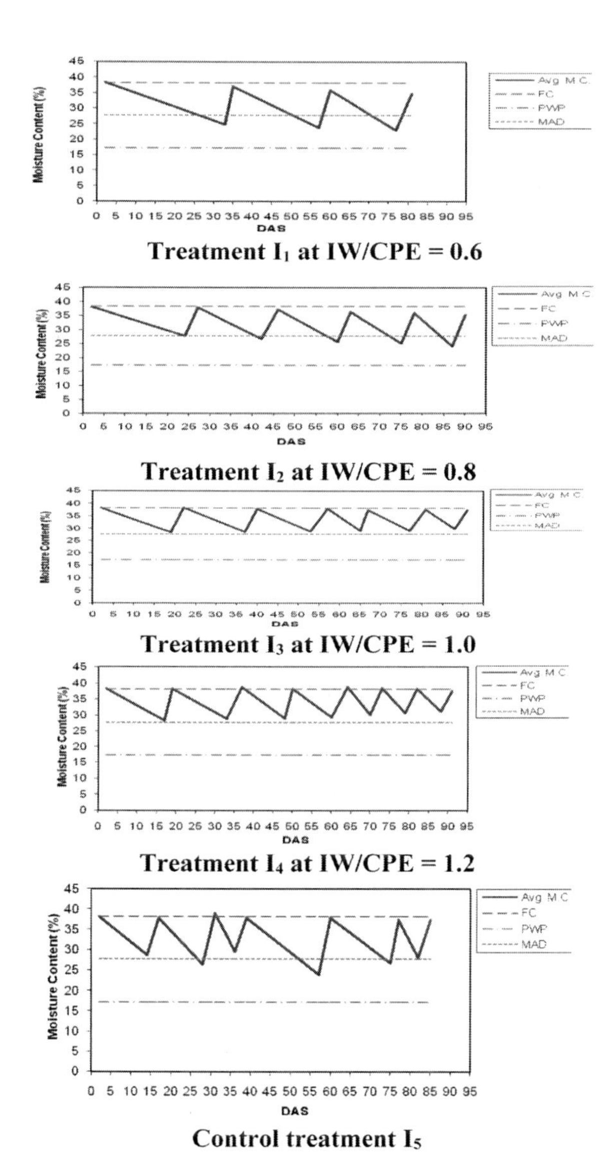

FIGURE 11.3 Soil moisture depletion pattern versus days after sowing (DAS), under different irrigation treatments.

TABLE 11.9 Moisture Availability Index Under Different Irrigation Scheduling Treatments, During the Growing Season

Treatment	Total water requirement, mm	Reference evapotranspiration (ETo), mm	MAI
I_1 (IW/CPE=0.6)	303	312.97	0.97
I_2 (IW/CPE=0.8)	450	312.97	1.44
I_3 (IW/CPE=1.0)	525	312.97	1.68
I_4 (IW/CPE=1.2)	600	312.97	1.92
I_5 (Control)	546.4	312.97	1.75

It is seen that MAI was more than unity for all the treatments except I_1. MAI was highest in I_4, which indicates the availability of sufficient moisture in the root zone throughout the season, followed by I_5, I_3, and I_2.

11.4.4 PERFORMANCE OF WHEAT

View of wheat crop at time of dough stage for different treatments is shown in Fig. 11.4. The Table 11.10 shows *plant height* at various growth stages and number of effective tillers per meter as influenced by different irrigation scheduling treatments.

Plant height of wheat (AKAW-4627) was different significantly due to irrigation scheduling at 30 DAS, 60 DAS and at harvest. At 30 DAS, highest plant height (36.90 cm) was significantly different in I_5 treatment compared to I_1 (32.38 cm). The lowest plant height was in I_1 irrigation scheduling. At 60 DAS and at harvest, significantly highest plant height of 91.25 cm and 92.33 cm was recorded in treatment I_4, respectively. The lowest plant height was recorded in I_1 irrigation scheduling treatment.

The *effective tiller per meter of row length* was influenced significantly by irrigation scheduling at harvest. Highest number of effective tillers was observed in I_4 (88.00), but was on par with I_3 (81.25) and I_5 (79.50). The lowest number of effective tillers was observed in I_1 (70.75), which were on par with I_2 (77.00).

FIGURE 11.4 View of wheat crop at time of dough stage for different treatments.

TABLE 11.10 Plant Height and Number of Effective Tillers Per Meter Under Different Irrigation Treatments

Treatment	Plant height at days after sowing, cm			Number of effective tillers per meter of row length
	30	60	At harvest	
I_1 (IW/CPE=0.6)	32.38	83.45	84.85	70.75
I_2 (IW/CPE=0.8)	34.46	84.05	86.00	77.00
I_3 (IW/CPE=1.0)	35.26	84.80	87.70	81.25
I_4 (IW/CPE=1.2)	36.09	91.25	92.33	88.00
I_5 (Control)	36.90	86.33	87.20	79.50
Mean	35.02	85.98	87.62	79.30
F – test	Sig.	Sig.	Sig.	Sig.
SE(m)+	0.705	1.564	1.261	3.016
CD at 5%	2.173	4.819	3.886	9.293
CV (%)	4.029	3.639	2.879	7.607

TABLE 11.11 Growth Characteristics Observed at Harvest

Treatment	Length of peduncle, cm	Length of earhead, cm	No. of grains/ earhead	Weight of 1000 grains, gm
I_1 (IW/CPE=0.6)	30.96	6.97	40.10	38.63
I_2 (IW/CPE=0.8)	31.84	7.10	42.85	42.33
I_3 (IW/CPE=1.0)	35.19	7.42	45.35	44.38
I_4 (IW/CPE=1.2)	35.28	7.50	47.50	45.30
I_5 – Control	32.35	7.29	43.75	43.00
Mean	33.12	7.26	43.91	42.73
'F' test	Sig.	Sig.	Sig.	Sig.
SE(m)+	0.644	0.111	1.500	0.848
CD at 5%	1.985	0.342	4.621	2.614
CV (%)	3.890	3.060	6.832	3.971

At harvest, Table 11.11 indicates length of peduncle, length of earhead, number of grains per earhead and weight of 1000 grains. Significant differences in *length of peduncle* (cm) were observed among different irrigation treatments. Significantly highest length of peduncle was observed in I_4 (35.28 cm) compared to I_1 (30.96 cm) and I_2 (31.84 cm) but was at par with I_3 (35.19 cm). Significantly lowest length of peduncle was observed in I_1 (30.96 cm). Significant differences in *length of earhead* (cm) were observed among different irrigation scheduling's. Significantly highest length of earhead was observed in I_4 (7.50 cm) compared to I_1 (6.97 cm) and I_2 (7.10 cm). Significantly lowest length of earhead was observed in I_1 (6.97 cm), however, it was on par with I_2 (7.10 cm).

A significant differences in *number of grain per earhead* were observed among different irrigation scheduling's. Significantly highest number of grains per earhead was observed in I_4 (47.50) but was at par with I_3 (45.35) and I_5 (43.75). Significantly lowest number of grains per earhead was observed in I_1 (40.10). There was significant influence of irrigation scheduling on *weight of 1000 grains of wheat*. Significantly highest 1000-grain weight (45.30 gm) was recorded in I_4 (IW/CPE =1.2) compared to I_1 (38.63) and I_2 (42.33), but was at par with I_5 (43.00 g). Significantly lowest 1000 grain weight (38.63 gm) was observed in I_1 irrigation scheduling treatment.

TABLE 11.12 Grain Yield (100-kg/ha) and Water Use Efficiency (100-kg/ha-cm)

Treatments	Grain yield, 100-kg/ha	Straw yield, 100-kg/ha	Consump-tiveuse, ha-cm	WUE, 100-kg per ha-cm
I_1 (IW/CPE=0.6)	32.15	78.08	30.30	1.06
I_2 (IW/CPE=0.8)	38.01	85.48	45.00	0.85
I_3 (IW/CPE=1.0)	40.35	89.52	52.50	0.77
I_4 (IW/CPE=1.2)	43.80	92.99	60.00	0.73
I_5 (Control)	38.40	86.12	54.64	0.70
Mean	38.54	86.44		
'F' test	Sig.	Sig.		
SE(m)+	0.508	1.099		
CD at 5%	1.565	3.386		
CV (%)	2.636	2.543		

11.4.5 GRAIN YIELD AND WATER USE EFFICIENCY

The Table 11.12 indicates yield attributes and WUE under different irrigation scheduling treatments. The irrigation scheduling influenced the grain yield of wheat significantly. The significantly highest *grain yield* (4380 kg/ha) was noticed in I_4 irrigation scheduling compared to I_1 (32.15 100-kg/ha), I_2 (3801 kg/ha) and I_5 (3840 kg/

ha) but was on par with I_3 (4035 kg/ha). But treatment I_5 (control) was at par with treatment I_2 (IW/CPE = 0.8). Significantly lowest (3215 kg/ha) grain yield was recorded in I_1 (IW/CPE = 0.6). Highest yield in I_4 and I_3 may be due to favorable soil moisture maintained in the root zone throughout the growing period. Grain yield recorded in treatment I_3 (IW/CPE = 1.0) was significantly higher than that in treatment I_5 (control) with saving of water of 4%.

The *straw yield* differed significantly due to irrigation scheduling. Significantly highest straw yield (9299 kg/ha) was recorded in I_4 compared rest of treatments. Similarly straw yield recorded in treatment I_5 (control) was at par with that of treatment I_2. Significantly lowest straw yield was observed in I_1 (7808 kg/ha).

WUE (100-kg/ha-cm) was increased progressively with decreasing number of irrigations. Highest WUE (1.06 × 100-kg/ha-cm) was observed in I_1 irrigation scheduling closely followed by I_2, I_3 and I_5 (0.85, 0.77, 0.73 and 0.70 × 100-kg/ha-cm), respectively. The lowest WUE was noticed in I_5 (0.70 100-kg/ha-cm) irrigation scheduling.

It was observed that water use in treatment I_4 was higher than treatment I_5. The WUE in I_4 was higher than I_5. It may be due to higher grain yield recorded in treatment I_4 compared to treatment I_5.

11.5 CONCLUSIONS

Following conclusions are drawn in this study:

1. Seasonal water requirement of wheat was highest (600 mm) under I_4 irrigation scheduling at IW/CPE = 1.2, followed by I_5 (Control with six irrigations: 546.50 mm), I_3 (IW/CPE=1.0, 525 mm) and I_2 (IW/CPE=0.8, 450 mm). It was lowest (303 mm) under I_1 irrigation scheduling at IW/CPE = 0.6.
2. Highest saving of water (45%) compared to control treatment was achieved in I_1 (IW/CPE=0.6), followed by I_2 (18%) and I_3 (4%), whereas 10% more water is required in I_4 compared to control treatment.
3. Favorable soil moisture was maintained in the irrigation scheduling treatments of IW/CPE=1.2 (I_4) and IW/CPE=1.0 (I_3) throughout the growing period and it was always maintained in allowable depletion regime. However, soil moisture was not adequate in irrigation scheduling at IW/CPE=0.6 (I_1). Whereas with irrigation scheduling treatments I_2 and I_5, soil moistures were slightly depleted below allowable limit.
4. The MAI was greater than unity for all the irrigation scheduling treatments expect I_1. This indicates the availability of sufficient moisture in the root zone throughout the season.
5. Number of effective tillers, length of earhead, number of grains per earhead and 1000-grain weight were highest in I_4 (IW/CPE = 1.2), but were at par with I_3 (IW/CPE = 1.0) and I_5 (Control with six irrigations). Length of pe-

duncle in I_4 was highest and at par with I_3. Plant height in I_4 treatment was significantly highest compared to all other treatments.

6. Irrigation scheduling at IW/CPE = 1.2 (I_4) recorded significantly highest grain yield and was superior than rest of the treatments, followed by treatments I_3 (IW/CPE=1.0), I_5 (Control-six irrigations) and I_2 (IW/CPE=0.8). Treatment I_1 (IW/CPE=0.6) recorded significantly lowest yield compared to all other treatments.
7. Highest straw yield was recorded in treatment I_4 (IW/CPE=1.2), which is significantly superior over other treatments and followed by treatment I_3 (IW/CPE=1.0). Similarly straw yield recorded in treatment I_5 (Control) was found at par with that of treatment I_2 (IW/CPE=0.8).
8. Highest water use efficiency was recorded in treatment I_1, which may due to lowest water use, followed by I_2, I_3, and I_4. However, lowest WUE was recorded in treatment I_5 (Control). Water use and WUE in treatment I_4 was higher than treatment I_5. It may be due to higher grain yield recorded in treatment I_4 compared to treatment I_5.

11.6 SUMMARY

A field experiment was conducted at Wheat Research Unit of Dr. PDKV, Akola during *rabi* season of 2010–2011 to study the effects of open pan evaporation based irrigation scheduling on growth and yield of wheat.

The total water requirement and number of irrigations were highest (600 mm and seven) under irrigation scheduling at I_4 (IW/CPE = 1.2) followed by I_5 (546.4 mm), I_3 (525 mm) and I_2 (450 mm). Highest Water use efficiency was recorded in treatment I_1 (IW/CPE = 0.6), which may be due to lowest water use followed by I_2, I_3, and I_4.

Soil moisture depletion patterns showed that favorable soil moisture was maintained in the treatment I_4 and I_3 throughout the growing period. The soil moisture was always maintained in allowable depletion regime. Soil moisture was inadequate in treatment I_1 as it was depleted below maximum allowable depletion before each irrigation. MAI was more than unity for all the treatments expect I_1. The MAI was highest in I_4, which indicates the availability of sufficient moisture in the root zone throughout the season followed by I_5, I_3, I_2.

Results indicated that the grain and straw yield were significantly higher 4380 kg/ha and 9299 kg/ha, respectively in I_4 irrigation scheduling using seven irrigations. The values of growth and yield parameters were also significantly higher in I_4 over I_1, I_2, I_3 and I_5 treatments.

KEYWORDS

- **benefit cost ratio**
- **carrot**
- **College of Agriculture Engineering and Technology Akola, CAET**
- **consumptive use**
- **critical difference**
- **crop coefficient**
- **crop water requirement**
- **crown root**
- **cumulative open pan evaporation, COPE**
- **Days after sowing, DAS**
- **electrical conductivity**
- **evaporation**
- **evapotranspiration**
- **fertilization**
- **field capacity**
- **grain yield**
- **India**
- **irrigation**
- **irrigation depth**
- **irrigation scheduling**
- **Maharashtra**
- **management allowable depletion, MAD**
- **moisture availability index**
- **nutrient uptake**
- **onion**
- **open pan evaporation**
- **permanent wilting point**
- **Punjabrao Deshmukh Krishi Vidyapeeth**
- **quintal, q (=100 kg)**
- **root distribution**
- **water extraction pattern**
- **water use efficiency**
- **wheat**
- **winter wheat**

REFERENCES

1. Alam, M. S., Mallik, S. A., Costa, D. J. (2010). Effect of irrigation on the growth and yield of carrot (*Daucus carota* spp. Sativus) in hill valley. *Bangladesh J. Agril. Res.*, 35(2):323–329.
2. Anonymous, (2009). Economic survey of Maharashtra (2006–07). Govt. of Maharashtra, pp. 81–89.
3. Bandyopadhyay, P. K. (1997). Effect of irrigation schedules on evapotranspiration and water use efficiency of winter wheat (*Triticum aestivum*). *Indian J. Agron.*, 42(1):90–93.
4. Batra, B. R., Rai, B. (1985). Effect of different levels of irrigation and fertilization on the growth, yield and quality of carrot for root and seed production. *Vegetable Sci.,* 17(2):127–139.
5. Bhoi, P. G., Goundaje, A. B., Magar, S. S. (1993). Response of HD-2278 wheat (*Triticum aestivum*) to irrigation and fertilizer level. *Indian J. Agric. Sci.,* 63(8):504–506.
6. Channagoudar, R. F., Janawade, A. D. (2006). Effect of different levels of irrigation and sulfur on growth, yield and quality of onion (*Allium cepa* L.). *Karnataka J. Agric. Sci.*, 19(3):489–492.
7. Chaudhary, T. N., Bhatnagar, V. K. (1980). Wheat root distribution water extraction pattern and grain yield as influenced by time and rate of irrigation. *Agricultural Water Mgmt.*, 3(2):115–124.
8. Chauhan, R. P. S. (1991). Effect of saline irrigation at different stages of growth on yield, nutrient uptake and water use efficiency of wheat (*Triticum aestivum*). *Indian J. Agric. Sci.,* 61(8):395–398.
9. Chavan, D. A., K. R. Pawar, (1987). Consumptive use of crop coefficient values for wheat. *J. Maharashtra Agric. Univ.,* 12(1):72–73.
10. Choudhary, P. N., V. Kumar, (1980). The sensitivity of growth and yield of dwarf wheat to water stress at tree growth stages. *Irrigation Sci.,* 1(4):223–231.
11. Deshmukh, S. V., Mulay, A. J., Bharad, G. M., Kale, H. B., Atale, S. B. (1992). Effect of depletion of irrigation at critical growth stages on grain yield of late-sown wheat (*Triticum aestivum*) in Vidarbha. *Indian J. Agron.*, 57(1):47–48.
12. Doorenbos, J., Pruitt, W. O. (1983). *Crop water requirements.* FAO Irrigation and Drainage Paper 24, Rome, Italy.
13. Gajri, P. R., Prihar, S. S., Chema, H. S., Kapoor, A. (1991). Irrigation and tillage effects of root development, water use and yield of wheat on coarse textured soils. *Irrig. Sci.,* 12:161–168.
14. Gill, K. S., Lenvain, J. S. (1992). Wheat response to open pan evaporation based irrigation scheduling on two Oxic Paleustafls. *Communications in soil science and Plant Analysis,* 23(9–10):1089–1103.
15. Megh R. Goyal, (Ed.), (2013). *Management of Drip/Trickle or Micro Irrigation.* Oakville–ON, Canada: Apple Academic Press Inc., pp. 1–408.
16. Megh R. Goyal, (Ed.), (2015). *Research Advances in Sustainable Micro Irrigation, volumes 1–10.* Oakville–ON, Canada: Apple Academic Press Inc.
17. Hallikeri, S. S., Halemani, H. L., Patil, V. C., Palled, Y. B., Patil, B. C., Katageri, I. S. (2009). Influence of sowing time and moisture regimes on growth, seed cotton yield and fiber quality of Bt-cotton. *Karnataka J. Agric. Sci.*, 22(5):985–991.
18. Hassan, V. A., Ogunlela, V. B., Sinha, T. D. (1986). Agronomic stress at various growth stages and seedling rates. *J. Agron. and Crop. Sci.*, 158(3):172–180.
19. Hosmani, M. H., Janawade, A. D. (2007). Influence of irrigation scheduling and sand application enhancing rabi groundnut (*Arachis hypogeae* L.) productivity of UKP command area of Karnataka. *Karnataka J. Agric. Sci.*, 20(3):457–461.
20. Imtiyaz, M., Mgadla, N. P., Chepete, B., Manase, S. K. (2000). Response of six vegetable crops to irrigation schedules. *Agricultural Water Management*, 45:331–342.

21. Kibe, A. M., Singh, S. (2003). Influence of irrigation, nitrogen and zinc on productivity and water use by late sown wheat (*Triticum aestivum*). *Indian J. Agron.*, 48(3):186–191.
22. Kumar, S., Imtiyaz, M., Kumar, A. (2009). Studying the feasibility of using Micro irrigation systems for vegetable production in a canal command area. *Irrigation and Drainage*, 58:86–95.
23. Mahmood, N., Akhtar, B., Saleem, M. (2002). Scheduling irrigation in winter grown at different seed rates. *Asian J. Pl. Sci.,* 1(2):136–137.
24. Maheshwar, J., Maragatham, N., Martin, G. J. (2007). Relatively simple irrigation scheduling and N application enhance the productivity of aerobic rice (*Oryza sativa* L.). *American J. Pl. Physiol.*, 2(4):261–268.
25. Maliwal, G. L., Patel, J. K., Kaswala, R. R., Patel, M. L., Bhatnagar, R., Patel, J. C. (2000). Scheduling of irrigation for wheat (*Triticum aestivum*) under restricted water supply in Narmada region. *Indian J. Agric. Sci.,* 70(2):90–92.
26. Mishra, R. K., Pandey, N., Bajpai, R. P. (1994). Influence of irrigation and nitrogen on yield and water pattern of wheat (*Triticum aestivum*). *Indian J. Agron.*, 39(4):560–564.
27. Paradkar, K. V. K., Raghuwanshi, R. K. S., Ranade, D. H. (1990). Depth and frequency of irrigation on performance of wheat grown on a moderately sodic clay soil. *J. Indian Soc. Soil Sci.*, 38:1–5.
28. Parihar, S. S., Tiwari, R. B. (2003). Effect of irrigation and nitrogen level on yield, nutrient uptake and water use of late sown wheat (*Triticum aestivum*). *Indian J. Agron.*, 48(2):103–107.
29. Patel, A. G., Patel, B. S., Patel, P. H. (2010). Effect of irrigation levels based on IW:CPE ratios and time of nitrogen application on yield and monetary return of frenchbean (*Phaseolus vulgaris* L.). *Legume Res.*, 33(1):42–45.
30. Patel, G. N., Patel, P. T., Patel, P. H. (2008). Yield, water use efficiency and moisture extraction pattern of summer groundnut as influenced by irrigation schedules, sulfur levels and sources. *SATe Journal/ejournal.icrisat.org.*, 6:1–4.
31. Phogat, B. S., Singh, V. R., Verma, R. S. (1987). Response of sugarcane varieties to irrigation. *Indian J. Agron.,* 32(1):77–79.
32. Prasad, U. K., Pandey, R. D., Prasad, T. N., Jha, A. K. (1987). Effect of irrigation and nitrogen on wheat. *Indian J. Agron.*, 32(4):310–313.
33. Prihar, S. S., Khera, K. L., Sandhu, K. S., Sandhu, B. S. (1976). Comparison of irrigation schedules based on pan evaporation and growth stages in winter wheat. *Agron. J.,* 68:650–653.
34. Ramamoorthy, K., Arthanari, M. P., Subbian, P. (2009). Effect of irrigation scheduling and method of irrigation on productivity and water economy in hybrid sunflower (*Helianthus annuus* L.). *Helia.*, 32(50):115–122.
35. Rathod, I. R., Vadodaria, R. P. (2010). Response of irrigation and weed management on productivity of wheat (*Triticum aestivum*) under middle Gujarat condition. *Pakistan J. Biological Science,* 7(3):346–349.
36. Rathore, A. L., Patel, S. L. (1991). Studies on nitrogen and irrigation requirement of late sown wheat. *Indian J. Agron.*, 36(2):184–187.
37. Reddy, M. G., Rao, Y. Y., Rao, K. S., Ramaseshaih, K. (1983). A preliminary study on scheduling irrigation with an evaporimeters. *Agric. Water Mgmt.*, 6(4):403–407.
38. Sharma, D. K. (1994). Effect of scheduling irrigation based on pan evaporation to wheat on sodic soils. *J. Indian Soc. Soil Sci.*, 42(3):346–351.
39. Singh, P. N., Mohan, S. C. (1994). Water use and yield response of sugarcane under different irrigation schedules and nitrogen levels in a subtropical region. *Agric. Water Mgmt.*, 26:253–264.

40. Singh, V., Bhunia, S. R., Chauhan, R. P. S. (2003). Response of late sown wheat (*Triticum aestivum*) to row spacing cum-population densities and levels of nitrogen and irrigation in north-western Rajasthan. *Indian J. Agron.,* 48(3):178–181.
41. Singh, V. P. N., Uttam, S. K. (1993). Performance of wheat (*Triticum aestivum*) cultivars under limited irrigation and fertility management. *Indian J. Agron.,* 38(3):386–388.
42. Solanki, R. M., Patel, R. G. (2000). Response of Lucerne to different moisture regimes, phosphate levels and sowing methods. *Indian J. Agric. Res.*, 34(3):160–163.
43. Thakur, R., Pal, S. K., Singh, M. K., Verma, U. N., Upsani, R. R. (2000). Response of late sown wheat (*Triticum aestivum*) to irrigation schedules. *Indian J. Agron.,* 45(3):586–589.
44. Wajid, A. K. H., Magsood, M., Ahmad, A., Hussain, A. (2007). Influence of drought on water use efficiency in wheat in semiarid regions of Punjab. *Pak. J. Bot.,* 42, 1703–1711.

APPENDIX I – LIST OF SYMBOLS

%	Percent
/	Per
°C	Degree Celsius
μ-mol/g	Micro molecular per gram
B:C	Benefit cost ratio
Ca	Calcium
cm	Centimeter
Cu	Coefficient of uniformity
CU	Consumptive water use
ds/m	Desi Siemen per meter
E	Evaporation
ha	Hectare
I_0	No irrigation
I_1	Irrigation at IW/CPE = 0.6
I_2	Irrigation at IW/CPE = 0.8
I_3	Irrigation at IW/CPE = 1.0
I_4	Irrigation at IW/CPE = 1.2
I_c	Irrigation at Control
K	Potassium
Kc	Crop coefficient
kg/ha	Kilogram per hectare
Mm/day	Millimeter per day
N	Nitrogen
PAN-E	Cumulative evaporation from US weather Bureau Class A pan less rain since previous irrigation
q/ha	Quintal per hectare, 1 q = 100 kg
R	Ratio
Rs	Rupees
t/ha	Tons per hectare
viz.	Namely

Y.	Economic yield
Zn/ha	Zinc per hectare
Zr	Root zone of soil layer
ρ	Bulk density

CHAPTER 12

DRIP IRRIGATION SCHEDULING BASED ON PARTIAL ROOT ZONE IN CITRUS

P. PANIGRAHI, R. K. SHARMA, and S. MOHANTY

CONTENTS

12.1 INTRODUCTION

Throughout world, water supply is a major constraint to crop production due to water demand for rapid industrialization and high population growth. The further scarcity of irrigation water for crop production should be checked for sustaining the food supply through efficient water conservation and management practices even in high rainfall areas. Moreover, the harvest per every drop of irrigation water should be enhanced while considering the best water use efficiency (WUE) associated with any crop.

In recent years, irrigated agriculture is shifting the paradigm of irrigation management from full to partial supply of water needs, which becomes a need in water scarce regions. Water scarcity in irrigation sector demands for the improvement in water use efficiency as a critical goal [11]. One of the most promising techniques is the use of partial root-zone drying (PRD) irrigation in crop production. The PRD is a practice of using irrigation to alternately wet and dry, two spatially prescribed parts of the plant root system to simultaneously maintain plant water status at optimum water potential and control transpiration without bringing a significant change in photosynthesis rate of leaves [8]. Achieving higher WUE in any crop can be possible by enhancing yield and/or reducing the water losses through deep percolation and evaporation from the field [12].

Past research has revealed the potential of PRD technique as a way of reducing water use in tree crops and vines with little or no impact on yield and fruit quality [5, 7, 8, 18]. Besides increased transpiration efficiency, another effect of PRD is the limitation of vegetative growth of the trees [3]. Overall, PRD strategy under deficit irrigation can produce optimum yield with higher WUE under limited water availability conditions.

Citrus is grown under irrigated conditions in northern region of India. The sub-optimum productivity with poor fruit quality is the major pomological constraint, affecting the economics of Kinnow production in this region. Limited irrigation water availability following faulty irrigation method (surface irrigation) is one of the major reasons of low productivity of citrus [14]. Micro irrigation is an efficient irrigation method compared to surface irrigation in Kinnow mandarin. Due to positive impact of drip irrigation on crop production with less water, the area under drip irrigation has been increased substantially in last few years in India [13, 15].

Irrigation scheduling with full water requirement of the crops including citrus is a common irrigation practice in India. The irrigation water shortage in arid and semiarid areas of India forces the orchard growers to impose deficit irrigation in crop production. PRD has been found successful in grape [3], peach [5], apple [9] and sweet orange [6]. However, there is not enough research related to the effects of PRD on citrus in India. Moreover, the studies in relation to the response of mandarin cultivars of citrus to PRD are very limited worldwide.

This chapter discusses the effects of PRD on citrus production, taking Kinnow as a test crop, under a semiarid subtropical conditions of India.

12.2 MATERIALS AND METHODS

12.2.1 EXPERIMENTAL SITE

During 2010, the study was conducted with 10-year-old 'Kinnow' mandarin (*Citrus reticulata* Blanco) trees budded on *Jatti Khatti* (*Citrus jambhiri* Lush) rootstock at New Delhi (latitude 28°38′23″ N, longitude 77°09′27″ E and at an elevation of 228.6 m above the mean sea level). The plant to plant spacing was 4 m, whereas row to row spacing was 5 m.

The soil at the experimental site varied from sandy loam (top 40 cm soil) to sandy clay loam (40–100 cm) with bulk density of 1.47–1.61 $g.cm^{-3}$. The irrigation water was free from salinity (EC, 1.15 dS m^{-1}), alkalinity (pH, 7.3) and sodicity (SAR, 4.4). The ground water contribution to plant water requirement is assumed to be negligible as water level in the nearby wells of the experimental plot was at 15–18 m depth from ground surface.

The experimental site encounters semiarid, subtropical climate with hot and dry summers. The hottest months of the year are May and June with mean daily temperature of 39°C, whereas January is the coldest month with mean temperature of 14°C. The mean annual rainfall at the site is 770 mm, out of which around 85% falls mainly during June to September.

12.2.2. EXPERIMENTAL LAYOUT

Two irrigation regimes *viz.*, 50% and 75% of the crop evapotranspiration (ETc) were imposed through PRD and traditional deficit irrigation strategy (DI), and compared with irrigating both sides of the root zone at 100% ETc (FI). Irrigation was switched over from one side to another side of the plant basin under PRD, when the available soil water at drying side was depleted by 40%, as suggested for citrus. The treatments were:

PRD_{50}: Irrigation at 50% ETc through PRD,

PRD_{75}: Irrigation at 75% ETc through PRD,

FI: Full irrigation (100% ETc),

The irrigation was continued from mid-January to June and mid-October to December in each year of experiment. Thirty-two Kinnow trees were selected for this experiment and four treatments except FI were imposed using randomized complete block design, with four replications per treatment and two trees per replication. The trees imposed under FI in RDI experiment were taken for comparison with PRD-irrigated trees.

12.2.3 IRRIGATION SCHEDULING AND CROP MANAGEMENT PRACTICES

Irrigation water was applied on alternate day using six on-line 8 lph pressure compensating emitters per tree mounted on two 12 mm diameter lateral pipes (3 emitters per lateral). The emitters were located at 1.0 m away from tree stem. The water quantity applied under FI was calculated based on 100% Class A pan evaporation rate for Kinnow mandarin in Delhi conditions, using the following formula [4]:

$$ETc = Kp \times Kc \times Ep \quad (1)$$

where, ETc = crop evapotranspiration (mm/day), Kp = pan coefficient (0.8), Kc = crop coefficient (0.85 for mature Kinnow tree), and Ep = 2-days cumulative pan evaporation (mm). The volume of water applied under FI was computed following the formula:

$$V_{id} = \pi\,(D^2/4) \times (ET_c - R_e)/E_i \quad (2)$$

where. V_{id} = irrigation volume (liter/tree) applied in each irrigation, D = mean tree canopy diameter measured in N-S and E-W directions (m), ETc = crop evapotranspiration (mm), R_e = effective rainfall depth (mm), and E_i = irrigation efficiency of drip system (90%). The effective rainfall during the experiment was worked out as the summation of changes in soil water content in root zone before and after rainfall and potential crop evapotranspiration for the day of rainfall.

The depth of mean daily water applied per tree in various months of the study years are presented in Table 12.1. The required amount of water to each irrigation treatment was regulated by adjusting the operating hours based on the actual discharge of the emitters from time to time. The flow of irrigation water in lateral pipes was controlled by lateral valves provided at the inlet end of lateral pipes.

TABLE 12.1 Irrigation Water (mm) Applied to Kinnow Mandarin Under Various Irrigation Treatments

Treatment	Months									
	Jan.	Feb.	Mar.	Apr.	May	June	Oct.	Nov.	Dec.	Total
PRD_{50}	8.2	12.5	66.4	94.0	99.9	105.9	35.6	29.7	21.8	474.0
PRD_{75}	12.3	18.7	99.6	140.9	149.8	158.8	53.4	44.5	32.7	710.7
FI_{100}	16.4	24.9	132.8	187.9	199.7	211.7	71.2	59.3	43.6	947.5

The NPK-based fertilizers (354 g N, 160 g P_2O_5 and 345 g K_2O per tree) as per recommendation were applied through drip irrigation system in monthly intervals from January to June. Intercultural operation and the plant protection measures against insect pests and diseases were adopted uniformly for all trees in the experimental block, following the recommendations given for Kinnow mandarin in Delhi region.

12.2.4 MEASUREMENTS AND ANALYSIS

Soil samples at 30 cm, 60 cm, 90 cm, 120 cm, and 150 cm distances from tree trunk along and in between the drip emitters were collected from 0–20 cm, 20–40 cm, 40–60 cm, 60–80 cm, and 80–100 cm depths during January of each year and analyzed for available macronutrients (N, P and K) and micronutrients (Fe, Mn, Cu and Zn) following the standard procedures [17]. Four tree basins from each treatment were taken for soil sampling.

For leaf nutrients (N, P, K, Fe, Mn, Cu and Zn) determination, 3- to 5- months old leaf samples (3rd and 4th leaf from tip of nonfruiting branches) were taken from each side of tree canopy, at a height of 1.5 m from ground surface during October and analyzed following the standard methods [17].

The mid-day leaf water potential was determined fortnightly taking two leaves per tree (sun-exposed) from the outer canopy using a pressure chamber (PMS instrument, Oregon, USA). For determination of stem water potential, two leaves per tree near to the trunk or a main scaffold branch was selected and covered with aluminum sheet and black polythene sheet to measure its potential at mid-day (12:00–13:00 PM). The leaves were enclosed in black polythene and aluminum sheet cover before 2 h of measurement for determination of both leaf and stem water potential. The water stress integral (S_ψ) under each treatment was calculated for midday leaf and stem water potential data, according to the equation [10]:

$$S\psi = Absolute\,value\,of \sum_{i=0}^{i=1} \{(\psi i, i+1) - c\}\, n \tag{3}$$

where, S_ψ = water stress integral (MPa day), $\psi_{i,\,i+1}$ = average midday leaf/stem water potential for any interval i and i+1 (MPa), c = maximum leaf/stem water potential measured during the study. and n = number of days in the interval.

Relative leaf water content (RLWC) and leaf water concentration (LWC) were determined for two leaves per tree (4 trees per treatment) using procedures by Bowman [2].

The measurement of net photosynthesis rate (P_n), stomatal conductance (g_s), and transpiration rate (T_r) of leaves was performed fortnightly from 9:00 am to 3:00 pm on a clear-sky day with a portable photosynthesis meter (LI-COR-6400, Lincoln, Nebraska, USA). Four mature leaves per tree (3rd or 4th leaf from tip of shoot) from exterior canopy position (one leaf in each North, South, East and West direction), and two trees per treatment were taken for this measurement. Leaf water use efficiency (LWUE) was defined as a ratio of P_n to T_r of leaves.

The vegetative growth of trees (tree height, stem height, canopy diameter, stock girth diameter and scion girth diameter) was measured annually by using a metric tape. Tree canopy volume was estimated as follows:

$$V_{pc} = 0.\ 5238\ H\ (D)^2 \tag{4}$$

where, V_{pc} = tree canopy volume (m^3), H = tree canopy height (difference between tree height and stem height, meter), and D = mean tree canopy spread diameter (North-South and East-West) in meter.

The number of fruits harvested from each tree was counted and the total weight was recorded, and the mean yield per tree under various treatments was estimated. Five fruits per tree were taken randomly for determination of fruit quality parameters, such as: juice percent, acidity, total soluble solids (TSS), ascorbic acid, sugars (total and reducing) using the methodology by Ranganna [16]. Water use efficiency (WUE) and irrigation water use efficiency (IWUE) were worked out as the fruit yield per total tree water use and fruit yield per unit quantity of irrigation water applied, respectively.

12.2.5 STATISTICAL ANALYSIS

The analysis of variance (ANOVA) for the data was done using SPSS statistical software, and separation of means was obtained using Duncan multiple range test (DMRT).

12.3 RESULTS AND DISCUSSION

12.3.1 AVAILABLE NUTRIENTS IN SOIL

The available N, P and K status in the soil under different irrigation strategies shows an increasing trend (Table 12.2). The maximum increase in the available nutrients was observed under FI, whereas the minimum was with PRD_{50}. The higher availability of N, P and K under FI was due to increased soil moisture in this treatment, which induced better nutrient concentration in soil water in the *rhizosphere*. The annual increase in available nutrients under the treatments suggests for both annual-soil nutrients based and yield-based fertilization strategies for Kinnow plants. The available micronutrients (Fe, Mn, Cu and Zn) in the soil showed a decreasing trend under all the irrigation strategies. The maximum decrease in available micronutrients was observed with FI and minimum was with PRD_{50}. However, the consistent reduction of micronutrients in soil suggests the application of appropriate quantity of micronutrients-based fertilizers to mandarin plants for sustaining higher yield for long run.

Table 12.2. Changes in Available N, P, K, Fe, Mn, Cu and Zn (mg kg^{-1} soil) in Soil Under Different Irrigation Treatments in Kinnow Mandarin

Treatments	Macro-nutrients			Micro-nutrients			
	N	P	K	Fe	Mn	Cu	Zn
PRD_{50}	$+3.44^{a}$	$+0.71^{a}$	$+3.90^{b}$	-0.82^{a}	-0.73^{a}	-0.17^{a}	-0.18^{a}
PRD_{75}	$+3.91^{b}$	$+0.81^{a}$	$+3.95^{d}$	-0.92^{a}	-0.92^{a}	-0.24^{a}	-0.20^{a}
FI_{100}	$+4.29^{c}$	$+0.93^{a}$	$+4.25^{e}$	-1.19^{b}	-1.06^{b}	-0.22^{a}	-0.23^{b}

Data in each column followed by a different letter are significantly different at $P < 0.05$, based on "separation by Duncan's multiple range test."

12.3.2 LEAF NUTRIENTS COMPOSITION

The leaf nutrient (N, P, K, Fe, Mn, Cu and Zn) analysis shows that all the nutrients except P and Cu were significantly affected by irrigation treatments (Table 12.3). The highest concentration of the nutrients was registered with FI, followed by PRD_{75}. The higher concentration of leaf-nutrients with fully irrigated trees was resulted by higher plant uptake with increased availability of such nutrients in root zone under FI. Among micronutrients, the magnitudes of all nutrients (Fe, Mn and Zn) were at par under PRD_{50} and PRD_{75}. The higher micronutrient concentration was observed with fully irrigated plants.

TABLE 12.3 Total N, P, K, Fe, Mn, Cu and Zn in Leaves (%, Dry Weight Basis) of 'Kinnow' Mandarin As Affected By Various Irrigation Treatments

Treatments	Macro-nutrients			Micro-nutrients			
	N	P	K	Fe	Mn	Cu	Zn
PRD_{50}	2.35^{a}	0.18^{a}	1.48^{a}	55.6^{a}	51.2^{a}	7.4^{a}	25.2^{a}
PRD_{75}	2.47^{b}	0.21^{a}	1.59^{b}	59.9^{a}	58.4^{a}	7.6^{a}	25.8^{a}
FI_{100}	2.69^{c}	0.22^{a}	1.64^{c}	62.6^{b}	61.5^{b}	8.2^{a}	26.9^{b}

Data in each column followed by a different letter are significantly different at $P < 0.05$, based on "separation by Duncan's multiple range test."

12.3.3 LEAF AND STEM WATER POTENTIAL, RELATIVE LEAF WATER CONTENT AND LEAF WATER CONCENTRATION

The midday-leaf (Ψ_l) and -stem water potential (Ψ_s), leaf water stress integral ($S\Psi_l$), and stem water stress integral ($S\Psi_s$) of the mandarin trees were affected significantly by irrigation treatments (Table 12.4). The mean Ψ_l and Ψ_s were higher under FI, followed by PRD_{75}. Earlier the similar responses of leaf and stem water potential to PRD was observed in citrus [6, 18].

The mean relative leaf water content (RLWC) and leaf water concentration (LWC) under different irrigation treatments were affected significantly under various irrigation treatments (Table 12.4). The highest value of RLWC and LWC were observed with fully irrigated trees, whereas the lowest values were observed with the trees under PRD_{50}.

TABLE 12.4 Mean Seasonal Mid-Day Leaf Water Potential and Content of Kinnow Mandarin as Affected By Partial Root Zone Drying

Treatments	**Leaf water contents**					
	Ψ_l	Ψ_s	$S\Psi_l$	$S\Psi_s$	RLWC	LWC
	MPa	**MPa**	**MPa-day**	**MPa-day**	**%**	**%**
PRD_{50}	-1.7^{b}	-1.1^{b}	48.4^{b}	25.8^{b}	87.8^{b}	71.1^{b}
PRD_{75}	-1.4^{d}	-0.9^{d}	31.5^{d}	21.1^{d}	90.4^{d}	75.1^{d}
FI_{100}	-1.2^{e}	-0.7^{e}	24.5^{e}	18.9^{e}	92.7^{e}	78.3^{e}

Data in each column followed by a different letter are significantly different at $P < 0.05$, based on "separation by Duncan's multiple range test."

12.3.4 LEAF PHYSIOLOGICAL PARAMETERS

The mean net photosynthesis rate (P_n), stomatal conductance (g_s), transpiration rate (T_r), and leaf water use efficiency ($LWUE = P_n/T_r$) under different irrigation treatments were significantly affected (Table 12.5). The maximum Pn value was registered with fully irrigated trees, followed by the trees under PRD_{75}. The lowest value of P_n was recorded under PRD_{50}. The g_s and T_r followed the same trend of P_n under different treatments. However, the LWUE was maximum in PRD_{50} treatment, whereas the minimum LWUE was in PRD_{75} treatment.

TABLE 12.5 Net Photosynthesis Rate (Pn, $\mu mol\ m^{-2}\ s^{-1}$), Stomatal Conductance (gs, $mmol\ m^{-2}\ s^{-1}$), Transpiration Rate (Tr, $mmol\ m^{-2}\ s^{-1}$) and Leaf Water Use Efficiency (LWUE) of Kinnow Mandarin Under Different Irrigation Treatments

Treatments	**Leaf physiological parameters**			
	Pn	**gs**	**Tr**	**LWUE**
PRD_{50}	3.11^{b}	20.13^{b}	1.43^{a}	2.17^{d}
PRD_{75}	3.20^{c}	23.13^{c}	1.79^{c}	1.78^{b}
FI_{100}	3.88^{d}	37.78^{e}	2.08^{e}	1.86^{c}

Data in each column followed by a different letter are significantly different at $P < 0.05$, based on "separation by Duncan's multiple range test."

12.3.5. PLANT VEGETATIVE GROWTH

The plant vegetative growth parameters (tree height, PH; stem girth diameter, SD; canopy diameter, CD and canopy volume, CV) were significantly affected by irrigation treatments (Table 12.6). The highest growth of the trees was observed with FI, followed by PRD_{75}. Previously, the similar findings of decrease in vegetative growth were observed with deficit-irrigated citrus [13].

TABLE 12.6 Tree Growth of Kinnow Mandarin Under Various Irrigation Treatments

Treatments	Tree growth parameters			
	TPH, cm	SD, mm	CD, cm	CV, m^3
DI_{50}	33.4^a	20.4^b	25.8^a	0.80^a
DI_{75}	36.2^b	22.5^d	31.3^d	0.83^b
PRD_{50}	32.5^a	19.7^a	25.3^a	0.79^a
PRD_{75}	35.9^b	22.0^c	30.9^c	0.80^b
FI_{100}	40.7^c	26.2^e	$48.^{7e}$	0.86^c

TPH: total tree height; SD: stem diameter; CD: canopy diameter;

CV: canopy volume.

Data in each column followed by a different letter are significantly different at $P < 0.05$, based on "separation by Duncan's multiple range test."

12.3.6 FRUIT YIELD AND WATER PRODUCTIVITY

The number of fruit dropped and harvested per tree, average fruit weight and total fruit yield in various treatments are presented in Table 12.7. The maximum number of fruits was dropped in PRD_{50}. The minimum fruit drop took place in FI. The fruit drop decreased with increase in irrigation regime under PRD. The number of fruit harvested under different treatments followed the reverse trend of fruit drop. However, the mean fruit weight recorded under PRD_{75} was maximum, followed by FI. The increased number of fruits with FI may be a reason for smaller fruits in this treatment. Both fruit number per tree and mean fruit weight decreased with decreasing irrigation regime from 75% ETc to 50% ETc with PRD. The highest fruit yield was recorded in FI, followed by PRD_{75}. The increased irrigation regime from 50% ETc to 75% ETc enhanced the fruit yield under PRD, resulting from less number of fruits with lower fruit weight under lower regime of irrigation. The IWUE and WUE were maximum under PRD_{50}. The higher IWUE in PRD_{50} was attributed to higher increase in fruit yield with comparatively less increase in irrigation water use under this treatment over other treatments.

TABLE 12.7 Fruit Drop (Number Fruit), Yield Harvested (Number of Fruits, Average Fruit Weight, Fruit Yield), Irrigation Water Use Efficiency (IWUE) and Water Use Efficiency (WUE) of Kinnow Mandarin Under Different Irrigation Treatments

Treatments	Yield parameters					
	No. fruits dropped	No. fruits harvested	Average fruit weight	Fruit yield	IWUE	WUE
	No./tree	No./tree	g	t/ha	t/ha-mm	t/ha-mm
PRD_{50}	148 (80, 48, 20)	703^{b}	160.7^{b}	56.48^{b}	0.119^{d}	0.062^{d}
PRD_{75}	100 (61, 28, 11)	755^{d}	163.0^{b}	58.73^{b}	0.082^{b}	0.053^{b}
FI_{100}	92 (64, 15, 13)	763^{d}	162.3^{b}	61.91^{b}	0.065^{a}	0.047^{a}

Data in each column followed by a different letter are significantly different at $P < 0.05$, based on "separation by Duncan's multiple range test."

TABLE 12.8 Fruit Quality Parameters (Fruit Size, Juice Content, TSS, Acidity, Vitamin-C, Reducible Sugar, Total Sugar) of Kinnow Mandarin Under Different Irrigation Treatments

Treatments	Fruit quality parameters					
	Juice content (%)	TSS (°Brix)	TA (%)	Ascorbic acid (mg/l)	Reducing sugar (mg/l)	Total sugar (mg/l)
PRD_{50}	44.3^{b}	11.4^{b}	0.83^{b}	123.6^{b}	61.7^{b}	69.3^{b}
PRD_{75}	47.9^{c}	11.1^{a}	0.80^{a}	111.9^{a}	49.2^{a}	63.2^{a}
FI_{100}	49.5^{c}	10.9^{a}	0.79^{a}	119.1^{b}	38.7^{a}	68.7^{a}

Data in each column followed by a different letter are significantly different at $P < 0.05$, based on "separation by Duncan's multiple range test."

12.3.7 *FRUIT QUALITY*

The fruit quality (juice content, total soluble solids, titratable acidity (TA), ascorbic acid content, reducible sugar and total sugar) were significantly affected by irrigation treatments (Table 12.8). Higher level of irrigation resulted in higher juice content in fruits, even under PRD. Earlier study also observed the juicy fruits under PRD in comparison to DI in citrus [6, 18]. The TSS of juice decreased with increase in irrigation regime under PRD. The TA increased with increase in irrigation from 50% ETc to 75%, etc. The ascorbic acid content of the fruits decreased from irrigation at 50% ETc to irrigation at 75% ETc, probably due to dilution effect of higher juice content on it under higher regime of irrigation. The total sugar (TS) and reducing sugar (RS) of fruits under different irrigation treatments followed the same trend of TSS. The highest TS and RS were observed in PRD_{50}. Moreover, the percent of reduction of reducing sugar with respect to total sugar content from irrigation at 100% ETc to irrigation at 75% ETc was higher than that from irrigation at 75% ETc to irrigation at 50% ETc treatment.

12.4 CONCLUSIONS

Water scarcity is one of the major causes of citrus decline in arid and semiarid regions. Drip irrigation has been found to be a potential water saving technique. PRD is a recently proposed water saving technique in irrigated agriculture. The present study was planned with a hypothesis that drip irrigation scheduling with PRD technique can save substantial amount of irrigation water over full irrigation (FI), besides improving the fruit yield and quality of citrus trees grown in a semiarid region. Keeping this in view, an experiment was conducted to study the response of citrus trees to PRD at 50% ET_C (PRD_{50}), PRD at 75% ET_C (PRD_{75}) and FI.

The higher vegetative growth and fruit yield were recorded with fully irrigated trees. However, the fruit yield in PRD_{75} was statistically at par with that in FI. The maximum irrigation water use efficiency was observed with PRD_{50} followed by PRD_{75}. The maximum net photosynthesis rate, stomatal conductance, and transpiration rate of leaves were registered with fully irrigated trees, followed by the trees under PRD_{75}. However, the maximum leaf water use efficiency was observed with PRD_{50}. The leaf nutrients content followed the same trend of vegetative growth under different treatments. Overall, these results reveal that partial root zone drying under drip irrigation scheduled at 50% ETc is a productive and potential water saving technique in citrus cultivation in arid and semiarid regions of Northern India. The adoption of such technique can bring more area under irrigation, resulting in higher production of quality citrus fruits.

12.5 SUMMARY

The maximum irrigation water use efficiency and water use efficiency were obtained from irrigation at 50% crop evapotranspiration under partial root zone drying, with some minor reduction in yield over that under full irrigation. It also produced the fruits with superior quality than that produced under full irrigation. The adoption of partial root-zone drip irrigation scheduled at 50% crop water requirement is a viable option against traditional full irrigation for citrus cultivation in sandy loam soils under similar agro-climatic conditions such as those found in Delhi region, India. The optimal NPK-fertigation strategy under partial root zone drying is suggested for drip-irrigated Kinnow mandarin.

KEYWORDS

- **acidity**
- **citrus**
- **drip irrigation**
- **fertigation**
- **fruit number**
- **fruit quality**
- **irrigation**
- **micro irrigation**
- **North India**
- **partial root zone drying, PRD**
- **TSS**
- **water use efficiency, WUE**
- **yield parameters**

REFERENCES

1. Anonymous, (1997). *Annual Progress Report for 1996–97*. Water Technology Centre, IARI, New Delhi.
2. Bowman, W. D. (1989). The relationship between leaf water status, gas exchange, and spectral reflectance in cotton leaves. *Remote Sensing of Environment,* 30:249–255.
3. Dry, P. R., Loveys, B. R., During, H. (2000). Partial drying of the root zone of grapes, I–Transient changes in shoot growth and gas exchange. *Vitis.,* 39:3–7.
4. Germana, C., Intrigliolo, F., Coniglione, L. (1992). Experiences with drip irrigation in orange trees. In: *Proceedings of the VII International Citrus Congress of the International Society of Citriculture. 8–13 March*. Acireale (Catania), Italy, pp. 661–664.
5. Goldhamer, D. A., Salinas, M., Crisosto, C., Day, K. R., Soler, M., Moriana, A. (2002). Effects of regulated irrigation and partial root zone drying on late harvest peach tree performance. *Acta Horticulturae,* 592:343–350.
6. Hutton, R. J., Loveys, B. R. (2011). A partial root zone drying irrigation strategy for citrus-Effects on water use efficiency and fruit characteristics. *Agric. Water Manage.*, 98:1485–1496.
7. Kang, S., Hu, X., Goodwin, I., Jerie, P. (2002). Soil water distribution, water use, and yield response to partial root zone drying under a shallow ground water table condition in a pear orchard. *Scientia Horticulture*, 92:277–291.
8. Kriedemann, P. E., Goodwin, I. (2001). *Regulated deficit irrigation and partial root zone drying–An overview of principles and applications*. Irrigation insights Bulletin 3: Land and Water Australia, pp. 102.
9. Leib, B. G., Capari, H. W., Redulla, C., Andrews, P. K., Jabro, J. J. (2006). Partial root zone drying and deficit irrigation of Fuji apples in a semiarid climate. *Irrig. Sci.*, 24:85–99.
10. Myers, B. J. (1988). Water stress integral–a link between short-term stress and long-term growth. *Tree Physiology*, 4:315–323.

11. Panigrahi, P., Huchche, A. D., Srivastava, A. K., Singh, S. (2008). Effect of drip irrigation and plastic mulch on performance of Nagpur mandarin (*Citrus reticulata* Blanco) grown in central India. *Indian Journal of Agricultural Sciences*, 78:1005–1009.
12. Panigrahi, P., Srivastava, A. K., Huchche, A. D. (2010). Optimizing growth, yield and water use efficiency (WUE) in Nagpur mandarin (*Citrus reticulate* Blanco) under drip irrigation and plastic mulch. *Indian Journal of Soil Conservation*, 38(1):42–45.
13. Panigrahi, P., Srivastava, A. K. (2011). Deficit irrigation scheduling for matured Nagpur mandarin (*Citrus reticulate* Blanco) trees of central India. *Indian Journal of Soil Conservation*, 39(2):149–154.
14. Panigrahi, P., Srivastava, A. K., Huchche, A. D. (2012). Effects of drip irrigation regimes and basin irrigation on Nagpur mandarin agronomical and physiological performance. *Agricultural Water Management,* 104:79–88.
15. Panigrahi, P., Srivastava, A. K., Huchche, A. D., Singh, S. (2012). Plant nutrition in response to drip versus basin irrigation in young Nagpur mandarin on Inceptisol. *Journal of Plant Nutrition*, 35:215–224.
16. Ranganna, R. (2001). *Handbook of Analysis and Quality Control for Fruit and Vegetable Products*. Second ed. Tata McGraw Hill, New Delhi.
17. Tandon, H. L. S. (2005). *Methods of Analysis of Soils, Plants, Waters, Fertilizers and Organic Manures*. Second ed. Fertilizer Development and Consultancy Organization, New Delhi, India.
18. Treeby, M. T., Henriod, R. E., Bevington, K. B., Milne, D. J., Storey, R. (2007). Irrigation management and rootstock effects on navel orange (*Citrus Sinensis Osbeck*) fruit quality. *Agric. Water Management,* 91:24–32.

CHAPTER 13

DRIP IRRIGATION SCHEDULING FOR *CITRUS RETICULATA* BLANCO

P. PANIGRAHI, A. K. SRIVASTAVA, and A. D. HUCHCHE

CONTENTS

13.1 INTRODUCTION

Nagpur mandarin (*citrus reticulata* var. Blanco) is commonly cultivated in India under furrow or basin method of irrigation [8, 10, 12]. Recently, water table is declining in bore wells creating water shortage in summer for sustaining the crop. Therefore, every year thousands of hectare of orchards is permanently wilted due to short of water, causing economical loss. Water management studies in Nagpur mandarin show that optimum soil water regime under drip irrigation can increase growth and yield [6, 9]. Plastic PE mulching has also proved its effectiveness in conserving the soil moisture and in increasing growth, yield and quality of different citrus cultivars [1, 3–5, 7]. However, information on the interactive effect of drip irrigation and mulching for Nagpur mandarin is meager. Therefore, this study was undertaken to evaluate the performance of drip irrigation in conjunction with plastic mulch in young Nagpur mandarin.

13.2 MATERIALS AND METHODS

During 2003–2007, the experiment was carried out at Research farm of National Research Center for Citrus, Nagpur – India. The Nagpur mandarin plants budded on rough lemon root stock under the study were one-year-old with 6 ´ 6 meter row spacing. The alternate day irrigation treatments imposed were:

T1 = Irrigation at 40% of pan evaporation with plastic mulch,
T2 = Irrigation at 60% of pan evaporation with plastic mulch,
T3 = Irrigation at 80% of pan evaporation with plastic mulch,
T4 = Irrigation at 100% of pan evaporation with plastic mulch, and
Control (C) = Basin irrigation at 50% depletion of available soil moisture.

The study consisted of randomized block design with three replications and three trees per replication. The black linear low-density poly ethylene (LLDPE) plastic mulch having thickness 100 micron was used. Mulching was 1.0 ´ 1.0 m^2 size polythene sheets on each tree basin keeping the tree at the center. The experimental soil type was clay loam with field capacity and permanent wilting point of 24.8% (weight basis) and 15.7% (weight basis), respectively. Recommended dose of fertilizers [13] and irrigation water through one dripper (4 lph)/tree were used.

The volume of water requirement was computed using the equation:

$$V = Ep \times Kc \times Kp \times Wp \times D \quad (1)$$

where, V = volume of water (liter/tree/day), Ep = cumulative pan evaporation for two consecutive days (mm), Kc = crop factor, Kp = pan factor, Wp = wetting factor, and D = canopy diameter observed at noon. The crop factor was taken as 0.6 and pan factor was 0.7 in winter and 0.8 in summer based on FAO-24 [2].

The moisture content at 0–15 cm depth was estimated by gravimetric method, whereas at 0–30 cm depth it was recorded by neutron moisture probe (Troxler model-4300) once in a week. The leaf physiological parameters such as photosynthesis rate (P), transpiration rate (E) and stomatal conductance (C) were recorded fortnightly and the seasonal pooled data was compared for different treatments. The vegetative growth parameters (tree height, stem height, canopy diameter, stock and scion girth) were measured and the incremental magnitudes under different treatments were compared. Analysis of leaf samples for macronutrients (N, P and K) and micronutrients (Fe, Mn, Cu and Zn) were also done under different levels of irrigation with mulch and control.

13.3 RESULTS AND DISCUSSION

13.3.1 IRRIGATION WATER

Drip irrigation scheduling based on cumulative pan evaporation on alternate day under plastic mulching from November to June (Table 13.1) indicates that the maximum water was applied at 100% of pan evaporation with mulch, which varied from 42.1 to 160.2 mm during different months. On the whole, the total amount of water applied was 262.34, 396.62, 524.68 and 655.85 mm at irrigation levels 40%, 60%, 80% and 100% of pan evaporation, respectively. The amount of water applied was maximum in the month of May and lowest in the month of December due to highest and lowest evaporation demand in these months, respectively.

TABLE 13.1 Irrigation Water Applied Under Different Irrigation Treatments With Plastic Mulch During November to June in Nagpur Mandarin

Treatment	Irrigation water applied, mm								
	Months								
	Nov.	Dec.	Jan.	Feb.	Mar.	Apr	May	June	Total
Irrigation at 40% evaporation + Mulch	18.42	16.84	17.24	20.31	41.68	51.25	64.1	32.5	262.34
Irrigation at 60% evaporation + Mulch	27.63	25.27	25.87	30.46	62.52	76.87	91.2	51.8	396.54
Irrigation at 80% evaporation + Mulch	36.84	33.69	34.50	40.62	83.32	83.36	128.16	69.1	509.59

TABLE 13.1 *(Continued)*

Treatment	Irrigation water applied, mm								
	Months								
	Nov.	Dec.	Jan.	Feb.	Mar.	Apr	May	June	Total
Irrigation at 100% evaporation+ Mulch	46.06	42.12	43.12	50.78	104.2	104.2	160.2	86.3	636.98
Irrigation at 50% ASMD through Basin irrigation	39.61	37.67	39.5	47.84	96.5	103	128.9	70.5	563.52

13.3.2 SOIL MOISTURE VARIATION

The soil moisture values at different depths (Table 13.2) indicates that the irrigation at 100% of pan evaporation and 80% of pan evaporation with mulch maintained the soil moisture nearly or more than field capacity during the study period. There were less fluctuations of soil moisture at 0.3 m depth in different months. But there was a significant effect of irrigation and mulch on soil moisture at 0.15 m depth, whereas at 0.3 m depth it was nonsignificantly affected.

TABLE 13.2 Average Soil Moisture Content (v/v) in Different Months at 15 cm and 30 cm Depths Under Different Irrigation Regimes With Plastic Mulch

Treatment	0.15 m				Mean	0.30 m				Mean
	Nov-Dec	Jan-Feb	Mar-Apr.	May-Jun.		Nov-Dec.	Jan-Feb	Mar-Apr.	May-Jun.	
Irrigation at 40% evaporation + Mulch	27.1	27.4	28.4	29.6	28.4	31.4	32.5	32.0	31.9	32.0
Irrigation at 60% evaporation + Mulch	28.6	28.6	29.6	29.8	29.4	31.7	32.6	32.1	32.0	32.2

TABLE 13.2 *(Continued)*

Treatment	0.15 m				Mean	0.30 m				Mean
	Nov-Dec	Jan-Feb	Mar-Apr.	May-Jun.		Nov-Dec.	Jan-Feb	Mar-Apr.	May-Jun.	
Irrigation at 80% evaporation + Mulch	30.5	30.1	31.0	31.4	30.9	32.2	33.2	31.9	32.6	32.6
Irrigation at 100% evaporation+ Mulch	32.3	34.0	34.1	34.9	33.5	33.1	33.1	32.1	33.0	33.0
Irrigation at 50% ASMD through Basin irrigation	25.4	25.1	27.3	27.9	26.5	31.5	32.5	32.1	32.1	32.1
CD (0.05)	—	—	—	—	0.82	—	—	—	—	NS

13.3.3 LEAF PHYSIOLOGY

The physiological parameters such as photosynthesis rate (P), transpiration rate (E) and stomatal conductance (C) were recorded by CO_2 gas analyzer CI-301PS (CID, Inc.) during December to March from 11:00 AM to 3:00 PM in one hour interval twice in a month (Table 13.3). It was observed that P was highest in IW/CPE = 0.6 with mulch in winter. Leaf water use efficiency (LWUE) in IW/CPE = 0.6 with mulch was highest in both winter and summer. It was also observed that all the parameters were affected significantly in both summer and winter, with exception to transpiration and stomatal conductance in summer.

TABLE 13.3 Photosynthesis Rate (P), Transpiration Rate (E), Stomatal Conductance (C) and Leaf Water Use Efficiency (LWUE) of Nagpur Mandarin Under Different Irrigation Regimes and Mulch in Winter and Summer

Treatments	P (μmol/m²/s)		E (m mol/m²/s)		C (mmol/m²/s)		LWUE	
	*Win.	+Sum.	Win.	Sum.	Win.	Sum.	Win.	Sum.
Irrigation at 40% evaporation + Mulch	3.931	3.794	2.342	2.434	69.6	46.1	1.609	1.630
Irrigation at 60% evaporation + Mulch	4.935	4.312	2.746	2.467	57.8	51.6	1.829	1.880
Irrigation at 80% evaporation + Mulch	3.923	3.938	2.753	2.723	77.4	71.9	1.484	1.418
Irrigation at 100% evaporation+ Mulch	2.021	2.163	2.004	2.736	38.1	74.8	0.896	0.980
Irrigation at 50% ASMD through Basin irrigation	3.152	1.712	2.324	1.891	76.1	31.5	1.123	0.961
CD (0.05)	0.39	0.62	0.28	NS	3.03	NS	0.18	0.09

Win.: winter; and Sum: summer

13.3.4 LEAF NUTRIENT COMPOSITION

Leaf samples were collected and analyzed for macronutrients (N, P and K) and micronutrients (Fe, Mn, Cu and Zn) under different levels of irrigation with mulch and basin irrigation (Table 13.4). A highly significant response was found in case of nitrogen, potassium and iron due to irrigation and mulching. The highest leaf – N (2.41%), – K (1.97%), – Fe (122.6 ppm) were observed in irrigation level at 60% pan evaporation under mulch, whereas lowest values were found at irrigation at 40% pan evaporation treatment. These results are in agreement with the findings of Shukla et al. [11].

TABLE 13.4 Leaf Nutrients of Nagpur Mandarin Under Different Irrigation Regimes With Plastic Mulch

Treatment	Macro-nutrients, %			Micro-nutrients, ppm			
	N	P	K	Fe	Mn	Cu	Zn
Irrigation at 40% evaporation + Mulch	1.23	0.073	1.22	99.6	48.5	11.2	10.2
Irrigation at 60% evaporation + Mulch	2.41	0.209	1.97	122.6	61.6	12.4	14.7
Irrigation at 80% evaporation + Mulch	1.77	0.167	1.77	105.7	58.5	15.8	15.5
Irrigation at 100% evaporation+ Mulch	1.67	0.075	1.26	101.3	49.3	12.3	22.9
Irrigation at 50% ASMD through Basin irrigation	1.22	0.092	1.24	99.8	49.66	8.5	10.3
CD (0.05)	0.4	NS	0.28	5.3	NS	NS	NS

13.3.5 TREE GROWTH

The tree growth parameters (average tree height, stock and scion girth and canopy spread) were recorded during July, to March (Table 13.5). It was observed that all the incremental growth parameters were highest in irrigation at 60% pan evaporation with mulch followed by irrigation at 80% pan evaporation with mulch. During October to December, none of the parameters were affected significantly, whereas during July to September except scion girth all the parameters, and during January to March none of the parameters except scion girth responded significantly to irrigation and mulch. In case of cumulative growth data during observation period, all the parameters except stock girth were significantly affected by different treatments. This may be due to optimum soil moisture supply and favorable soil temperature under mulch, which resulted in better availability and uptake of nutrients by the plants.

TABLE 13.5 Incremental Vegetative Growth of Nagpur Mandarin Under Different Irrigation Regimes With Plastic Mulch

Treatments	Tree height (m)	Stock girth (mm)	Scion girth (mm)	Canopy volume (m^3)
Irrigation at 40% evaporation + Mulch	0.48	42	40	0.582
Irrigation at 60% evaporation + Mulch	0.62	52	49	0.988
Irrigation at 80% evaporation + Mulch	0.53	48	45	0.661
Irrigation at 100% evaporation+ Mulch	0.45	40	38	0.503
Irrigation at 50% ASMD through Basin irrigation	0.43	36	36	0.451
CD (0.05)	0.06	ns	6.3	0.07

13.4 SUMMARY

A field experiment was carried out during 2003–07 at Nagpur to study the effects of alternate day irrigation levels (40% of pan evaporation, 60% of pan evaporation, 80% of pan evaporation and 100% of pan evaporation) under black linear low density polythene mulch (100 micron) for Nagpur mandarin. The results were compared with conventional irrigation (Basin irrigation at 50% depletion of available soil moisture). It was found that the irrigation at 60% of pan evaporation under plastic mulch gave the best growth along with 29.63% of irrigation water saving compared to control.

This study concludes that optimum drip irrigation scheduling (60% of pan evaporation) along with plastic mulch improved the vegetative growth of young Nagpur mandarin trees and enhances nutrient uptake efficiency, besides conserving a good quantum of irrigation water against basin irrigation. Thus, adoption of drip irrigation under plastic mulch is the most suitable approach for cultivation of Nagpur mandarin in Central India.

The study also reveals that there was a significant effect of drip irrigation and mulch on soil moisture at 15 cm depth, whereas at 30 cm depth it was nonsignificantly affected. It was also observed that all the physiological parameters (photosynthesis rate (P), transpiration rate (E) and stomatal conductance (C)) were affected significantly in both summer and winter with the exception of transpiration and stomatal conductance in summer. Analysis of leaf samples for macronutrients (N, P and K) and micronutrients (Fe, Mn, Cu and Zn) under different levels of irrigation with mulch and control indicated a highly significant response in case of N, K and Fe that is correlated well with all the growth parameters.

KEYWORDS

- **air temperature**
- **citrus**
- **crop water stress index, CWSI**
- **drip irrigation**
- **India**
- **infrared thermometry**
- **irrigation scheduling**
- **leaf nutrients**
- **macro nutrients**
- **micro irrigation**
- **micro nutrients**
- **mulching**
- **Nagpur**
- **Nagpur mandarin**
- **plastic mulching**
- **soil moisture**
- **tree canopy**
- **vegetative growth**
- **Vertisols**
- **water productivity**
- **water stress**
- **water use efficiency, WUE**

REFERENCES

1. Barua, P., Barua, H. K., Borah, A. (2000). Plant growth and yield of Assam lemon as influenced by different drip irrigation levels and plastic mulch. *Annals of Biology*, 16(1):17–20.
2. Doorenbos, J., Pruitt, W. O. (1977). *Guidelines for predicting crop water requirements*. Irrigation and Drainage paper 24, FAO, United Nations, Rome, Italy.
3. Ghali, M. H., Nakhlla, F. G. (1996). Evaluation of perforated polyethylene mulch on loamy sand soil under drip irrigated orange trees, part II–Soil thermal regime and moisture, root distribution and tree productivity. *Annals of Agricultural Science* Moshtohor., 34(3):1099–1116.
4. Lal, H., Samra, J. S., Arora, Y. K. (2003). Kinnow mandarin in Doon Valley, part II- Effect of irrigation and mulching on water use, soil temperature, weed population and nutrient losses. *Indian Journal of Soil Conservation*, 31(3):281–286.
5. Mohanty, S., Sonkar, R. K., Marathe, R. A. (2002). Effect of mulching on Nagpur mandarin cultivation in drought prone region of Central India. *Indian Journal of Soil Conservation*, 30(3):286–289.
6. Panigrahi, P., Huchche, A. D., Srivastava, A. K., Singh, S. (2008). Effect of drip irrigation and plastic mulch on performance of Nagpur mandarin (*Citrus reticulata* Blanco) grown in central India. *Indian Journal of Agricultural Sciences*, 78:1005–1009.
7. Panigrahi, P., Srivastava, A. K., Huchche, A. D. (2010). Optimizing growth, yield and water use efficiency (WUE) in Nagpur mandarin (*Citrus reticulate* Blanco) under drip irrigation and plastic mulch. *Indian Journal of Soil Conservation*, 38(1):42–45.
8. Panigrahi, P., Srivastava, A. K. (2011). Deficit irrigation scheduling for matured Nagpur mandarin (*Citrus reticulate* Blanco) trees of central India. *Indian Journal of Soil Conservation*, 39(2):149–154.
9. Panigrahi, P., Srivastava, A. K., Huchche, A. D. (2012). Effects of drip irrigation regimes and basin irrigation on Nagpur mandarin agronomical and physiological performance. *Agricultural Water Management,* 104:79–88.
10. Panigrahi, P., Srivastava, A. K., Huchche, A. D., Singh, S. (2012). Plant nutrition in response to drip versus basin irrigation in young Nagpur mandarin on Inceptisol. *Journal of Plant Nutrition*, 35:215–224.
11. Shukla, A. K., Pathak, R. K., Tiwari, R. P., Nath, V. (2000). Influence of Irrigation and mulching on plant growth, leaf nutrient status of aonla under sodic soil. *J. Appl. Hort.*, 2(1):37–38.
12. Singh, S. (1999). Citrus Industry of India: Current status and future Strategies. *Intensive Agriculture*, January–February issue:3–7.
13. Srivastava, A. K., Singh, S. (1997). *Nutrients management in Nagpur mandarin and acid Lime*. Extension Bulletin 5, National Research Center for Citrus, Nagpur–India.

CHAPTER 14

MICRO-IRRIGATION AND FERTIGATION SCHEDULING IN CITRUS

P. PANIGRAHI and A. K. SRIVASTAVA

CONTENTS

14.1 INTRODUCTION

Nagpur mandarin (*Citrus reticulate* var. Blanco) is commercially grown in around 0.185 million-ha of Central India [21]. The productivity of the crop (9 tons/ha) is too low as compared to productivity of mandarin varieties (25–30 tons/ha) in other countries like USA, Spain, Brazil and China. One of the major causes of low productivity of citrus in India is shortage of irrigation water in critical growth stages of the crop. The crop is conventionally cultivated with surface irrigation using ground water and conventional fertilizer application methods [13, 15]. Low water use efficiency (WUE) and fertilizer use efficiency (FUE) under surface irrigation are the two major drawbacks of surface irrigation systems [16, 17]. Moreover, the substantial loss of nutrients from plant rhizosphere through surface runoff and deep percolation under traditional methods of fertilization and irrigation causes regional water pollution, which is a threat to human life [6]. Thus, the strategy, which allows judicious use of water as well as nutrients in concurrence with plant demand, is likely to impart an improvement in quality of citrus production. Drip irrigation has higher WUE and FUE, resulting in improved productivity in different citrus varieties of the world [3, 7, 9, 14, 20].

In recent years, drip irrigation is gradually gaining popularity among the citrus growers of India [12, 21]. The growers are more interested in application of fertilizers through drip irrigation, as it saves manpower over traditional fertilization methods. However, the information regarding optimal fertilizer and irrigation scheduling through drip, and it's efficacy over conventional irrigation and fertilization methods, is meager in citrus cultivation in India. This lack of information keeps the orchard growers in stack to adopt the fertigation technology in citrus.

Therefore, this chapter discusses the interactive effects of micro-irrigation and fertilizers and compares these results with conventional band placement of fertilizers under basin irrigation, in Nagpur mandarin grown under hot subhumid tropical climate of central India.

14.2 MATERIALS AND METHODS

The field experiment was conducted at experimental farm of National Research Centre for Citrus, Nagpur (21°08′45″ N, 79°02′15″E and 340 m above mean sea level) during 2006–2007 and 2007–2008 in a 12-year-old Nagpur mandarin (*Citrus reticulata* var. Blanco) plants budded on rough lemon (*Citrus Jambhiri* Lush) root stock with spacing of 6 × 6 m^2. The experimental soil was clay loam (31.65% sand, 23.6% silt and 44.8% clay) with a field capacity (at −0.33 bar) and permanent wilting point (at −15.00 bar) of 29.26% (v/v) and 18.5% (v/v), respectively. The bulk density of soil was 1.18 $g.cm^{-3}$. The mean available N, P, K, Fe, Mn, Cu, and Zn in the upper 30 cm of soil were of 115, 10.0, 144, 18.2, 9.4, 1.1 and 0.72 mg/kg, respectively. The average daily Class-A pan evaporation rate varied from 2 mm/day in month of December to as high as 12 mm/day in May at the experimental site. Rainfall of 10 to 15 mm/day was observed during irrigation seasons (Nov to Jun). The water level in the well near the experimental site was around 15 m deep from ground surface.

The irrigation treatments were:

Drip irrigation at 50% of daily pan evaporation (Ep): I_1,

Drip irrigation at 75% Ep: I_2,

Drip irrigation at 100% Ep: I_3, and

Basin irrigation at 50% depletion of available soil water, control: I_4.

Irrigation was applied using four pressure compensated on-line drippers/plant of 8 lph, placed at 1.0 m away from plant stem and basin (ring of 1.2 m in radius) irrigation at 50% depletion of available soil water at 0–0.30 m soil profile. Irrigation quantity for different drip irrigation treatments was calculated using the formula [4]:

$$V = [S \times Kp \times K_C \times (E - ER)]/r \quad (1)$$

Where, V = the irrigation volume (liters/tree), S = the tree canopy area (m^2), Kp = the pan coefficient (0.7), Kc = the crop coefficient (0.6) [1], E = the daily Class A pan evaporation rate (mm/day), ER = the effective rainfall (mm/day), and r = the water application efficiency of irrigation system (~90%). Under basin irrigation, water was supplied through flexible hosepipe, when the soil moisture at 30 cm depth attained 50% of available soil water (23.9%, v/v). Water quantity applied in basin irrigation method was computed using the equation:

$$V = (FC - RSM) \times d \times A \quad (2)$$

where, V = volume of irrigation water (m^3), FC = field capacity (v/v, %), RSM = required soil water level » 23.9 (v/v, %), d = depth of effective root zone (0.30 m), and A = mean canopy area of tree. No runoff during irrigation cycle was observed in the orchard, assuming effective rainfall » rainfall.

The fertilizers applied under each drip irrigation treatment were at:

25% of recommended dose of fertilizer (RDF), F_1,

50% RDF, F_2, and

75% RDF, F_3.

RDF was taken as N: P_2O_5: K_2O = 600 g: 200 g: 100 g annually, as suggested for bearing Nagpur mandarin [22]. Water soluble form of urea phosphate (N: P: K = 18:44:0) and murate of potash (N: P: K = 0:0:60) were used for supplying required quantity of P and K, respectively. Some quantity of N was also applied through urea phosphate and the rest amount was supplied through urea (N: P: K= 46:0:0). Monthly fertigation was done using fertilizer injection pump during November to June, with equal splits of annually required fertilizers as estimated in various treatments. The circular band placement of granular fertilizers (urea, single super phosphate and murate of potash) at one meter radius from tree stem under basin irrigation (control) was performed three times in a year.

As a whole, ten treatments (I_1F_1, I_1F_2, I_1F_3, I_2F_1, I_2F_2, I_2F_3, I_3F_1, I_3F_2, I_3F_3 and control) were imposed in split plot design (SPD), with four replications and three adjacent trees in a row per replication. The orchard floor was kept cleaned and all the experimental trees were grown under uniform cultural and management practices.

The soil water content was monitored twice in a week at 0.30 m, and 0.60 m depths by neutron moisture meter (Troxler model-4300, USA). The indexed leaf samples (2nd and 4th leaf from tip of branches) surrounding the trees at a height of 1.5 m to 1.8 m from the ground were collected at the end of irrigation seasons and nutrient (N, P, K, Fe, Mn, Cu, and Zn) analysis was done as per the standard procedure. The vegetative growth parameters (tree height, stem height, canopy width, and stem (stock and scion) girth) were measured for all trees and their polled annual incremental magnitudes were compared. The canopy volume was calculated as follows {10}:

$$\text{Canopy volume} = 0.5233\ H\ W^2 \tag{3}$$

where, H = (tree height–stem height), and W = the canopy.

The weight of total fruits from each tree for various treatments was recorded and 5 fruits per tree were taken randomly for determination of fruit quality parameters (juice percent, acidity and total soluble solids). The total soluble solid (TSS) and acidity were measured using standard methods [18]. All the data generated were subjected to analysis of variance (ANOVA) and the critical difference (CD) at P = 5% was obtained using standard method [5].

14.3 RESULTS AND DISCUSSION

14.3.1 *IRRIGATION WATER QUANTITY*

The daily water applied under various irrigation treatments (Table 14.1) was lowest in December (12.5 to 41 $L.day^{-1}.plant^{-1}$) and highest in May (75.6 to 165.5 $L.day^{-1}.plant^{-1}$). The basin irrigation consumed higher quantity of water (41 to 165.5 $L.day^{-1}.plant^{-1}$) over drip irrigation (12.5 to 151.2 $L.day^{-1}.plant^{-1}$) during both the years of study. Overall, the water requirement under drip irrigation were 2798, 4196 and 5595 $m^3\ ha^{-1}\ year^{-1}$ under 50%, 75% and 100% Ep irrigation regimes, respectively, compared to 6340 $m^3ha^{-1}year^{-1}$ in basin irrigation. Earlier studies also demonstrated 40% reduction of water consumption in Verna lemon in Spain [19], 30% in Thompson Navel orange in Chile [7] and 35% in Kinnow in North India [20] under drip over conventional basin irrigation.

TABLE 14.1 Mean Daily Irrigation Water Applied (liters.day–1.plant–1) Under Different Irrigation Treatments In Citrus Orchard

Treatment	Daily irrigation water applied, liters.day–1.plant–1								TWA,
	Months								m3
	Nov.	Dec.	Jan.	Feb.	Mar.	Apr.	May	Jun.	ha–1 yr–1
I1	21.8	12.5	19.6	31.7	47.6	61.8	75.6	59.3	2798
I2	32.8	19.0	29.4	47.6	71.4	92.6	113.4	88.9	4196
I3	43.7	25.0	39.2	63.4	95.2	123.5	151.2	118.6	5595
+BI+BPF (Control)	50.5	41.0	45.2	82.3	107.8	134.6	165.5	132.4	6340

Note: Mean data during 2006–2008;
TWA: Total water applied, I1: Drip irrigation (DI) at 50% Class A pan evaporation (Ep), I2: DI at 75% Ep, I3: DI at 100% Ep; +BI, I4: Basin irrigation and BPF = Band placement of fertilizers.

14.3.2 SOIL WATER VARIATION

The mean monthly soil water variation observed at 0.30 and 0.60 m depths during irrigation periods indicated that all the drip irrigation regimes (except 50% Ep) with fertigation showed significantly higher soil water content (25.8–28.0%, v/v) compared to basin irrigation with band placement of fertilizer (24.2–26.7%, v/v) at 0.30 m depth (Fig. 14.1). The soil water content at 0.30 m depth increased invariably in all the treatments during January–February due to some un-seasonal rains (10 to 15 mm) in these months. The fluctuation of soil water content between two measurements in a week under basin irrigation was observed to be wider than that under any of the drip irrigation treatments. It was due to higher rate of evaporation from larger wetted surface area under basin irrigation, as reported by Cohen [2]. Among different drip irrigation regimes, the range of soil water depletion at 0.30 m depth was progressively increased with increasing irrigation level, indicating the higher rate of evapotranspiration (ET) of the trees under higher level of irrigation, even under low volume irrigation system. However, the soil water fluctuation under drip irrigation treatments was almost nil at 0.60 m depth, suggesting the confinement of effective root zone of drip-irrigated trees within top 0.30 m soil profile (Fig. 14.1b). The higher soil water content at 0.60 m depth under basin irrigation indicated the percolation of irrigation water from 0 to 0.30 m soil profile under basin irrigation. The fluctuation of soil water content at 0.60 m depth in basin irrigation was relatively higher during April to June than November to March, probably due to higher percolation caused by higher quantity of irrigation water supply during summer months (April to June) under basin irrigation. The higher soil water depletion was observed in higher levels of fertilizer with any irrigation regime. It might be due to increased plant ET caused by higher vegetative growth of the plants under higher fertilizer levels.

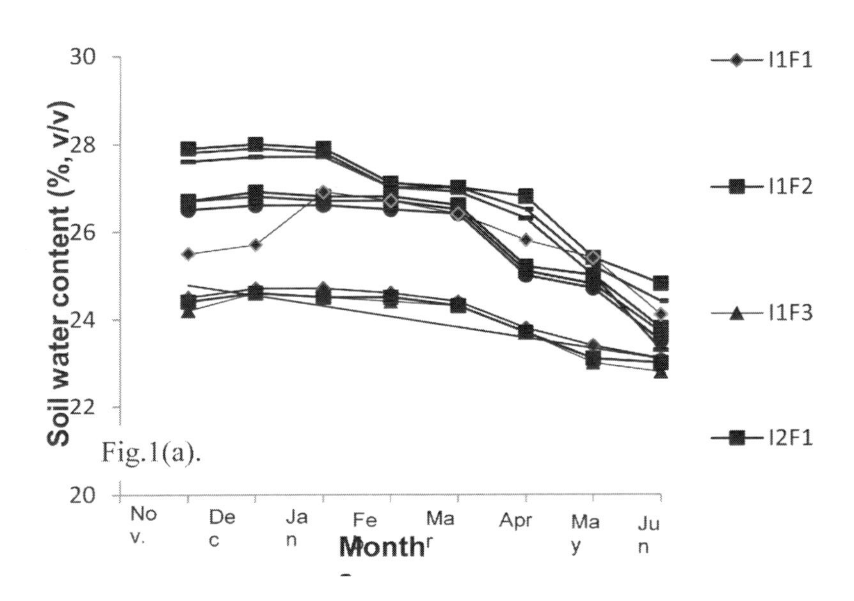

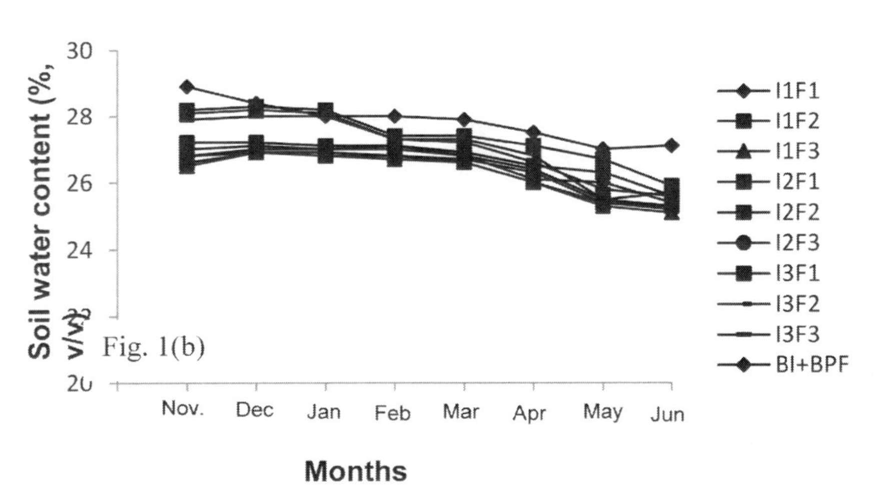

FIGURE 14.1 Soil water content at 0–0.30 m and 0.30–0.60 m depths, under various irrigation treatments during different months in Nagpur mandarin.

14.3.3 LEAF NUTRIENTS COMPOSITION

The various irrigation and fertigation treatments were observed to produce a significant response on leaf nutrient composition (Table 14.2). Increase in irrigation regime from 50% Ep (I_1) to 75% Ep (I_2) under drip irrigation enhanced the leaf – N content in corresponding fertilizers levels. The further increase in irrigation to 100% Ep (I_3) reduced leaf – N content. However, the leaf – N content increased with increase in fertilizer levels under each irrigation regime. Overall, the highest leaf – N (2.15%) was registered under I_2F_3 and lowest under I_1F_1 (1.65% N) versus 1.98% leaf – N in BI with BPF. Leaf – K content followed the similar pattern of response as leaf – N. The higher leaf N and K under optimal irrigation and fertilizer level was also observed earlier in 'Thompson Navel' orange [7] and Valencia orange [3]. Leaf – P content increased with increasing both irrigation and fertilizer levels applied through drip-irrigation, with highest magnitude under I_3F_3. The response of irrigation and fertigation in relation to leaf – P corroborates the prefindings [11] in citrus in Florida. Among micro -nutrients (Fe, Mn, Cu, Zn), only Fe concentration was observed to be significantly higher in I_3F_3 (113.99 ppm) over other treatments.

TABLE 14.2 Leaf Nutrients Composition Under Different Irrigation and Fertigation Strategy in Nagpur Mandarin.

Treatment N		Macronutrients, %			Micronutrients. Ppm.			
		P	K	Fe	Mn	Cu	Zn	
I1	F1	1.65	0.071	1.16	79.52	28.87	13.00	15.03
	F2	1.88	0.081	1.34	86.52	31.90	13.76	15.27
	F3	1.93	0.081	1.55	92.69	32.70	13.23	15.28
I2	F1	1.99	0.080	1.76	98.06	29.90	14.47	15.43
	F2	2.12	0.082	1.78	99.53	29.97	15.17	15.84
	F3	2.15	0.084	1.87	101.67	33.10	15.10	16.63
I3	F1	1.97	0.084	1.67	108.46	29.93	16.00	15.50
	F2	2.08	0.108	1.71	109.68	30.77	15.80	15.89
	F3	2.09	0.110	1.78	113.99	29.77	16.37	15.50
+BI+BPF = I4		1.98	0.081	1.64	13.52	31.90	13.76	15.27
	I	0.02	0.006	0.07	1.2	NS	NS	NS
#CD0.05	F	0.04	0.008	0.08	NS	NS	NS	NS
	IxF	0.12	0.01	0.10	0.8	NS	NS	NS

14.3.4 PLANT GROWTH

The annual incremental vegetative growth parameters (plant height, stock girth, scion girth, canopy volume) under different treatments were recorded for two years and the mean values are presented in Table 14.3. The trend of increase in all the growth parameters in relation to irrigation and fertigation was similar in both the years of observation. The mean annual increase varied in the range 0.25–0.55 m, 3.1–5.6 cm, 2.9–5.4 cm and 6.23–9.98 m^3, for in plant height, stock girth, scion girth, and canopy volume, respectively, under drip-fertigation treatments (except I_1F_1) over 0.22 m (plant height), 3.0 m (stock girth), 2.8 cm (scion girth), and 5.96 m^3 (canopy volume) under BI with BPF. However, among different growth parameters, only plant height and canopy volume showed a significant response to irrigation and fertigation individually as well as in combination. All the growth parameters were increased with increasing irrigation level from 50% Ep to 100% Ep under drip irrigation with corresponding fertilizer dose (except I_3F_3). The similar response of plant growth parameters to fertigation levels was also observed within any irrigation treatments. However, the highest magnitude of all incremental growth parameters was recorded under drip irrigation at 75% Ep with 75% RDF. The higher plant growth under drip-fertigation over BI with BPF was also observed earlier in Valencia orange [8] and in Thompson Navel orange [7].

TABLE 14.3 Annual Incremental Vegetative Growth of the Nagpur Mandarin Plants Under Different Irrigation and Fertigation Strategies

Treatment		Plant height, m	Stock girth, cm	Scion girth, cm	Canopy volume, m3
I1	F1	0.20	2.8	2.6	5.68
	F2	0.26	3.1	2.9	6.23
	F3	0.30	3.1	3.0	7.44
I2	F1	0.25	3.4	2.9	6.91
	F2	0.37	3.8	3.7	9.67
	F3	0.55	5.6	5.4	9.98
I3	F1	0.37	3.9	3.4	9.32
	F2	0.46	4.0	3.7	9.73
	F3	0.48	4.4	4.2	9.79
+BI+BPF = I4		0.22	3.0	2.8	5.96
	I	0.02	NS	NS	0.41
CD0.05	F	0.009	NS	NS	0.26
	IxF	0.024	NS	NS	0.73

14.3.5 FRUIT YIELD AND QUALITY

Fruit yield was changed significantly in response to irrigation and fertigation levels individually and in combination (Table 14.4). The highest fruit number (523) and fruit weight (110.3 g) was recorded in I_2F_3 followed by I_3F_3. The average annual fruit yield was increased with increase in fertilizer level under each drip-irrigation regime. However, the annual fruit yield was observed to increase from 8.5 t/ha in I_1F_1 to as high as 16.03 t/ha in I_2F_3 against 10.0 t/ha in BI with BPF. The higher fruit yield under optimal drip-fertigation over surface irrigation with conventional fertilization was also observed earlier in various citrus cultivars [3, 7, 9].

All the fruit quality parameters (juice content, TSS and acidity) varied significantly with irrigation and fertigation levels. The fruits having highest juice content (41.8%) and TSS (10.2 °Brix) and lowest acidity (0.82%) were harvested in I_2F_3. The BI with BPF produced the fruits with 38.4% juice content, 0.84% acidity and 9.4 °Brix TSS. The higher TSS and lower acidity in fruits under optimal water supply and fertilization through drip system over surface irrigation with broadcasting method of fertilization were also observed earlier in Valencia orange [8] and 'Thompson Navel' orange [7].

TABLE 14.4 Yield and Fruit Quality As Affected By Irrigation and Fertigation in Nagpur Mandarin

Treatment No. of fruits/ plant		**Yield parameters**			**Fruit quality parameters**		
		Average Fruit weight, g	**Total yield ton/ha**	**Juice %**	**Acidity %**	**T.S.S °Brix**	
I1	F1	364	84.0	8.50	37.4	0.85	9.4
	F2	384	87.2	9.30	38.6	0.83	9.5
	F3	421	89.6	10.48	38.7	0.83	9.5
I2	F1	409	89.8	10.21	38.4	0.81	9.6
	F2	447	98.5	12.24	39.3	0.82	9.6
	F3	523	110.3	16.03	41.8	0.82	10.2
I3	F1	411	96.2	11.00	39.9	0.84	10.1
	F2	468	102.6	13.35	40.2	0.85	10.0
	F3	518	109.5	15.77	40.4	0.86	9.7
+BI+BPF = I4		380	94.6	10.00	38.4	0.84	9.4
	I	5.6	9.8	0.82	1.08	0.006	0.62
CD0.05	F	4.2	5.6	0.67	0.87	0.008	0.44
	IxF	6.8	10.2	0.97	1.12	0.01	0.87

14.4 CONCLUSIONS

A study was conducted to optimize the combine use of irrigation water and fertilizers through drip irrigation for citrus trees at Nagpur, Maharashtra, India. The irrigation at 50% (I_1), 75% (I_2), and 100% (I_3) of daily Class A pan evaporation rate (Ep) along with 25% (F_1), 50% (F_2) and 75% (F_3) of RDF were applied through drip emitters to 11 year-old mandarin plants. The band placement of fertilizer (BPF) at 100% RDF under basin irrigation (BI) was taken as control for comparison.

All the irrigation and fertilizer interacted treatments (except I_1F_1) imposed through drip system produced higher plant growth and fruit yield, with better quality fruits over BI with BPF. The highest fruit yield (16.03 t/ha) with superior quality fruits (41.8% juice content, 10.2 °Brix TSS and 0.82% acidity) was recorded under I_2F_3. Leaf nutrients (N, P, K, Fe, Cu, Zn and Mn) analysis indicated that I_2F_3 registered significantly higher leaf-N (2.15%) and K (1.87%), whereas I_3F_3 produced higher P (0.11%) and Fe (113.99 ppm) as compared with BI with BPF (1.98% N, 0.081% P, 1.64% K, and 93.52 ppm Fe).

The application of irrigation and fertilizers through drip system is found to be a potential water and fertilizer saving technique in Nagpur mandarin. The optimal drip irrigation regime (75% of Class A pan evaporation) in combination with 75% of RDF saved around 50% water and 25% fertilizer over basin irrigation with band placement of fertilization. Drip-fertigation also enhanced the fruit yield to the tune of 60% and improved the fruit qualities (Juice percent, TSS, acidity) over basin irrigation with band fertilization method in Nagpur mandarin. The higher productivity with superior quality fruits using less water and fertilizers under drip-fertigation warrants its adoption in Nagpur mandarin orchards of central India. This can help in bringing more area under irrigation, resulting in large increase in production of citrus with prolonged orchard longevity.

The overall results of this study demonstrated that the application of optimum quantity of water and fertilizers (I_2F_3) through drip irrigation saved 50% of water and 25% of fertilizers, respectively, besides producing 60% higher fruit yield with better quality fruits over BI with BPF.

14.5 SUMMARY

Efficient use of water and nutrients is essential for sustainable production of citrus. Drip irrigation and application of fertilizers through this irrigation system has been found as one of the potential practices to save water and nutrients in different crops. The present study was conducted to evaluate effects of irrigation levels and fertilizers through drip irrigation on citrus trees at Nagpur, Maharashtra, India. All the irrigation and fertilizer interacted treatments (except I_1F_1) imposed through drip system produced higher plant growth and fruit yield, with better quality fruits over BI with BPF. The highest fruit yield (16.03 t/ha) with superior quality fruits (41.8% juice content, 10.2 °Brix TSS and 0.82% acidity) was recorded under I_2F_3. Leaf nutrients (N, P, K, Fe, Cu, Zn and Mn) analysis indicated that I_2F_3 registered significantly higher leaf-N (2.15%) and K (1.87%), whereas I_3F_3 produced higher P (0.11%) and Fe (113.99 ppm) as compared with BI with BPF (1.98% N, 0.081% P, 1.64% K, and 93.52 ppm Fe). The overall results of this study demonstrated that the application of optimum quantity of water and fertilizers (I_2F_3) through drip irrigation saved 50% and 25% of water and fertilizers, respectively, besides producing 60% higher fruit yield with better quality fruits over BI with BPF.

KEYWORDS

- **citrus**
- **drip irrigation**
- **fertigation**
- **fruit acidity**
- **fruit quality**
- **India**
- **Nagpur mandarin**
- **soil moisture**
- **TSS**
- **vegetative growth**
- **water productivity**

REFERENCES

1. Allen, R. G., Pereira, L. S., Raes, D., Smith, M. (1998). *Crop evapotranspiration guidelines for computing crop water requirements*. FAO Irrigation and Drainage Paper 56 by FAO, Rome, Italy.
2. Cohen, M. (2001). Drip irrigation in lemon orchards. *International Water and Irrigation,* 21(1):18–21.
3. Duenhas, L. H., Villas, Bôas R. L., Souza Cláudio, M. P., Oliveira, M. V. A. M., Dalri, A. B. (2005). Yield, fruit quality and nutritional status of Valencia Orange under fertigation and conventional fertilization. *J. Brazilian Association of Agricultural Engineering,* 25(1):154–160.
4. Germanà, C., Intrigliolo, F., Coniglione, L. (1989). Experiences with drip irrigation in orange trees. *In: Proc. Int. Soc. Citriculture,* held during 31 July–4 August at Royal Botanical Garden, Kew, pp. 661–664.
5. Gomez, K. A., Gomez, A. A. (1984). *Statistical procedures for agricultural research.* John Wiley and Sons, Inc., New York.
6. Hanson, B. R., Nek, J. I., Hopmans, J. W. (2006). Evaluation of urea–ammonium–nitrate fertigation with drip irrigation using numerical modeling. *Agric. Water Management*, 86:102–113.

7. Holzapfel, E. A., Lopez, C., Joublan, J. P., Matta, R. (2001). The effect of water and fertigation on canopy growth and yield on Thompson Navel oranges. *Chilean J. Agric. Res.*, 61(1):51–60.
8. Koo, R. C. J., Smjstrala, A. G. (1984). Effect of trickle irrigation and fertigation on fruit production and fruit quality of Valencia orange. *Proceedings of Florida State Horticultural Society*, 97:8–10.
9. Morgan, K. T., Wheaton, T. A., Castle, W. S., Parsons, L. R. (2009). Response of young and maturing citrus trees grown on a sandy soil to irrigation scheduling, nitrogen fertilizer rate, and nitrogen application method. *Hort Science*, 44(1):145–150.
10. Obreza, T. A. (1991). Young Hamlin orange tree fertilizer response in south-west Florida. *Proceedings of Florida State Horticultural Society*, 103:12–16.
11. Obreza, T. A., Schumann, A. (2010). Keeping water and nutrients in the Florida citrus tree. *Hort Technology*, 20(1):67–73.
12. Panigrahi, P., Huchche, A. D., Srivastava, A. K., Singh, S. (2008). Effect of drip irrigation and plastic mulch on performance of Nagpur mandarin (*Citrus reticulata* Blanco) grown in central India. *Indian Journal of Agricultural Sciences*, 78:1005–1009.
13. Panigrahi, P., Srivastava, A. K., Huchche, A. D. (2010). Optimizing growth, yield and water use efficiency (WUE) in Nagpur mandarin (*Citrus reticulate* Blanco) under drip irrigation and plastic mulch. *Indian Journal of Soil Conservation*, 38(1):42–45.
14. Panigrahi, P., Srivastava, A. K. (2011). Deficit irrigation scheduling for matured Nagpur mandarin (*Citrus reticulate* Blanco) trees of central India. *Indian Journal of Soil Conservation*, 39(2):149–154.
15. Panigrahi, P., Srivastava, A. K., Huchche, A. D. (2012). Effects of drip irrigation regimes and basin irrigation on Nagpur mandarin agronomical and physiological performance. *Agricultural Water Management,* 104:79–88.
16. Panigrahi, P., Srivastava, A. K., Huchche, A. D., Singh, S. (2012b). Plant nutrition in response to drip versus basin irrigation in young Nagpur mandarin on *Inceptisol. Journal of Plant Nutrition*, 35:215–224.
17. Phene, C. J. (1995). Research trends in microirrigation. In: *Proc. 5th International Micro Irrigation Congress On Micro-Irrigation For A Changing World: Conserving Resource Preserving The Environment,* Lamm F R (Ed.). ASABE, Orlando, Florida.
18. Ranganna, R. (2001). *Handbook of Analysis and Quality Control for Fruit and Vegetable Products.* 2nd edition, Tata McGraw Hill, pp. 860.
19. Sáchez Blanco, M. J., Torrecillas, A., León, A., Amor, F. D. (1989). Growth of Verna lemons under different irrigation regimes. *Adv. Horti. Sci.*, 3(3):109–111.
20. Singh, R., Kumar, S., Kumar. A. (2001). Studies on performance of drip irrigated Kinnow orchard in south-west Punjab-A case study. In: *Micro-irrigation* (Ed. H. P. Singh, S. P. Kaushish, A. Kumar, T. S. Murthy, J. C. Samuel). CBIP. pp. 794.
21. Singh, S., Srivastava, A. K. (2004). Citrus industry of India and overseas. *In: Advances in Citriculture* by Singh, S., Shivankar, V. J., Srivastava, A. K., Singh, I. P. (Eds). Jagmander Book Agency, New Delhi. pp. 8–67.
22. Srivastava, A. K., Singh, S. (1997). *Nutrients management in Nagpur mandarin and acid lime.* Extension Bulletin No. 5, NRC for Citrus, Nagpur, Maharashtra, India.

CHAPTER 15

INTEGRATING MICRO-IRRIGATION WITH RAINWATER HARVESTING IN CITRUS

P. PANIGRAHI, A. D. HUCHCHE, and A. K. SRIVASTAVA

CONTENTS

In this chapter: One Rs. (Indian rupees) = 0.01571 US$.

15.1 INTRODUCTION

Water scarcity is one of the major causes of low productivity and decline of citrus orchards in tropical and subtropical regions of the world. In India, it is third important fruit crop after banana and mango. Nagpur mandarin (*Citrus reticulata* Blanco) is mostly grown on Vertisol of gently sloppy lands, which is characterized with producing abundant runoff during monsoon on one hand and soil moisture shortage for sustaining the crop in post monsoon period on the other hand [4, 10]. Although, some in-situ runoff conservation measures for different crops such as lemon [3] and sweet orange [1] were advocated for better growth and production, yet such information is lacking for Nagpur mandarin. Moreover, rainwater harvesting in tank and recycling it in orchards is one of the potential options for enhancing productivity of citrus in water scarce regions.

The citrus in central India is basically irrigated by basin or furrow irrigation method. Proper irrigation water management by optimum use of available water resources along with water resource development in the region is quite necessary. Water management studies in Nagpur mandarin show that optimum soil water regime under drip irrigation can increase its growth and yield [8, 9]. Also plastic polythene mulching [5, 6] and deficit irrigation [7] have proved the effectiveness in conserving the soil moisture and increasing the growth, yield and quality of different citrus cultivars. However, the information on rainwater harvesting and recycling it through drip irrigation and mulching for citrus is meager. Keeping this in view, a study was undertaken to evaluate the performance of rainwater harvesting and recycling the harvested water through drip irrigation in Nagpur mandarin under plastic mulch. The impact of rainwater harvesting on groundwater of the study site was also studied.

15.2. MATERIALS AND METHODS

During 2006–2009, the experiment was carried out at research farm of National Research Center for Citrus, Nagpur – India on one-year-old Nagpur mandarin trees budded on rough lemon root stock with 6 ´ 6 m^2 tree spacing. The treatments were:

T_1: continuous bunding,

T_2: continuous trenching,

T_3: staggered trenching between rows, and

control (C): without any soil and water conservation measures.

The randomized block design was used with seven replications in block size of 36×18 m^2 on a slope of 4.2%. The soil type was clay loam with field capacity and permanent wilting point of 24.8% (weight basis) and 15.7% (weight basis), respectively. Runoff was measured through multislot divisor and well-stirred runoff samples were collected for estimation of sediment yield and nutrients loss after each rainfall under different treatments. Runoff sample analysis consisted of alkaline $KMNO_4$ distillation for available N, $NaHCO_3$ (pH 8.3) extractable-P as Olsen-P, 1N neutral NH_4OAc – K [11]. The moisture content at 0–30 cm depth was recorded each week by neutron moisture probe (Troxler model-4300) in various treatments.

A water harvesting tank of size $35 \times 35 \times 3$ m^3 was constructed in 2005. Prior to construction of the tank, the groundwater level in the well was recorded. The trees were irrigated by groundwater during initial years (2003–2005). The irrigation systems in the orchards were traditional surface irrigation and drip irrigation with and without mulch and compared with rain-fed treatment. After construction of the tank, the harvested water was used at the best level of drip irrigation (60% of pan evaporation) with black plastic mulch (100 micron thickness). Mulching was done with a 1.0 ´ 1.0 m^2 size polythene sheets on each tree basin keeping the tree at the center in one ha of Nagpur mandarin. The harvested water also recharged the groundwater and water from wells was used for irrigation purpose after drying of tank during May and June. The volume of water required was computed using the equation:

$$V = Ep ´ Kc ´ Kp ´ Wp ´ D \qquad (1)$$

Where, V = volume of water (liter/plant/day), Ep = cumulative pan evaporation for two consecutive days (mm), Kc = crop factor, Kp = pan factor, Wp = wetting factor, and D = canopy diameter observed at noon. The crop factor was taken as 0.6 and pan factor as 0.7 in winter and 0.8 in summer [2].

Recommended dose of fertilizers was applied. The vegetative growth parameters (tree height, stem height, canopy diameter, stock and scion girth) were measured and their incremental magnitudes under different treatments were compared.

The fruits were harvested from each tree and the weight was measured to estimate the yield in different treatments. Five fruits per tree were taken randomly for determination of fruit quality (juice percent, acidity and total soluble solids) parameters. Juice was extracted manually by juice extractor and the percent was estimated on weight basis with respect to fruit weight. The total soluble solid (TSS) was determined by digital refractometer and acidity was measured by volumetric titration with standardized sodium hydroxide, using phenolphthalein as an internal indicator. The economics of citrus cultivation under different treatments was determined with indices such as net return and benefit-cost ratio (BCR).

The data were subjected to analysis of variance (ANOVA), and significance of the data within the treatments was determined using SAS-9.2 statistical software.

15.3 RESULTS AND DISCUSSION

15.3.1 RUNOFF, SOIL AND NUTRIENT CONSERVATION

The mean annual rainfall, runoff and soil loss data under different treatments indicated that the maximum runoff (38.15%) and soil loss (4.98 t/ha) occurred in control, whereas the minimum (runoff 27.3%; soil loss 3.74 t/ha) was under continuous trenching, followed by continuous bunding (Table 15.1). The runoff and soil loss, occurred under staggered trenching, were 10 and 6% lower over control. Continuous trenching conserved the maximum runoff (28.4%) and soil (24.9%) among the conservation measures over control due to higher runoff conservation in trenches between the rows.

TABLE 15.1 Runoff, Soil Loss, and Nutrient Loss Under Drip Irrigation With Different Soil and Water Conservation Measures in Nagpur Mandarin

Treatment	Runoff, mm	Soil loss, t/ha/yr	Nutrient loss, kg/ha		
			N	P	K
Drip irrigation + Mulch + Continuous bunding	263 (28.8)***	4.11	0.75	0.15	1.24
Drip irrigation + Mulch + Continuous trenching	249 (27.3)	3.74	0.62	0.13	1.09
Drip irrigation + Mulch + Staggered trenching	313 (34.3)	4.67	0.87	0.17	1.57
Without conservation measure (Control)	348 (38.15)	4.98	1.08	0.24	2.08

** ARF, Annual Rainfall, Mean ARF = 912 mm;

*** Numbers in parenthesis indicate runoff as % of mean annual rainfall.

The analysis of runoff samples under different treatments for available N, P and K (Table 15.1) showed that nutrient loss was maximum in control (1.08 kg N/ha, 0.24 kg P/ha and 2.08 kg K/ha), and lowest in continuous trenching (0.62 kg N/ha, 0.13 kg P/ha and 1.09 kg K/ha) followed by continuous bunding. The lowest nutrient loss under continuous trenching was attributed to the lowest soil loss. Due to heavy loss of upper fertile soil through runoff, the nutrient concentration in eroded soil was invariably higher than the original soil, irrespective of the treatments.

15.3.2 SOIL MOISTURE VARIABILITY AND GROUNDWATER RECHARGE

The mean monthly moisture content at 0–30 cm soil profile revealed that the soil moisture status was improved considerably in various conservation treatments with drip irrigation over control (Table 15.2). Among different treatments, the highest soil moisture content (24.55–30.52%, v/v) was observed under continuous trenching followed by continuous bunding (24.25–28.33%, v/v). The moisture content under staggered trenching was 23.43–26.92% (v/v) in various months. The higher moisture content in continuous trenching was due to maximum rainwater conservation during the rainy period. The moisture content under various conservation measures and control was reduced with time, except during the month of February, due to some unseasonal rainfall (11 mm) in the month. This was due to more consumptive use by trees under increased soil moisture content under various conservation treatments. Moreover, the moisture content under different treatments did not vary significantly at the initial period (October) of observation. But during the period between November and February, the moisture content under various conservation measures was significantly higher over control.

The groundwater level in the wells located in the orchard was increased by 1.5–2.3 m after construction of water harvesting tanks compared to water level before construction.

TABLE 15.2 Soil Water Content (%, v/v) at 0.30 m Depth Under Different Soil and Water Conservation Measures in Nagpur Mandarin

	Month								
Treatment	Oct.	Nov.	Dec.	Jan.	Feb.	Mar.	Apr.	May	June
Drip irrigation + Mulch + Continuous bunding	28.33	28.26	27.94	26.76	28.24	27.45	26.45	26.15	24.25
Drip irrigation + Mulch + Continuous trenching	30.52	30.31	28.86	28.45	30.37	27.75	26.57	26.35	24.55
Drip irrigation + Mulch + Staggered trenching	24.36	24.18	23.85	23.74	25.69	26.92	25.88	24.20	23.43
No conservation measure (Control)	23.63	23.38	21.84	21.54	23.46	25.88	23.85	22.80	22.33
**CD (P=0.05)	NS	1.92	2.21	2.40	2.97	NS	NS	NS	NS

**CD = Critical difference at 5% probability, NS = Not significant.

15.3.3 VEGETATIVE GROWTH, YIELD, FRUIT QUALITY AND ECONOMICS

The incremental growth of vegetative parameters (tree height, canopy volume, and stem girth) showed that all the parameters were significantly higher under various conservation measures over control (Table 15.3). The highest magnitude of the incremental growth parameters was observed in continuous trenching. Similarly, all the conservation measures produced higher fruit yield (7–29%) with better fruit quality over control. The highest fruit yield (9.60 kg/tree) was observed in continuous trenching. Quality assessment of fruits showed that the juice contents (40.42%) and TSS (10.10 °Brix) were significantly higher under continuous trenching treatment. The higher vegetative growth and fruit yield with better fruit quality in various conservation measures was due to better availability of soil moisture to mandarin trees during flowering and fruiting stages during the postmonsoon period.

TABLE 15.3 Incremental Vegetative Growth, Fruit Yield and Quality of Nagpur Mandarin Under Various Soil and Water Conservation Measures

	Vegetative growth			Fruit yield			Fruit quality		
Treatment	**PH**	**SG**	**CV**	**No. of fruits/ tree**	**Fruit weight**	**Total yield/ tree**	**Juice**	**Acidity**	**TSS**
	m	**cm**	**m³**	**No.**	**g**	**kg/ plant**	**%**	**%**	**°Brix**
Drip irrigation + Mulch + Continuous bunding	0.28	2.23	0.759	65	136.7	8.88	39.33	0.83	10.00
Drip irrigation + + Mulch + Continuous trenching	0.35	2.4	0.846	69	139.2	9.60	40.42	0.82	10.10
Drip irrigation + + Mulch + Staggered trenching	0.24	1.8	0.563	59	135.4	7.98	37.66	0.84	9.98
Without conservation measure (Control)	0.19	1.35	0.402	55	135.2	7.43	35.42	0.86	9.94
CD (P = 0.05)	0.08	0.3	0.031	2.5	NS	0.31	3.6	NS	0.05

CD = Critical difference at 5% probability.

TABLE 15.4 Economics of Citrus Cultivation Under Rainwater Harvesting and Drip Irrigation

Treatments	Yield tons/ha	Gross return × 10^5 Rs.	Net return × 10^5 Rs.	BCR
RWHT + CT + Drip irrigation + Mulch	13.0	2.4	1.8	2.0
Rain-fed system	8.5	1.5	1.1	1.7

RWHT = rainwater harvesting tank, CT = continuous trenching; DI = drip irrigation.

The citrus production (Table 15.4) under continuous trenching, water harvesting tank, drip irrigation and mulch was more economical (net return of Rs. 1.8 × 10^5 and BCR of 2.0) compared to that under rain-fed condition (net return of Rs. 1.1 × 10^5 and BCR of 1.7)

15.4 CONCLUSIONS

Citrus is extensively grown in central India. The crop is basically irrigated by bore well or dug well through conventional surface irrigation method in this region. For the last few years, water level in wells has declined alarmingly creating water shortage for sustaining the crop. Thus, keeping this problem in view, various in-situ rainwater conservation treatments (continuous trenching, continuous bunding, staggered trenching between the rows across the slope (4.2%) and control (without any soil and water conservation treatment)) were evaluated in one-year-old Nagpur mandarin at Nagpur, India.

The continuous trenching produced the best response conserving 38% runoff, 32.28% soil, 32.44% N, 27.67% P, and 28.95% K over control, besides 15.7% higher fruit yield with better fruit quality. Moreover, rainfall runoff from 3.2 hectare of land was harvested in a tank size of 35×35×3 m^3 and recycled at the best level of irrigation (60% of pan evaporation) through drip irrigation with black plastic mulch of 100 micron thickness in one ha of Nagpur mandarin. Over all, by conjunctive use of both continuous trenching and tank-based drip irrigation with black plastic mulch, the tree growth and fruit yield were enhanced up to 45% over basin irrigation without any conservation measure. These studies suggested the combine adoption of continuous trenching and tank-based drip irrigation with black plastic mulch to reduce the water scarcity for inducing better growth, yield and health of Nagpur mandarin central India.

15.5 SUMMARY

Drip irrigation with continuous trenching was found to be a superior soil and water conservation technique for cultivation of Nagpur mandarin. The method warrants its' adoption in mandarin orchards of Central India and elsewhere having similar agro-pedological conditions. Moreover, the citrus cultivation under continuous trench, drip irrigation and plastic mulch using water from rainwater harvesting tank was found more productive and economical compared to rain-fed citriculture in central India. The technique therefore suggested to be adopted in the study region for improving the quality production of citrus without bringing any sizeable reduction in soil fertility.

The continuous trenching produced 15.7% higher fruit yield with better fruit quality. Over all, by conjunctive use of both continuous trenching and tank-based drip irrigation with black plastic mulch, the tree growth and fruit yield were enhanced up to 45% over basin irrigation without any conservation measure. These studies suggested the combine adoption of continuous trenching and tank-based drip irrigation with black plastic mulch.

KEYWORDS

- **Central India**
- **citrus**
- **drip irrigation**
- **FAO**
- **fruit quality**
- **fruit yield**
- **India**
- **irrigation scheduling**
- **Mahatma Phule Agricultural University**
- **Nagpur mandarin**
- **National Research Center for Citrus**
- **rain water harvesting**
- **soil moisture**
- **water productivity**
- **water scarcity**

REFERENCES

1. Arora, Y. K., Mohan, S. C. (1985). Water harvesting and water management for fruit crops in waste Lands, Third National Workshop on arid zone fruit research held at Mahatma Phule Agricultural University, Rahuri, India, July 5–8. *Technical Document*, 17:104–112.
2. Doorenbos, J., Pruitt, W. O. (1977). *Guidelines for predicting crop water requirements*. Irrigation and Drainage Paper 24, FAO, United Nations, Rome, Italy.
3. Ghosh, S. P. (1982). *Water harvesting for fruit orchards in Dehra Dun valley*. Proceedings of International Symposium on hydrological aspects of mountainous watersheds at Roorke, India, November 4–6, pp. 31–33.
4. Huchche, A. D., Srivastava, A. K., Ram, L., Singh, S. (1999). Nagpur mandarin orchard efficiency in central India. Hi-Tech Citrus Management. Proceedings of International Symposium on citriculture held at National Research Center for Citrus, Nagpur, Maharashtra, India, November 23–27, pp. 24–28.
5. Panigrahi, P., Huchche, A. D., Srivastava, A. K., Singh, S. (2008). Effect of drip irrigation and plastic mulch on performance of Nagpur mandarin (*Citrus reticulata* Blanco) grown in central India. *Indian Journal of Agricultural Sciences*, 78:1005–1009.
6. Panigrahi, P., Srivastava, A. K., Huchche, A. D. (2010). Optimizing growth, yield and water use efficiency (WUE) in Nagpur mandarin (*Citrus reticulate* Blanco) under drip irrigation and plastic mulch. *Indian Journal of Soil Conservation*, 38(1):42–45.
7. Panigrahi, P., Srivastava, A. K. (2011). Deficit irrigation scheduling for matured Nagpur mandarin (*Citrus reticulate* Blanco) trees of central India. *Indian Journal of Soil Conservation*, 39(2):149–154.
8. Panigrahi, P., Srivastava, A. K., Huchche, A. D. (2012). Effects of drip irrigation regimes and basin irrigation on Nagpur mandarin agronomical and physiological performance. *Agricultural Water Management,* 104:79–88.
9. Panigrahi, P., Srivastava, A. K., Huchche, A. D., Singh, S. (2012). Plant nutrition in response to drip versus basin irrigation in young Nagpur mandarin on Inceptisol. *Journal of Plant Nutrition*, 35:215–224.
10. Singh, S., Srivastava, A. K. (2004). Citrus industry of India and overseas. In: Singh, S., Shivankar, V. J., Srivastava, A. K., Singh, I. P. (Eds.), *Advances in Citriculture*. Jagmander Book Agency, New Delhi, India, pp. 8–67.
11. Tandon, H. L. S. (1998). *Methods of Analysis of Soils, Plant, Water and Fertilizers*, Fertilizer Development and Consultation Organization, New Delhi, India, pp. 42–44.

APPENDICES

(Modified and reprinted with permission from: Megh R. Goyal (2012). Appendices. Pages 317–332. In: *Management of Drip/Trickle or Micro Irrigation,* edited by: Megh R. Goyal. New Jersey, USA: Apple Academic Press Inc.)

APPENDIX A

CONVERSION SI AND NON-SI UNITS

To convert the Column 1 in the Column 2	Column 1	Column 2	To convert the Column 2 in the Column 1
	Unit	Unit	
Multiply by	SI	Non-SI	Multiply by

LINEAR

0.621	kilometer, km (10^3 m)	miles, mi	1.609
1.094	meter, m	yard, yd	0.914
3.28	meter, m	feet, ft	0.304
3.94×10^{-2}	millimeter, mm (10^{-3})	inch, in	25.4

SQUARES

2.47	hectare, he	acre	0.405
2.47	square kilometer, km^2	acre	4.05×10^{-3}
0.386	square kilometer, km^2	square mile, mi^2	2.590
2.47×10^{-4}	square meter, m^2	acre	4.05×10^{-3}
10.76	square meter, m^2	square feet, ft^2	9.29×10^{-2}
1.55×10^{-3}	mm^2	square inch, in^2	645

CUBICS

9.73×10^{-3}	cubic meter, m^3	inch-acre	102.8
35.3	cubic meter, m^3	cubic-feet, ft^3	2.83×10^{-2}
6.10×10^4	cubic meter, m^3	cubic inch, in^3	1.64×10^{-5}

2.84×10^{-2} — liter, L (10^{-3} m^3) bushel, bu —— 35.24
1.057 —— liter, L liquid quarts, qt —— 0.946
3.53×10^{-2} — liter, L cubic feet, ft^3 —— 28.3
0.265 —— liter, L gallon —— 3.78
33.78 —— liter, L fluid ounce, oz —— 2.96×10^{-2}
2.11 —— liter, L fluid dot, dt —— 0.473

WEIGHT

2.20×10^{-3} — gram, g (10^{-3} kg) pound, —— 454
3.52×10^{-2} — gram, g (10^{-3} kg) ounce, oz —— 28.4
2.205 —— kilogram, kg pound, lb —— 0.454
10^{-2} —— kilogram, kg quintal (metric), q —— 100
1.10×10^{-3} — kilogram, kg ton (2000 lbs), ton —— 907
1.102 —— mega gram, mg ton (US), ton —— 0.907
1.102 —— metric ton, t ton (US), ton —— 0.907

YIELD AND RATE

0.893 —— kilogram per hectare pound per acre —— 1.12
7.77×10^{-2} — kilogram per cubic meter pound per fanega —— 12.87
1.49×10^{-2} — kilogram per hectare pound per acre, 60 lb —— 67.19
1.59×10^{-2} — kilogram per hectare pound per acre, 56 lb —— 62.71
1.86×10^{-2} — kilogram per hectare pound per acre, 48 lb —— 53.75
0.107 —— liter per hectare galloon per acre —— 9.35
893 —— ton per hectare pound per acre —— 1.12×10^{-3}
893 —— mega gram per hectare pound per acre —— 1.12×10^{-3}
0.446—— ton per hectare ton (2000 lb) per acre —— 2.24
2.24 —— meter per second mile per hour —— 0.447

SPECIFIC SURFACE

10 —— square meter per kilogram square centimeter per gram —— 0.1
10^3 ——square meter per kilogram square millimeter per gram ——10^{-3}

PRESSURE

9.90 —	megapascal, MPa	atmosphere —	0.101
10 —	megapascal	bar —	0.1
1.0 —	megagram per cubic meter	gram per cubic centimeter —	1.00
2.09×10^{-2} —	pascal, Pa	pound per square feet —	47.9
1.45×10^{-4} —	pascal, Pa	pound per square inch —	6.90×10^{3}

CONVERSION SI AND NON-SI UNITS

To convert the Column 1 in the Column 2	Column 1	Column 2	To convert the Column 2 in the Column 1
	Unit	Unit	
Multiply by	SI	Non-SI	Multiply by

TEMPERATURE

1.00 (K-273)—	Kelvin, K	centigrade, °C —	1.00 (C+273)
(1.8 C + 32)—	centigrade, °C	Fahrenheit, °F —	(F–32)/1.8

ENERGY

9.52×10^{-4} —	Joule J	BTU —	1.05×10^{3}
0.239 —	Joule, J	calories, cal —	4.19
0.735 —	Joule, J	feet-pound —	1.36
2.387×10^{5} —	Joule per square meter	calories per square centimeter —	4.19×10^{4}
10^{5} —	Newton, N	dynes —	10^{-5}

WATER REQUIREMENTS

9.73×10^{-3} —	cubic meter	inch acre —	102.8
9.81×10^{-3} —	cubic meter per hour	cubic feet per second —	101.9
4.40 —	cubic meter per hour	galloon (US) per minute —	0.227
8.11 —	hectare-meter	acre-feet —	0.123
97.28 —	hectare-meter	acre-inch —	1.03×10^{-2}
8.1×10^{-2} —	hectare centimeter	acre-feet —	12.33

CONCENTRATION

1 ———	centimol per kilogram	milliequivalents per 100 grams ———	1
0.1 ———	gram per kilogram	percents ———	10
1 ———	milligram per kilogram	parts per million ———	1

NUTRIENTS FOR PLANTS

2.29 ———	P	P_2O_5 ———	0.437
1.20 ———	K	K_2O ———	0.830
1.39 ———	Ca	CaO ———	0.715
1.66 ———	Mg	MgO ———	0.602

NUTRIENT EQUIVALENTS

		Conversion	Equivalent
Column A	**Column B**	**A to B**	**B to A**
N	NH3	1.216	0.822
	NO3	4.429	0.226
	KNO3	7.221	0.1385
	Ca(NO3)2	5.861	0.171
	(NH4)2SO4	4.721	0.212
	NH4NO3	5.718	0.175
	(NH4)2 HPO4	4.718	0.212
P	P2O5	2.292	0.436
	PO4	3.066	0.326
	KH2PO4	4.394	0.228
	(NH4)2 HPO4	4.255	0.235
	H3PO4	3.164	0.316
K	K2O	1.205	0.83
	KNO3	2.586	0.387
	KH2PO4	3.481	0.287
	Kcl	1.907	0.524
	K2SO4	2.229	0.449
Ca	CaO	1.399	0.715

	Ca(NO3)2	4.094	0.244
	CaCl2 × 6H2O	5.467	0.183
	CaSO4 × 2H2O	4.296	0.233
Mg	MgO	1.658	0.603
	MgSO4 × 7H2O	1.014	0.0986
S	H2SO4	3.059	0.327
	(NH4)2 SO4	4.124	0.2425
	K2SO4	5.437	0.184
	MgSO4 × 7H2O	7.689	0.13
	CaSO4 × 2H2O	5.371	0.186

APPENDIX B

PIPE AND CONDUIT FLOW

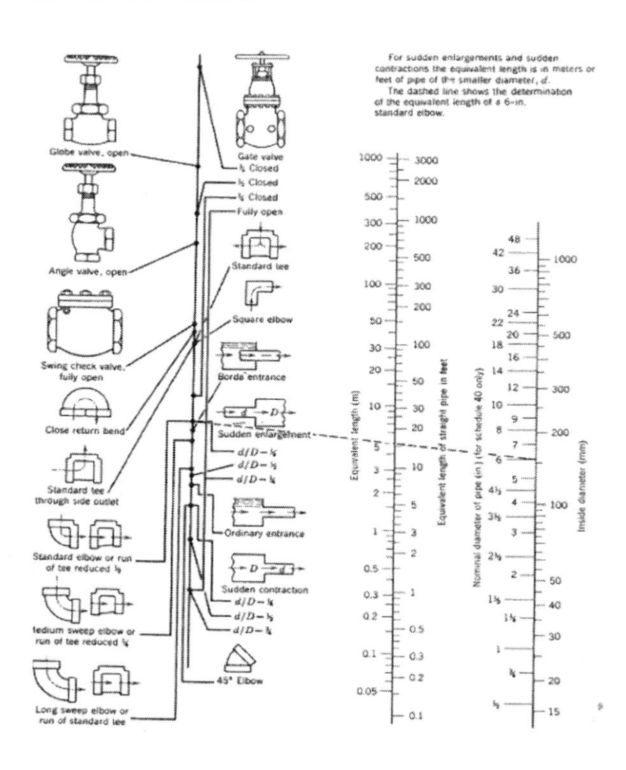

APPENDIX C

PERCENTAGE OF DAILY SUNSHINE HOURS: FOR NORTH AND SOUTH HEMISPHERES

Latitude	Jan	Feb	Mar	Apr	May	Jun	Jul	Aug	Sep	Oct	Nov	Dec
NORTH												
0	8.50	7.66	8.49	8.21	8.50	8.22	8.50	8.49	8.21	8.50	8.22	8.50
5	8.32	7.57	8.47	3.29	8.65	8.41	8.67	8.60	8.23	8.42	8.07	8.30
10	8.13	7.47	8.45	8.37	8.81	8.60	8.86	8.71	8.25	8.34	7.91	8.10
15	7.94	7.36	8.43	8.44	8.98	8.80	9.05	8.83	8.28	8.20	7.75	7.88
20	7.74	7.25	8.41	8.52	9.15	9.00	9.25	8.96	8.30	8.18	7.58	7.66
25	7.53	7.14	8.39	8.61	9.33	9.23	9.45	9.09	8.32	8.09	7.40	7.52
30	7.30	7.03	8.38	8.71	9.53	9.49	9.67	9.22	8.33	7.99	7.19	7.15
32	7.20	6.97	8.37	8.76	9.62	9.59	9.77	9.27	8.34	7.95	7.11	7.05
34	7.10	6.91	8.36	8.80	9.72	9.70	9.88	9.33	8.36	7.90	7.02	6.92
36	6.99	6.85	8.35	8.85	9.82	9.82	9.99	9.40	8.37	7.85	6.92	6.79
38	6.87	6.79	8.34	8.90	9.92	9.95	10.1	9.47	3.38	7.80	6.82	6.66
40	6.76	6.72	8.33	8.95	10.0	10.1	10.2	9.54	8.39	7.75	6.72	7.52
42	6.63	6.65	8.31	9.00	10.1	10.2	10.4	9.62	8.40	7.69	6.62	6.37
44	6.49	6.58	8.30	9.06	10.3	10.4	10.5	9.70	8.41	7.63	6.49	6.21
46	6.34	6.50	8.29	9.12	10.4	10.5	10.6	9.79	8.42	7.57	6.36	6.04
48	6.17	6.41	8.27	9.18	10.5	10.7	10.8	9.89	8.44	7.51	6.23	5.86
50	5.98	6.30	8.24	9.24	10.7	10.9	11.0	10.0	8.35	7.45	6.10	5.64
52	5.77	6.19	8.21	9.29	10.9	11.1	11.2	10.1	8.49	7.39	5.93	5.43
54	5.55	6.08	8.18	9.36	11.0	11.4	11.4	10.3	8.51	7.20	5.74	5.18
56	5.30	5.95	8.15	9.45	11.2	11.7	11.6	10.4	8.53	7.21	5.54	4.89
58	5.01	5.81	8.12	9.55	11.5	12.0	12.0	10.6	8.55	7.10	4.31	4.56
60	4.67	5.65	8.08	9.65	11.7	12.4	12.3	10.7	8.57	6.98	5.04	4.22
SOUTH												
0	8.50	7.66	8.49	8.21	8.50	8.22	8.50	8.49	8.21	8.50	8.22	8.50
5	8.68	7.76	8.51	8.15	8.34	8.05	8.33	8.38	8.19	8.56	8.37	8.68
10	8.86	7.87	8.53	8.09	8.18	7.86	8.14	8.27	8.17	8.62	8.53	8.88
15	9.05	7.98	8.55	8.02	8.02	7.65	7.95	8.15	8.15	8.68	8.70	9.10
20	9.24	8.09	8.57	7.94	7.85	7.43	7.76	8.03	8.13	8.76	8.87	9.33
25	9.46	8.21	8.60	7.74	7.66	7.20	7.54	7.90	8.11	8.86	9.04	9.58

30	9.70	8.33	8.62	7.73	7.45	6.96	7.31	7.76	8.07	8.97	9.24	9.85
32	9.81	8.39	8.63	7.69	7.36	6.85	7.21	7.70	8.06	9.01	9.33	9.96
34	9.92	8.45	8.64	7.64	7.27	6.74	7.10	7.63	8.05	9.06	9.42	10.1
36	10.0	8.51	8.65	7.59	7.18	6.62	6.99	7.56	8.04	9.11	9.35	10.2
38	10.2	8.57	8.66	7.54	7.08	6.50	6.87	7.49	8.03	9.16	9.61	10.3
40	10.3	8.63	8.67	7.49	6.97	6.37	6.76	7.41	8.02	9.21	9.71	10.5
42	10.4	8.70	8.68	7.44	6.85	6.23	6.64	7.33	8.01	9.26	9.8	10.6
44	10.5	8.78	8.69	7.38	6.73	6.08	6.51	7.25	7.99	9.31	9.94	10.8
46	10.7	8.86	8.90	7.32	6.61	5.92	6.37	7.16	7.96	9.37	10.1	11.0

APPENDIX D

PSYCHOMETRIC CONSTANT (Γ) FOR DIFFERENT ALTITUDES (Z)

$$\gamma = 10^{-3}\,[(C_p.P) \div (\varepsilon.\lambda)] = (0.00163) \times [P \div \lambda]$$

γ, psychrometric constant [kPa C–1]
cp, specific heat of moist air = 1.013 [kJ kg–1 °C–1]
P, atmospheric pressure [kPa].
ε, ratio molecular weight of water vapor/dry air = 0.622
λ, latent heat of vaporization [MJ kg–1] = 2.45 MJ kg–1 at 20 °C.

Z (m)	γ **kPa/°C**	z **(m)**	γ **kPa/°C**	z **(m)**	γ **kPa/°C**	z **(m)**	γ **kPa/°C**
0	0.067	1000	0.060	2000	0.053	3000	0.047
100	0.067	1100	0.059	2100	0.052	3100	0.046
200	0.066	1200	0.058	2200	0.052	3200	0.046
300	0.065	1300	0.058	2300	0.051	3300	0.045
400	0.064	1400	0.057	2400	0.051	3400	0.045
500	0.064	1500	0.056	2500	0.050	3500	0.044
600	0.063	1600	0.056	2600	0.049	3600	0.043
700	0.062	1700	0.055	2700	0.049	3700	0.043
800	0.061	1800	0.054	2800	0.048	3800	0.042
900	0.061	1900	0.054	2900	0.047	3900	0.042
1000	0.060	2000	0.053	3000	0.047	4000	0.041

APPENDIX E

SATURATION VAPOR PRESSURE [E_S] FOR DIFFERENT TEMPERATURES (T)

Vapor pressure function = e_s = [0.6108]*exp{[17.27*T]/[T + 237.3]}							
T °C	**e_s kPa**	**T °C**	**e_s kPa**	**T °C**	**e_s kPa**	**T °C**	**e_s kPa**
1.0	0.657	**13.0**	1.498	**25.0**	3.168	**37.0**	6.275
1.5	0.681	**13.5**	1.547	**25.5**	3.263	**37.5**	6.448
2.0	0.706	**14.0**	1.599	**26.0**	3.361	**38.0**	6.625
2.5	0.731	**14.5**	1.651	**26.5**	3.462	**38.5**	6.806
3.0	0.758	**15.0**	1.705	**27.0**	3.565	**39.0**	6.991
3.5	0.785	**15.5**	1.761	**27.5**	3.671	**39.5**	7.181
4.0	0.813	**16.0**	1.818	**28.0**	3.780	**40.0**	7.376
4.5	0.842	**16.5**	1.877	**28.5**	3.891	**40.5**	7.574
5.0	0.872	**17.0**	1.938	**29.0**	4.006	**41.0**	7.778
5.5	0.903	**17.5**	2.000	**29.5**	4.123	**41.5**	7.986
6.0	0.935	**18.0**	2.064	**30.0**	4.243	**42.0**	8.199
6.5	0.968	**18.5**	2.130	**30.5**	4.366	**42.5**	8.417
7.0	1.002	**19.0**	2.197	**31.0**	4.493	**43.0**	8.640
7.5	1.037	**19.5**	2.267	**31.5**	4.622	**43.5**	8.867
8.0	1.073	**20.0**	2.338	**32.0**	4.755	**44.0**	9.101
8.5	1.110	**20.5**	2.412	**32.5**	4.891	**44.5**	9.339
9.0	1.148	**21.0**	2.487	**33.0**	5.030	**45.0**	9.582
9.5	1.187	**21.5**	2.564	**33.5**	5.173	**45.5**	9.832
10.0	1.228	**22.0**	2.644	**34.0**	5.319	**46.0**	10.086
10.5	1.270	**22.5**	2.726	**34.5**	5.469	**46.5**	10.347
11.0	1.313	**23.0**	2.809	**35.0**	5.623	**47.0**	10.613
11.5	1.357	**23.5**	2.896	**35.5**	5.780	**47.5**	10.885
12.0	1.403	**24.0**	2.984	**36.0**	5.941	**48.0**	11.163
12.5	1.449	**24.5**	3.075	**36.5**	6.106	**48.5**	11.447

APPENDIX F

SLOPE OF VAPOR PRESSURE CURVE (Δ) FOR DIFFERENT TEMPERATURES (T)

$\Delta = [4098.\ e°(T)] \div [T + 237.3]^2$
$= 2504\{\exp[(17.27T) \div (T + 237.2)]\} \div [T + 237.3]^2$

T °C	Δ kPa/°C	T °C	Δ kPa/°C	T °C	Δ kPa/°C	T °C	Δ kPa/°C
1.0	0.047	13.0	0.098	25.0	0.189	37.0	0.342
1.5	0.049	13.5	0.101	25.5	0.194	37.5	0.350
2.0	0.050	14.0	0.104	26.0	0.199	38.0	0.358
2.5	0.052	14.5	0.107	26.5	0.204	38.5	0.367
3.0	0.054	15.0	0.110	27.0	0.209	39.0	0.375
3.5	0.055	15.5	0.113	27.5	0.215	39.5	0.384
4.0	0.057	16.0	0.116	28.0	0.220	40.0	0.393
4.5	0.059	16.5	0.119	28.5	0.226	40.5	0.402
5.0	0.061	17.0	0.123	29.0	0.231	41.0	0.412
5.5	0.063	17.5	0.126	29.5	0.237	41.5	0.421
6.0	0.065	18.0	0.130	30.0	0.243	42.0	0.431
6.5	0.067	18.5	0.133	30.5	0.249	42.5	0.441
7.0	0.069	19.0	0.137	31.0	0.256	43.0	0.451
7.5	0.071	19.5	0.141	31.5	0.262	43.5	0.461
8.0	0.073	20.0	0.145	32.0	0.269	44.0	0.471
8.5	0.075	20.5	0.149	32.5	0.275	44.5	0.482
9.0	0.078	21.0	0.153	33.0	0.282	45.0	0.493
9.5	0.080	21.5	0.157	33.5	0.289	45.5	0.504
10.0	0.082	22.0	0.161	34.0	0.296	46.0	0.515
10.5	0.085	22.5	0.165	34.5	0.303	46.5	0.526
11.0	0.087	23.0	0.170	35.0	0.311	47.0	0.538
11.5	0.090	23.5	0.174	35.5	0.318	47.5	0.550
12.0	0.092	24.0	0.179	36.0	0.326	48.0	0.562
12.5	0.095	24.5	0.184	36.5	0.334	48.5	0.574

APPENDIX G

NUMBER OF THE DAY IN THE YEAR (JULIAN DAY)

Day	Jan	Feb	Mar	Apr	May	Jun	Jul	Aug	Sep	Oct	Nov	Dec
1	1	32	60	91	121	152	182	213	244	274	305	335
2	2	33	61	92	122	153	183	214	245	275	306	336
3	3	34	62	93	123	154	184	215	246	276	307	337
4	4	35	63	94	124	155	185	216	247	277	308	338
5	5	36	64	95	125	156	186	217	248	278	309	339
6	6	37	65	96	126	157	187	218	249	279	310	340
7	7	38	66	97	127	158	188	219	250	280	311	341
8	8	39	67	98	128	159	189	220	251	281	312	342
9	9	40	68	99	129	160	190	221	252	282	313	343
10	10	41	69	100	130	161	191	222	253	283	314	344
11	11	42	70	101	131	162	192	223	254	284	315	345
12	12	43	71	102	132	163	193	224	255	285	316	346
13	13	44	72	103	133	164	194	225	256	286	317	347
14	14	45	73	104	134	165	195	226	257	287	318	348
15	15	46	74	105	135	166	196	227	258	288	319	349
16	16	47	75	106	136	167	197	228	259	289	320	350
17	17	48	76	107	137	168	198	229	260	290	321	351
18	18	49	77	108	138	169	199	230	261	291	322	352
19	19	50	78	109	139	170	200	231	262	292	323	353
20	20	51	79	110	140	171	201	232	263	293	324	354
21	21	52	80	111	141	172	202	233	264	294	325	355
22	22	53	81	112	142	173	203	234	265	295	326	356
23	23	54	82	113	143	174	204	235	266	296	327	357
24	24	55	83	114	144	175	205	236	267	297	328	358
25	25	56	84	115	145	176	206	237	268	298	329	359
26	26	57	85	116	146	177	207	238	269	299	330	360
27	27	58	86	117	147	178	208	239	270	300	331	361
28	28	59	87	118	148	179	209	240	271	301	332	362
29	29	(60)	88	119	149	180	210	241	272	302	333	363
30	30	—	89	120	150	181	211	242	273	303	334	364
31	31	—	90	—	151	—	212	243	—	304	—	365

APPENDIX H

STEFAN-BOLTZMANN LAW AT DIFFERENT TEMPERATURES (T)

$[\sigma^*(T_K)^4] = [4.903 \times 10^{-9}]$, MJ K^{-4} m^{-2} day^{-1}
Whççere: $T_K = \{T[°C] + 273.16\}$

T	$\sigma^*(T_K)^4$	T	$\sigma^*(T_K)^4$	T	$\sigma^*(T_K)^4$
Units					
°C	MJ m^{-2} d^{-1}	**°C**	MJ m^{-2} d^{-1}	**°C**	MJ m^{-2} d^{-1}
1.0	27.70	**17.0**	34.75	**33.0**	43.08
1.5	27.90	**17.5**	34.99	**33.5**	43.36
2.0	28.11	**18.0**	35.24	**34.0**	43.64
2.5	28.31	**18.5**	35.48	**34.5**	43.93
3.0	28.52	**19.0**	35.72	**35.0**	44.21
3.5	28.72	**19.5**	35.97	**35.5**	44.50
4.0	28.93	**20.0**	36.21	**36.0**	44.79
4.5	29.14	**20.5**	36.46	**36.5**	45.08
5.0	29.35	**21.0**	36.71	**37.0**	45.37
5.5	29.56	**21.5**	36.96	**37.5**	45.67
6.0	29.78	**22.0**	37.21	**38.0**	45.96
6.5	29.99	**22.5**	37.47	**38.5**	46.26
7.0	30.21	**23.0**	37.72	**39.0**	46.56
7.5	30.42	**23.5**	37.98	**39.5**	46.85
8.0	30.64	**24.0**	38.23	**40.0**	47.15
8.5	30.86	**24.5**	38.49	**40.5**	47.46
9.0	31.08	**25.0**	38.75	**41.0**	47.76
9.5	31.30	**25.5**	39.01	**41.5**	48.06
10.0	31.52	**26.0**	39.27	**42.0**	48.37
10.5	31.74	**26.5**	39.53	**42.5**	48.68
11.0	31.97	**27.0**	39.80	**43.0**	48.99
11.5	32.19	**27.5**	40.06	**43.5**	49.30
12.0	32.42	**28.0**	40.33	**44.0**	49.61
12.5	32.65	**28.5**	40.60	**44.5**	49.92
13.0	32.88	**29.0**	40.87	**45.0**	50.24
13.5	33.11	**29.5**	41.14	**45.5**	50.56
14.0	33.34	**30.0**	41.41	**46.0**	50.87

14.5	33.57	**30.5**	41.69	**46.5**	51.19
15.0	33.81	**31.0**	41.96	**47.0**	51.51
15.5	34.04	**31.5**	42.24	**47.5**	51.84
16.0	34.28	**32.0**	42.52	**48.0**	52.16
16.5	34,52	**32.5**	42.80	**48.5**	52.49

APPENDIX I

THERMODYNAMIC PROPERTIES OF AIR AND WATER

1. Latent Heat of Vaporization (λ)

$$\lambda = [2.501-(2.361 \times 10^{-3})\ T]$$

where: λ = latent heat of vaporization [MJ kg^{-1}]; and T = air temperature [°C].

The value of the latent heat varies only slightly over normal temperature ranges. A single value may be taken (for ambient temperature = 20 °C): λ = 2.45 MJ kg^{-1}.

2. Atmospheric Pressure (P)

$$P = P_o\ [\{T_{Ko}-\alpha(Z-Z_o)\} \div \{T_{Ko}\}]^{(g/(\alpha.R))}$$

where: P, atmospheric pressure at elevation z [kPa]

P_o, atmospheric pressure at sea level = 101.3 [kPa]

z, elevation [m]

z_o, elevation at reference level [m]

g, gravitational acceleration = 9.807 [m s^{-2}]

R, specific gas constant == 287 [J kg^{-1} K^{-1}]

α, constant lapse rate for moist air = 0.0065 [K m^{-1}]

T_{Ko}, reference temperature [K] at elevation z_o = 273.16 + T

T, means air temperature for the time period of calculation [°C]

When assuming P_o = 101.3 [kPa] at z_o = 0, and T_{Ko} = 293 [K] for T = 20 [°C], above equation reduces to:

$$P = 101.3[(293-0.0065Z)\ (293)]^{5.26}$$

3. Atmospheric Density (ρ)

$$\rho = [1000P] \div [T_{Kv}\ R] = [3.486P] \div [T_{Kv}], \text{ and } T_{Kv} = T_K[1-0.378(e_a)/P]^{-1}$$

where: **ρ,** atmospheric density [kg m^{-3}]

R, specific gas constant = 287 [J kg$^{-1\ K-1}$]

$T_{Kv,}$ virtual temperature [K]

$T_{K,}$ absolute temperature [K]: T_K = 273.16 + T [°C]

$e_{a,}$ actual vapor pressure [kPa]

T, mean daily temperature for 24-hour calculation time steps.

For average conditions (e_a in the range 1–5 kPa and P between 80–100 kPa), T_{Kv} can be substituted by: $T_{Kv} \approx 1.01\ (T + 273)$

4. Saturation Vapor Pressure Function (e_s)

e_s = [0.6108]*exp{[17.27*T]/[T + 237.3]}

where: e_s, saturation vapor pressure function [kPa]

T, air temperature [°C]

5. Slope Vapor Pressure Curve (Δ)

$\Delta = [4098.\ e°(T)] \div [T + 237.3]^2$

$= 2504\{\exp[(17.27T) \div (T + 237.2)]\} \div [T + 237.3]^2$

where: **Δ,** slope vapor pressure curve [kPa C^{-1}]

T, air temperature [°C]

e°(T), saturation vapor pressure at temperature T [kPa]

In 24-hour calculations, **Δ** is calculated using mean daily air temperature. In hourly calculations T refers to the hourly mean, T_{hr}.

6. Psychrometric Constant (γ)

$\gamma = 10^{-3}\ [(C_p.P) \div (\varepsilon.\lambda)] = (0.00163) \times [P \div \lambda]$

where: γ, psychrometric constant [kPa C^{-1}]

c_p, specific heat of moist air = 1.013 [kJ $kg^{-1\ °C-1}$]

P, atmospheric pressure [kPa]: Eqs. (2) or (4)

ε, ratio molecular weight of water vapor/dry air = 0.622

λ, latent heat of vaporization [MJ kg^{-1}]

7. Dew Point Temperature (T_{dew})

When data is not available, T_{dew} can be computed from e_a by:

$T_{dew} = [\{116.91 + 237.3Log_e(e_a)\} \div \{16.78–Log_e(e_a)\}]$

where: T_{dew}, dew point temperature [°C]

e_a, actual vapor pressure [kPa]

For the case of measurements with the Assmann psychrometer, T_{dew} can be calculated from:

$$T_{dew} = (112 + 0.9T_{wet})[e_a \div (e°\ T_{wet})]^{0.125}–[112–0.1T_{wet}]$$

8. Short Wave Radiation on a Clear-Sky Day (R_{so})

The calculation of R_{so} is required for computing net long wave radiation and for checking calibration of pyranometers and integrity of R_{so} data. A good approximation for R_{so} for daily and hourly periods is:

$$R_{so} = (0.75 + 2 \times 10^{-5}\ z)R_a$$

where: *z*, station elevation [m]

$R_{a,}$ extraterrestrial radiation [MJ m^{-2} d^{-1}]

Equation is valid for station elevations less than 6000 m having low air turbidity. The equation was developed by linearizing Beer's radiation extinction law as a function of station elevation and assuming that the average angle of the sun above the horizon is about 50°.

For areas of high turbidity caused by pollution or airborne dust or for regions where the sun angle is significantly less than 50° so that the path length of radiation through the atmosphere is increased, an adoption of Beer's law can be employed where P is used to represent atmospheric mass:

$$R_{so} = (R_a)\exp[(-0.0018P) \div (\mathbf{K_t}\sin(\Phi))]$$

where: K_t, turbidity coefficient, $0 < K_t < 1.0$ where $K_t = 1.0$ for clean air and $K_t = 1.0$ for extremely turbid, dusty or polluted air.

P, atmospheric pressure [kPa]

Φ, angle of the sun above the horizon [rad]

R_a, extraterrestrial radiation [MJ m^{-2} d^{-1}]

For hourly or shorter periods, Φ is calculated as:

$\sin\Phi = \sin\varphi \sin\delta + \cos\varphi \cos\delta \cos\omega$

where: φ, latitude [rad]

δ, solar declination [rad] (Eq. (24) in Chapter 3)

ω, solar time angle at midpoint of hourly or shorter period [rad]

For 24-hour periods, the mean daily sun angle, weighted according to R_a, can be approximated as:

$\sin(\Phi_{24}) = \sin[0.85 + 0.3\,\varphi \sin\{(2\pi J/365)–1.39\}–0.42\,\varphi^2]$

where: Φ_{24}, average Φ during the daylight period, weighted according to R_a [rad]

φ, latitude [rad]

J, day in the year

The Φ_{24} variable is used to represent the average sun angle during daylight hours and has been weighted to represent integrated 24-hour transmission effects on 24-hour R_{so} by the atmosphere. Φ_{24} should be limited to >0. In some situations, the estimation for R_{so} can be improved by modifying to consider the effects of water vapor on short wave absorption, so that: $R_{so} = (K_B + K_D)\,R_a$ where:

$K_B = 0.98\exp[\{(–0.00146P) \div (K_t \sin\Phi)\}–0.091\{w/\sin\Phi\}^{0.25}]$

where: K_B, the clearness index for direct beam radiation

K_D, the corresponding index for diffuse beam radiation

$K_D = 0.35–0.33\,K_B$ for $K_B > 0.15$

$K_D = 0.18 + 0.82\,K_B$ for $K_B < 0.15$

R_a, extraterrestrial radiation [MJ m^{-2} d^{-1}]

K_t, turbidity coefficient, $0 < K_t < 1.0$ where $K_t = 1.0$ for clean air and $K_t = 1.0$ for extremely turbid, dusty or polluted air.

P, atmospheric pressure [kPa]

Φ, angle of the sun above the horizon [rad]

W, perceptible water in the atmosphere [mm] = $0.14\,e_a\,P + 2.1$

e_a, actual vapor pressure [kPa]

P, atmospheric pressure [kPa]

APPENDIX J

PSYCHROMETRIC CHART AT SEA LEVEL

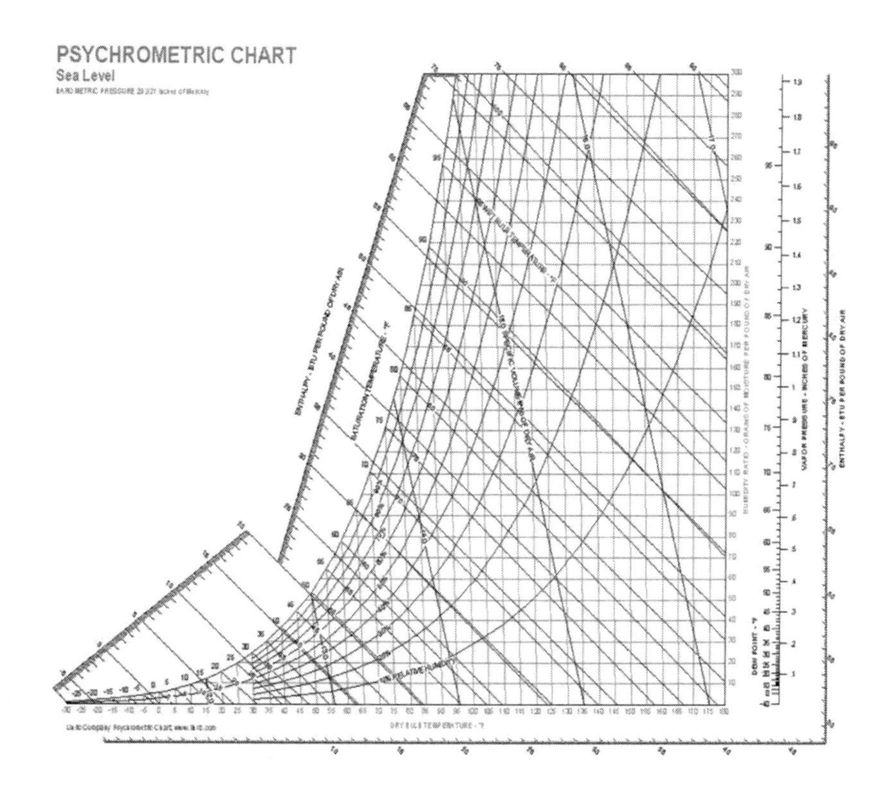

APPENDIX K

DESIGN TABLES AND CHARTS FOR MICRO IRRIGATION

[From: <http://www.jains.com/Designtechnical/design.htm>]

The http://www.jains.com/Designtechnical/design.htm lists *design tables* and *charts* for: Units and Conversion; Design Data; Friction Loss in Jain Tough Hose 12 mm OD; Friction Loss in Jain Tough Hose 16 mm OD; Friction Loss in Jain Tough Hose 20 mm OD; Friction Loss in J- Turbo Aqura®/J-Turbo Line® 12 mm OD; Friction Loss in J- Turbo Aqura®/J-Turbo Line® 16 mm OD; Friction Loss in J- Turbo Aqura®/J-Turbo Line® 20 mm OD; Friction Loss in Twin-Wall™- Deluxe/Twin-Wall™- BTF/Jain Turbo Slim® Drip Tape 16 mm (5/8″) ID; Friction Loss in Twin-Wall™- Marathon/Jain Turbo Slim® Drip Tape 22 mm (7/8″) ID;

Flow Nomogram for Polyethylene Pipes; Flow Diagram for PVC Pressure Pipes & Quick Fix™ Pipes; Guidelines for Climate Control of Temperature and Humidity in Greenhouses; General Guidelines for Maintenance of Drip irrigation system; General Maintenance of Guidelines for Filters; Guidelines for use of Drip Tapes; General Guidelines for Drip Irrigation Systems.

Selected tables, graphs, charts, and nomographs are included in this book for the information purpose only. The inclusion of such information does not imply endorsement or preference to the Jain Irrigation Products.

1. Filter screen size

Mesh No.	Opening Size		
	Inch	**mm**	**Micron**
80	0.0069	0.177	177
100	0.0059	0.149	149
120	0.0049	0.125	125
150	0.0041	0.105	105
200	0.0037	0.094	74

2. Design Data

Allowable design stress for PE Pressure Pipes

If PE pipes are used to convey fluids of excessive temperature then the allowable design stress or the pressure-class ratings should be appropriately reduced in accordance with following graph.

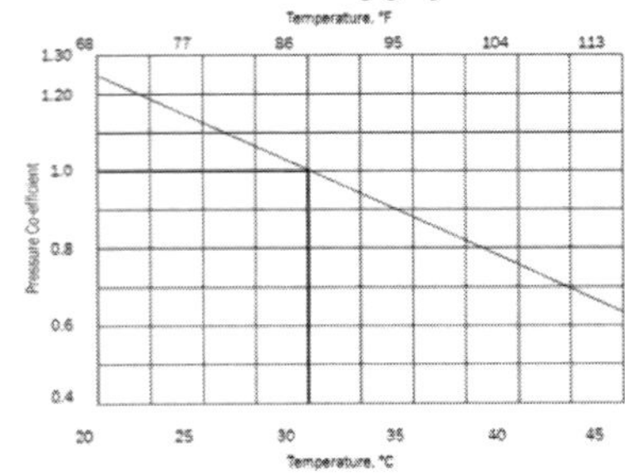

Source : IS:4984 - Indian Standard for HDPE Pipes.

Allowable design stress for uPVC Pressure Pipes

If uPVC pipe or uPVC valve is used to convey fluids of excessive temperature then the allowable design stress or the pressure class ratings should be appropriately reduced in accordance with following graph.

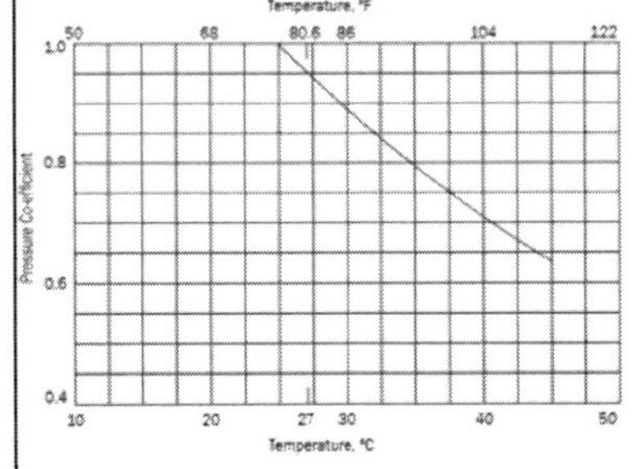

Source : IS:4985 - Indian Standard for uPVC Pipes.

Potential Evapotranspiration Rate (PET) for different Climate

Climate	Temperature (°C)	Humidity (%)	PET (mm per day)
Cool humid	≤18	50 - 85	2.54 - 3.80
Cool dry	18 - 25	≤50	3.80 - 5.00
Moderate	28 - 36	50 - 65	5.00 - 6.35
Hot humid	36 - 40	50 - 85	6.35 - 8.90
Hot dry	40 - 45	≤50	8.90 - 10.00
Hot dry desert	45 - 50	≤30	10.00 - 11.50

Solvent Cement Requirement for uPVC Pipes

Pipe Size mm	No. of Joints* per Liter	Pipe Size mm	No. of Joints * per Liter
20	330	125	45
25	270	140	36
32	225	160	27
40	180	180	20
50	135	200	15
63	125	225	12
75	103	250	9
90	79	280	6
110	54	315	5
* Each Joint represents one socket in a Fitting.			

Frictional Head Losses in Valves and Fittings

3. Friction Loss in Jain Tough Hose 12 mm OD

Note:

- Curves are labeled with specific discharge rate (SDR) in lph/meter.
- The friction loss has been calculated on the basis of the mean ID with dimensions and tolerances for pressure class-2 & for 0% ground slope.
- Based on the Hazen Williams flow equation.

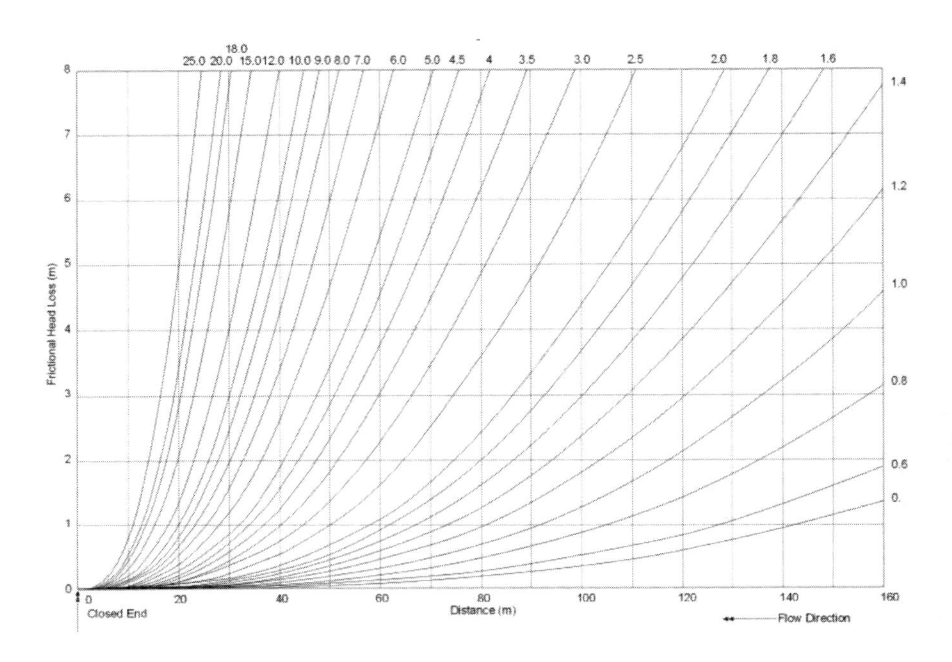

4. Friction Loss in Jain Tough Hose 16 mm OD, Note: See under section K3

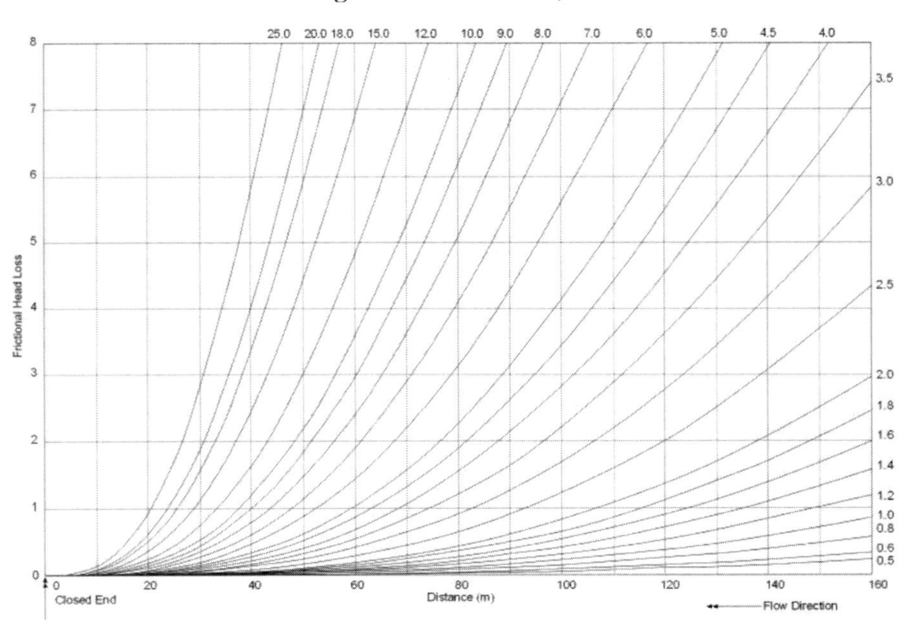

5. Friction Loss in Jain Tough 20 mm OD, Note: See under section K3

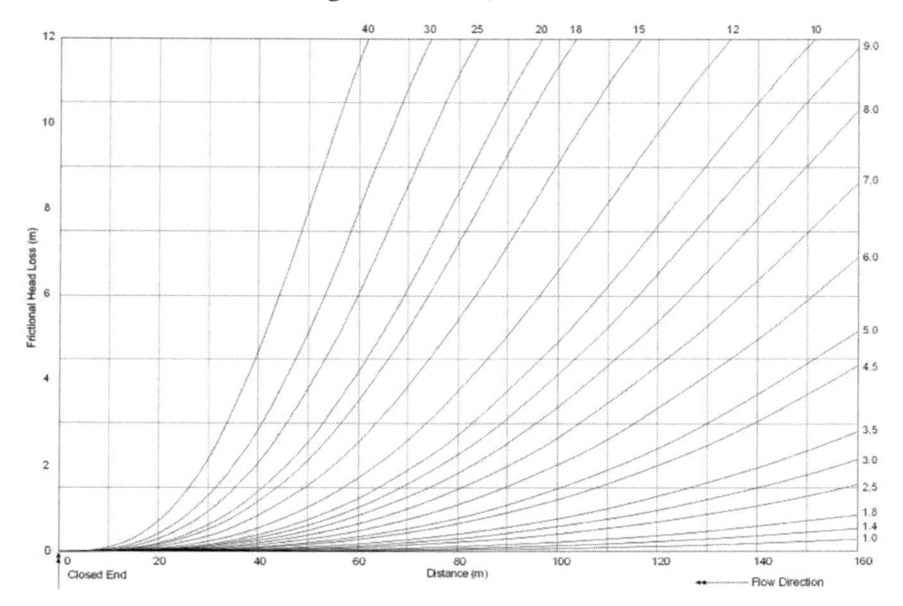

6. Friction Loss in J-Turbo Aqura®/J-Turbo Line® 12 mm OD, Note: See under section K3

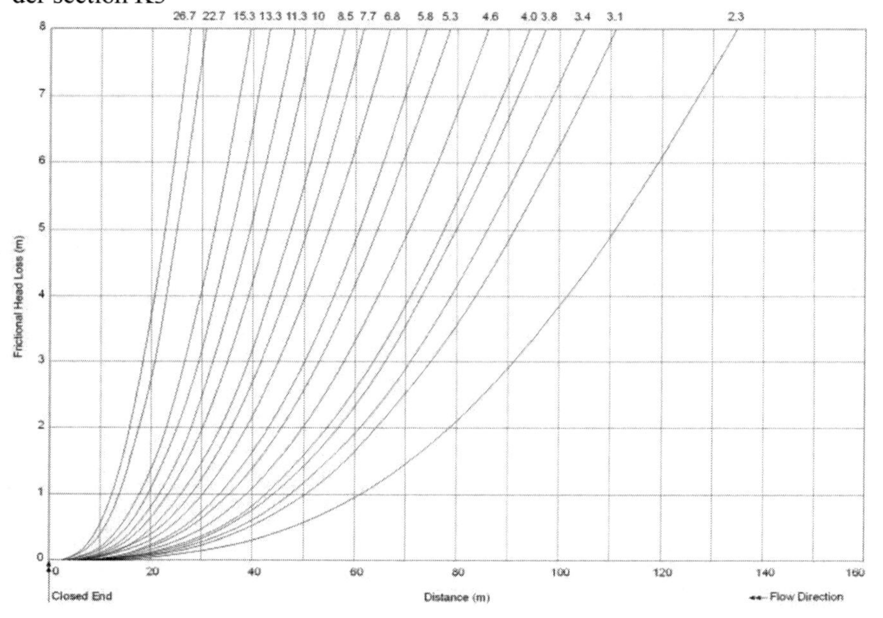

7. Friction Loss in J-Turbo Aqura®/J-Turbo Line® 16 mm OD, Note: See under section K3

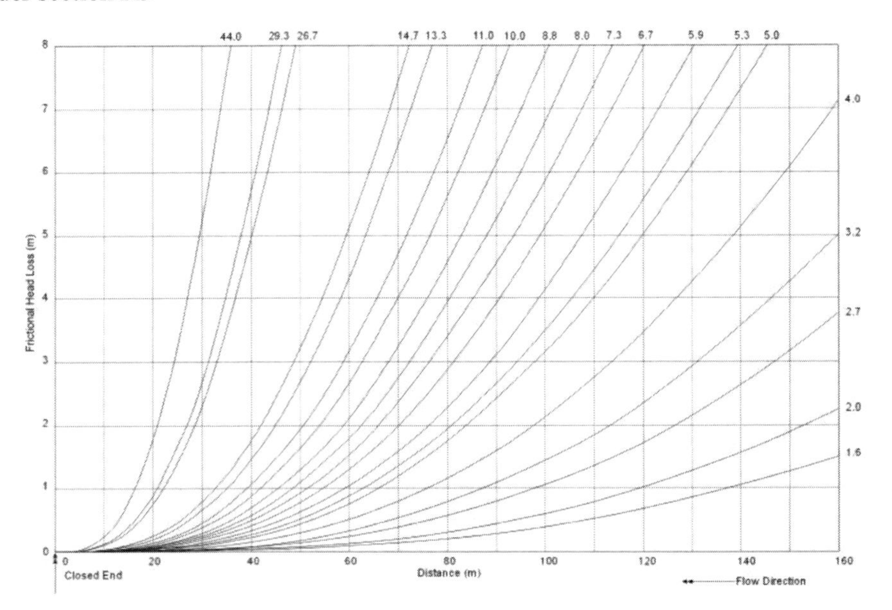

8. Friction Loss in J-Turbo Aqura®/J-Turbo Line® 20 mm OD, Note: See under section K3

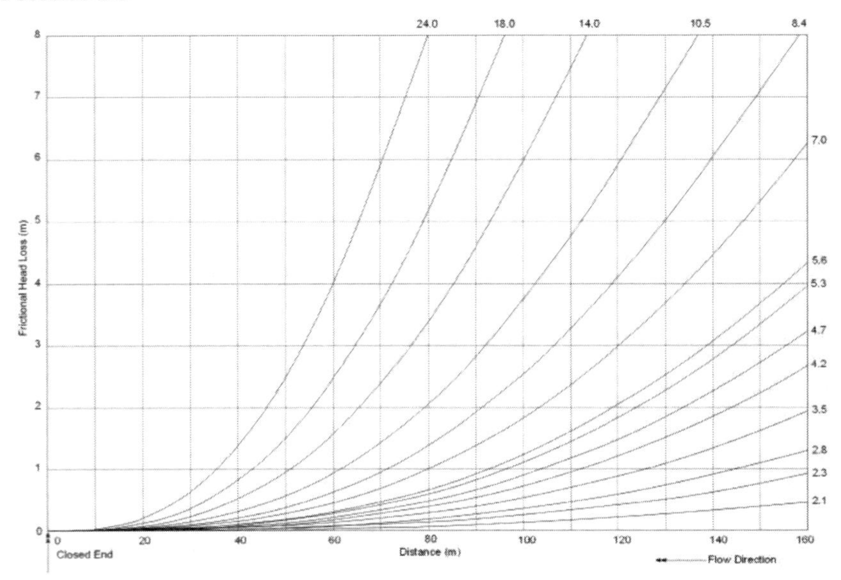

9. Frictional Loss Twin-Wall™ – BTF/Deluxe Drip Tape 16 mm ID, Note: See under section K3

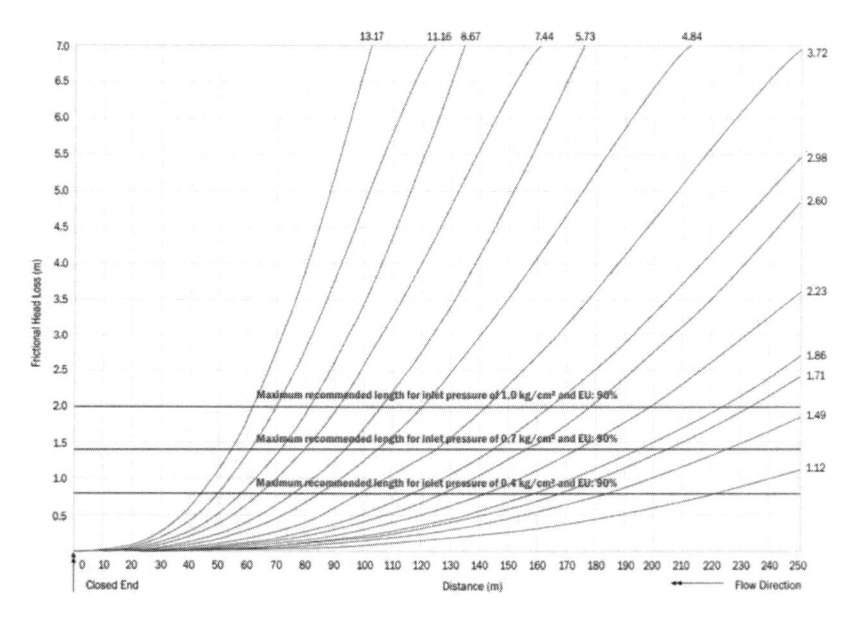

10. Frictional Loss Twin-Wall™ – BTF/Deluxe Drip Tape 22 mm ID, Note: See under section K3

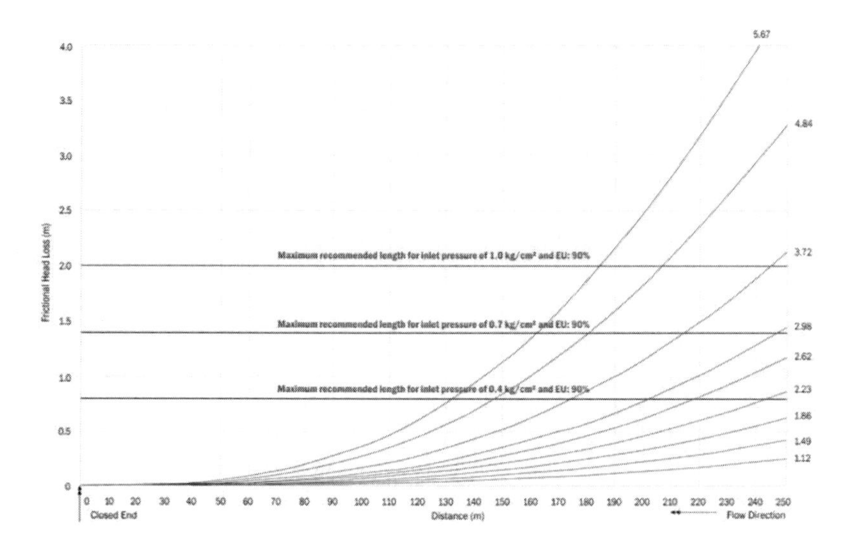

11. Flow Nomogram for Polyethylene Pipes

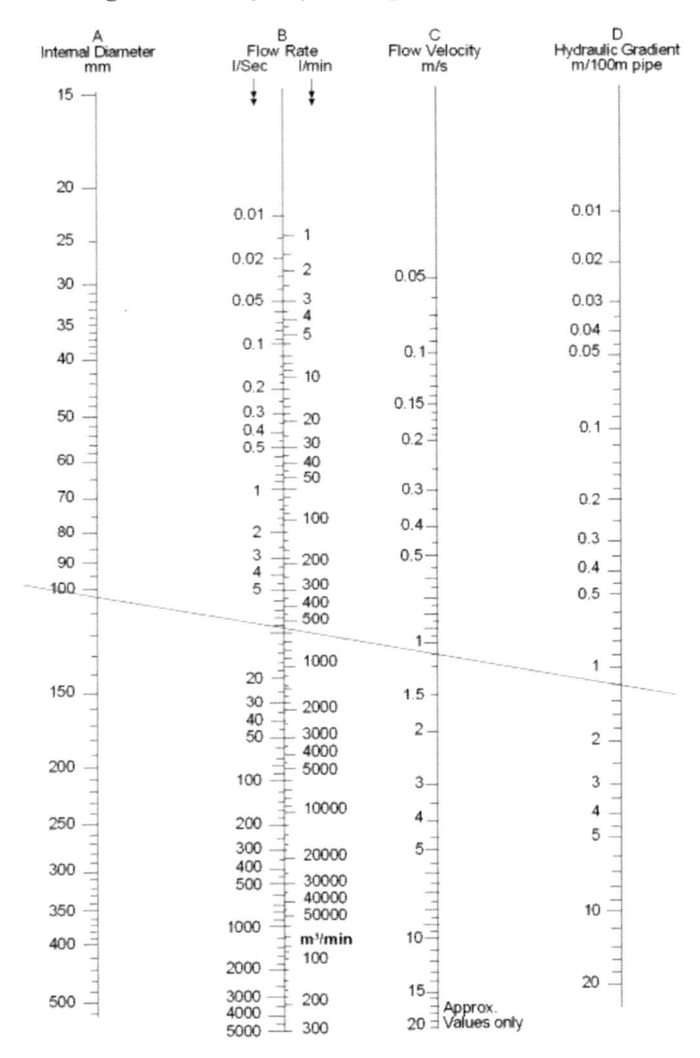

Example: Given – Flow 10 lps, Pipe size 110 mm OD (104 mm ID), Velocity 1.2 m/s. Then – Hydraulic Gradient 1.2 m/100 m.

To use the above FLOW NOMOGRAM at least two values out of A, B, C, D should be known. Joining the two values on lines and extending the line henceforth will give the desired values.

12. Flow Diagram for PVC Pressure Pipes & Quick Fix™ Pipes

Note: (i) The Diagram has been calculated on the basis of the mean ID with dimensions and tolerances to IS 4985; (ii) Based on the Hazen Williams Flow Formula.

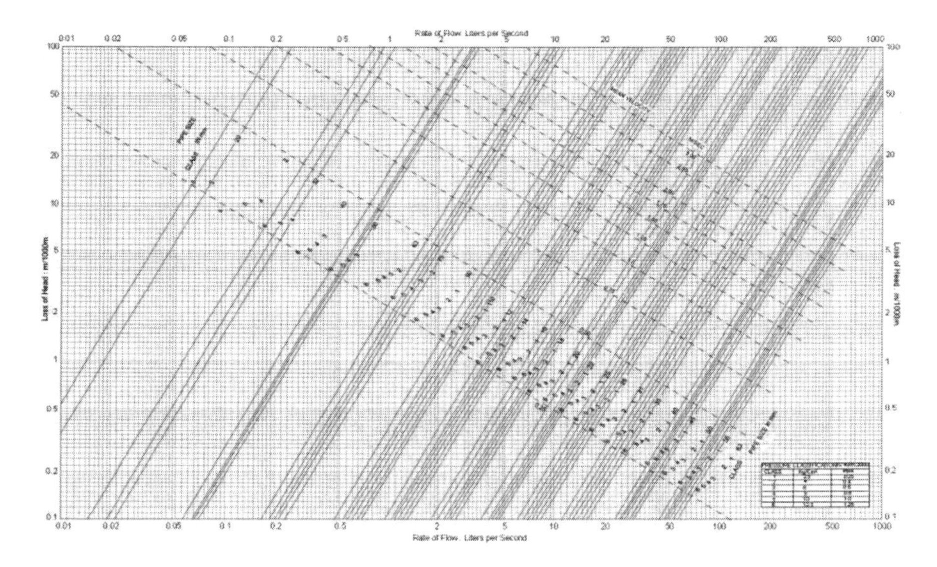

13. Guidelines for Climate Control of Temperature and Humidity in Greenhouses

- Climate control in a greenhouse is based on the principle of the exchange of energy between the air and the fog moisture supplied by the NaanDanJain Fogger.
- One calorie is the amount of heat necessary to raise the temperature of 1 cm3 of water by 1 °C.
- The conversion of water from liquid to vapor absorbs heat from the surrounding air, at a rate of 590 calories/1 gram of evaporated water (1 cm3). This process lowers the air temperature.
- Efficient installation and operation can reduce the temperature in the greenhouse by 4 to 6 °C, depending on the environmental conditions.
- Efficiency of the cooling system depends on two environmental factors:
  - External temperature
  - External humidity

Essential conditions for efficient cooling with the NaanDanJain Fogger are:

- An efficient ventilation system, which continuously introduces external dry air into the greenhouse to replace the humid air.
- Pulse operation of the fogging system, to minimize the amount of water that may settle on the foliage.

What is the importance of the droplet size created by the NaanDanJain Fogger?

With a 7 l/h nozzle at 4 kg/cm², the average fog droplet is 90 microns. These droplets evaporate without wetting the leaves or greenhouse floor.

Spacing Design

Type	Fogger Configuration	Nozzle	Flow Rate	Spacing		Precipitation Rate mm/h
				Between Laterals	On the line	
Fogger	T	7.0 x 2	14	3.0	1.5	3.1
	Cross	7.0 x 4	28	3.0	3.0	3.1
Super Fogger	T	6.5 x 2	13	3.0	1.5	2.9

- The foggers should be installed in as high as possible, with minimal height being 3.0 m. The nozzles should be situated perpendicularly to the main line.
- Operating pressure: 4.0 kg/cm² at the fogger.

How much water is needed to cool a greenhouse?

- A fog precipitation of 3 to 3.5 mm/h is suitable in the majority of cases (3 mm/h = 30 m³/h per ha.)

Water Consumption (meter cube per hectare per 10 h/day)

Nozzle size l/hr	Model configuration	Net precipitation, (l/hr/m² for various spacing's & pulses)					
		3.0 x 1.5 m			3.0 x 3.0		
Fogger							
		1:5*	1:10	1:15	1:5	1:10	1:15
Blue 7.0 × 2	T	0.62	0.31	0.21			
Blue 7.0 × 4	cross				0.62	0.31	0.21
Water consumption m³/ha/day (10 h)		62	31	21	62	31	21
Super Fogger							
6.5 × 2	T	0.58	0.29	0.19	* On: Off pulse ratio Example: ratio 1:5 The system will operate only fifth of the time so the actual precipitation will be only 20%, or 10% for ratio 1:10		
Water consumption m³/ha/day (10 h)		58	29	19			

- Cooling

- The pulse duration should be as short as possible, between 1 to 3 sec. (Under certain conditions, longer durations can be considered).
- The interval between fogging pulses should be based on the factors of humidity, temperature levels, greenhouse ventilation and height of the foggers.

Initial guide (pulse of 2.0 sec)

- Fine tuning is recommended to be done according to local conditions.

Outside relative humidity (%)	Interval (Sec.)	Replenish moisture (l/ha)
30	15	4100
40	25	2500
50	25	2500
60	30	2100

Cooling and Humidifying cannot be conducted simultaneously.

Humidification

- In the event that the humidity level needs to be increased then the ventilation system must be shut down. The duration of the fogging process should be as short as possible (1.0 second).
- The intervals between fogging pulses will vary according to the minimal relative humidity required.
- During the morning, when temperatures rise and humidity decreases, the humidity sensor will cause the fogging system to operate.

Humidity	Interval	Duration
30 – 40%	60 sec	1 sec
40 – 50%	90 sec	1 sec
50 – 60%	120 sec	1 sec

Spraying

- In the event that the humidity level needs to be increased then the ventilation system must be shut down. The duration of the fogging process should be as short as possible (1.0 second).

Water Quality

- In order to avoid clogging by carbonates or the accumulation of salt deposits on the leaves, it is recommended to avoid irrigation with hard or saline water. Rainwater, soft water or osmosis-treated water is the most suitable.

Optional Nozzles

Color	**blue**	**orange**	**red**	**black**
Flow rate l/h at 4.0 bar	7.0	14.0	21.0	28.0

14. Recommended Concentration of free Chlorine in water (ppm = mg/liter) for various purpose

Purpose of Chlorination	**Application Method**	**Concentration ppm (mg/l) at head control**	**ppm (mg/l) level at end of the system (dripper)**	**Remarks: 2HOCl (35%Cl) injection l/hr/10 m³ flow**
Algae prevention	Continuous	1–10 max.	0.5–1.0 max	0.029–0.29
Algae & bacteria killing	Intermittent	10–20 for 20 min	0.5–1.0 max	0.29–0.58
Dissolving organic matter	Hyper-chlorination	50–500 for 5 min	5–10 max	2.9–29
Oxidation of iron	Continuous	0.6 mg/l/ppm of Iron impurities	0.5–1.0 max	0.022 /ppm for Iron impurities
Oxidation of manganese	Continuous	0.6 mg/l/ppm of Mn. impurities	0.5–1.0 max	0.022/ppm for Mn. impurities
Sulfur impurities	Intermittent	0.6 mg/l/ppm of H2S impurities	0.5–1.0 max	0.022/ppm for Sulfur impurities

15. Where to contact for design of micro irrigation systems

Jalgaon – Head Office Jain Plastic Park, Jalgaon – 425 001. CIN: L29120 MH-1986PLC042028. Tel: +91 257 225 8011/22, Fax: +91 257 225 8111/22	Ontario, California Jain Irrigation Inc. P.O. Box 3760 1941 S. Vineyard Ave., Unit 6 Ontario, CA 91761 Tel: +1 800 828 9919, 909 395 5200, Fax: +1 800 777 6162, 909 395 5201, Website: www.jainirrigationinc.com
USA	
Columbus, Ohio Jain (Americals) Inc. 1819 Walcutt Road, Suite 1 Columbus, Ohio 43228 USA Tel: +1 614–850–9400 Website: www.jainamericas.com	Fresno California Jain Irrigation Inc. 2851 E. Florence Ave. Fresno, CA 93721 Tel: +1 800 695 7171, 559 485 7171 Fax: +1 888 434 3747, 559 485 7623, Website: www.jainirrigationinc.com
Watertown, New York, USA Jain Irrigation Inc. 740 Water St. Watertown, NY 13601 Tel: +1 800 242 7467, Fax: +1 866 329 2427, Website: www.jainirrigationinc.com	Winter Haven, Florida, USA Jain Irrigation Inc. P.O. Box 3546 3857 W. Lake Hamilton Dr. Winter Haven, FL 33881 Tel: +1 800 848 8153, 863 294 1900, Fax: +1 800 533 6421 863 299 6421, Website: www.jainirrigationinc.com

INDEX

D

E

I

J

K

P

Q

R

S

T

U

V

W

Y

Z